SOLAR ENERGY APPLICATIONS IN THE TROPICS

SOLAR ENERGY APPLICATIONS IN THE TROPICS

Proceedings of a Regional Seminar and Workshop on the Utilization of Solar Energy in Hot Humid Urban Development, held at Singapore, 30 October - 1 November, 1980

Edited by **BILL B. P. LIM**

National University of Singapore, Faculty of Architecture and Building

D. REIDEL PUBLISHING COMPANY

Dordrecht, Holland / Boston, U.S.A. / London, England

Library of Congress Cataloging in Publication Data

CIP

Regional Seminar and Workshop on the Utilization of Solar Energy in Hot
Humid Urban Development (1980 : Singapore)
Solar energy applications in the tropics.

Includes index.
1. Solar energy–Tropics–Congresses. I. Lim, Bill B. P., 1936–
II. Title.
TJ810.R43 1980 333.79'23'0913 82–16697
ISBN-13: 978-94-009-7938-3 e-ISBN-13: 978-94-009-7936-9
DOI: 10.1007/978-94-009-7936-9

Published by D. Reidel Publishing Company
P.O. Box 17, 3300 AA Dordrecht, Holland

Sold and distributed in the U.S.A. and Canada
by Kluwer Boston Inc.,
190 Old Derby Street, Hingham, MA 02043, U.S.A.

In all other countries, sold and distributed
by Kluwer Academic Publishers Group,
P.O. Box 322, 3300 AH Dordrecht, Holland

D. Reidel Publishing Company is a member of the Kluwer Group

CONTENTS

PREFACE

Renewable energy sources such as solar energy were advocated even before the energy crisis in 1973. Subsequent development in solar energy has been remarkable. Indeed it has been one of the most studied and researched topics in recent years.

Much of the technology in the utilization of solar energy has been associated with housing and industrial applications, and, to a smaller extent, with commercial use. The nature of the utilization has been related to the philosophy of independence or autonomy in that each housing or industrial unit is made self-sufficient by the installation of solar energy equipment. The general aim of the development of solar energy has been to substitute existing energy sources by solar energy especially among the developed countries in the temperate regions. In developing countries, the application of solar technology has been more acceptable in rural areas where conventional infrastructure is under-developed.

A different direction is necessary for the development of solar energy in urban areas. Buildings are inter-related, and have been dependent on the conventional infrastructure. The consumption of energy is many times more than that in the rural areas. A new approach to urban development must be generated, and studies should be made on the feasibility and viability of using solar energy for urban settlements.

The Regional Seminar and Workshop on Solar Energy Application in the Tropics attempts to introduce the topic of solar energy utilization, in general, to a wide audience which has not been familiar with these issues, and to highlight some of the problems associated with urban application.

Speakers at the Seminar came from the United States, Australia, Japan, France, Malaysia and Thailand as well as Singapore. The attendance was also international, and some 180 participants deliberated actively over 2½ days from 30 October to 1 November 1980.

The Seminar was partially financed by UNESCO, chiefly for travelling and accommodation expenses of overseas speakers and official delegates from 3 Asean countries, namely, Philippines, Indonesia and Thailand. The proceeds of this publication will be donated to the National University of Singapore for research purposes.

As this was the first international seminar on solar energy held in Singapore, it was felt desirable to publish the papers and proceedings

Bill B. P. Lim (ed.), Solar Energy Applications in the Tropics, ix–x.

in book form. The authors of the papers kindly consented to give copyrights to the National University of Singapore.

I wish to thank members of the Organising Committee for their untiring work during the preparation of the Seminar and their chairmanship in the Sessions. The collation and the preparation of the manuscripts was ably carried out by the Dean's Office of the Faculty of Architecture and Building, National University of Singapore. The Seminar was efficiently conducted by Mrs. Gloria Lim of the Executive Communication Services who also transcribed the proceedings. My thanks also go to Mr. Michael Brown for proof-reading and correction, and Miss Mary Tan for typing the manuscripts to camera-reading form.

It is hoped that this book will help architects, engineers, developers in all sectors of the building industry to understand the implication of solar technology in buildings. For this reason, many comments and questions, though elementary in nature, are included in the proceedings. It is unfortunate that many speakers' names were not recorded during the discussions. Consequently, all names except those of panel members were omitted. For editing purposes, the discussion sessions were abbreviated and in some cases speeches were paraphrased. In so doing, the printed text may not present the precise speeches made, but the essential parts are hopefully retained. I appreciate contributions made by all speakers during the discussion and workshop sessions, especially panel members who vetted the drafts.

Bill B. P. Lim
Chairman of the Organising Committee and Editor

November 1981

CONTRIBUTORS

Contributors were speakers at plenary sessions and authors of working papers for the Seminar and Workshop. They are listed according to the order of sessions of the Seminar and Workshop.

Prof. Jeffrey Cook	M.Arch., M.R.A.I.C., A.I.A.	Department of Planning, College of Architecture, Arizona State University, USA.
Prof. K. R. Rao	Ph.D. Roor., M.Sc., B.Sc.And., F.Inst.P.Lond., M.I.S.E. F.I.E.(Ind)	Department of Building, Science, National University of Singapore, Republic of Singapore.
Dr. Norman Sheridan	Ph.D., F.I.E.(Aust)	Department of Mechanical Engineering, University of Queensland, Australia.
Assoc. Prof. H. B. Robert Exell	M.A., D.Phil.(Oxon)	Division of Energy Technology, Asian Institute of Technology, Thailand.
Assoc. Prof. Bong Tet Yin	Ph.D., M.S. Kentucky, Ir Bandung, M.I.E.S.	Department of Mechanical and Production Engineering, National University of Singapore, Republic of Singapore.
Dr. Chiam Joon Tong	M.Sc. Tech.,Ph.D., M.I.E.S., P.Eng.(S'pore)	Consultant in Energy Management.
Assoc. Prof.Donald G.S. Chuah	B.Sc (S'pore), M.Sc., Ph.D (U.S.M.), M.Inst.P. (U.K., F.Inst.P.(M) M.I.E.E.E. (U.S.A.) Chevalier de l' Ordere des Palmes Academiques (France)	School of Physics, Universiti Sains Malaysia, Penang, Malaysia

Prof. Ken-ichi Kimura	Ph.D.(Waseda), B.Arch.	Department of Architecture, Waseda University, Tokyo, Japan.

WORKING PAPERS

Mr. Toshihiro Ishibashi	(read by Mr T. Hiramatsu)	Airconditioning Research and Development Laboratories, Yazaki Shigen Co. Ltd., Japan.
Mr. Don Hampton		U.N.D.P./U.N.E.S.C.O. Advisor, Department of Building, Ngee Ann Technical College, Republic of Singapore.

ORGANISING COMMITTEE

(National University of Singapore)

Prof. Bill B. P. Lim (Chairman)	Ph.D. B.Arch.Syd., Dip.T.C.P.Syd., F.R.A.I.A., M.R.I.B.A., M.S.I.A., M.S.I.P.	Department of Building Science, National University of Singapore, Republic of Singapore.
Prof. K. R. Rao	Ph.D.Roor., M.S.C., B.Sc.And., F.Inst.P.Lond., M.I.S.E.S.	Department of Building Science, National University of Singapore, Republic of Singapore.
Assoc. Prof. Bong Tet Yin	Ph.D., M.S.Kentucky, Ir Bandung, M.I.E.S.	Department of Mechanical and Production Engineering, National University of Singapore, Republic of Singapore.

Assoc. Prof. Kooi Pang Shyan	D.Phil.Oxon, M.Sc. Tech. Manc., BS Nat. Taiwan, M.I.E.E.	Department of Electrical Engineering, National University of Singapore, Republic of Singapore.
Dr. Goh Thong Ngee	Ph.D.Wis., B.E.Sask., M.I.E.S., M.A.I.I.E., M.A.S.M.E., M.S.M.E.	Department of Industrial and Systems Engineering, National University of Singapore, Republic of Singapore.
Mr. S. Sujarittanonta	M.Arch.Texas, B.Arch. Chulalongkorn, Dip.I.C.H.P.H. Bouwcentrum, Dip. Econ.Dev.N.I.D.A., Dip.Arch.Chulalongkorn, M.A.S.A.	School of Architecture, National University of Singapore, Republic of Singapore

WELCOMING ADDRESS

Prof. Bill B. P. Lim, Chairman, Organising Committee

Minister of State for Education, Mr Chai Chong Yii, distinguished guests, colleagues, ladies and gentlemen:

On behalf of the National University of Singapore, I would like to welcome all of you to the Regional Seminar and Workshop on Solar Energy Applications in the Tropics. The gathering today is to me most significant for at least two reasons.

First, the seminar is international. The participants come from not only within the region of ASEAN but also from as far away as the United States, Japan, Australia and Europe. This indicates that the energy issue is now fully recognised by many countries. Energy conservation and utilisation of alternative sources of energy have become a common platform of co-operation and interaction among nations. We now have a common goal to work for.

This is enforced by the leadership given by UNESCO which has spearheaded many projects on energy. The Organising Committee is very grateful for the support which UNESCO has given. A grant has been provided mainly to cover the travelling and subsistence expenses of speakers and official delegates from the ASEAN countries. It is hoped that this international gathering will set the pace for many conferences to come in the region at which the latest development and knowledge in energy conservation and the exploitation of other energy sources will be exchanged.

Second, we have gathered today professionals from a wide variety of disciplines. This demonstrates the far-reaching consequences of the energy issue which is inter-disciplinary in nature. It is more than just a matter of technology; proper use of energy irrespective of its form can only be made through the concerted effort of the community at large. Solutions can only be found after the necessary integration of all professionals involved in the various stages of the production, supply and control. The energy problem can be tackled only by addressing ourselves collectively to it. Through the co-operation of professionals, we will also be able to develop better understanding and appreciation of each other's role and of the contribution which each professional can make to the society as a whole.

Bill B. P. Lim (ed.), Solar Energy Applications in the Tropics, 1.

OPENING REMARKS

Dr U Kuruppu, Programme Specialist in Science and Technology, Regional Office for Science and Technology for South East Asia, UNESCO.

Prof. Lim, Mr. Chai Chong Yii, Honourable Minister of State for Education, distinguished guests, participants, ladies and gentlemen:

It is a great honour and a privilege for me to be here with you today and to bring you greetings and good wishes from the Director-General of UNESCO, and the Director-General of the Regional office, Science and Technology for South-east Asia. UNESCO is most thankful to the Government of Singapore and the National University of Singapore for orgánising this Seminar. Our special thanks go to you Mr Minister, for being here with us today despite your many commitments. Your presence is an encouragement to all of us and will inspire us to greater efforts. To Prof Lim goes much credit for the initiative in organising this seminar and workshop. It is a timely topic of great concern to the whole world, but especially for those in the tropics. UNESCO is privileged to be associated with this venture as the sponsor. I would like here to express our great appreciation to Prof. Lim, the Organising Committee and the staff for the excellent arrangements they have made and the very good facilities provided.

Development of alternative energy sources is one of the most critical areas for action at the present time. The forward march of science and technology has made mankind more and more energy dependent. This was all right when energy supplies were thought to be plentiful. But the oil crisis of 1973 gave the world a rude shock and marked the end of an era - the era of cheap energy. For the past seven years, the world has become aware that the supply of energy is of concern to the whole of mankind. It is vital that measures be taken immediately to conserve the dwindling supplies of costly fuel and at the same time to develop alternative sources of energy. The problem is doubly complicated for the developing countries for, to expand their economies and to enhance per capita income, they need to increase energy consumption, but at the same time they can hardly meet the present energy bills, let alone future increases. Hence, it is not surprising that many scientists, institutes, governments, non-governmental organisations and international organisations are interested in the quest for alternative sources of energy.

The presence of a large gathering of distinguished scientists and engineers here today is an indication of the great interest in one such source, namely solar energy. This augurs well for the future of

Bill B. P. Lim (ed.), Solar Energy Applications in the Tropics, 3–5.

mankind. However in our eagerness to search for alternative sources of energy, there might be unnecessary duplication and waste of energy. Therefore, it is appropriate to sound a word of caution and to request everyone, be they individuals, institutes, non-governmental organisations or international agencies to keep to their own fields of competence. The usual tendency is to expand in wider and wider areas but this might result in spreading resources too thinly and in unnecessary duplication. UNESCO's field of competence is the training of personnel, development of educational institutions and research and development. Applied to the field of energy, it could be the stimulation of research activities and development of innovative technologies up to pilot project stage.

But that alone is not enough. The innovations must filter down to the masses and be of actual use to them. Hence there must be tests of economic viability, commercial production coupled with marketing and sales promotion. There should be adequate publicity to make people energy-conscious and aware of present innovations, so that they may be interested in new possibilities of improving their working conditions and standard of living. To achieve this, the efforts of scientists, engineers, economists, planners, marketing and sales personnel, science writers, mass media people, extension workers and government officials must be harnessed.

Permit me at this stage to state in brief UNESCO's activities in the field of energy. Taken globally, there are three main areas of action. First, there is the promotion and development of conventional and non-conventional small-scale sources of energy for rural and dispersed people. This will be concerned with sources such as wood or coal, bio-mass, wind and low-cost applications of solar energy. In addition to renewable sources of energy, attention will be paid to the utilisation of low grade coal and other traditional fuel in developing countries. Application of indigenous technology to small-scale energy systems suitable for rural or remote areas will be encouraged.

Second, there is the development of a world-wide system for information on new and renewable sources of energy. As a follow-up to the U.N. conference on Science and Technology development as part of preparation of U.N. Conference on new renewable sources of energy, UNESCO will provide support for the development of a world-wide system for information, compilation, evaluation, processing and dissemination in the field of new renewable sources of energy. The emphasis will be on improving the access of users in developing countries to information on small-scale services. Full advantage will be taken of the possibilities offered by existing information systems and new arrangement will be established in areas in which existing system do not provide services of this kind. It is hoped that this system will promote dissemination of information and help to avoid unnecessary duplication.

Third, there is the strengthening of scientific programmes of operation in the general field of energy. The development of regional and sub-regional linkages, networks and associations in the field of energy will be promoted and supported in various parts of the world. This will be done in co-operation with interested organisations of the U.N. system, including the Regional Economic Commission and other international, regional and national organisations. Activities will include courses and workshops, exchanges of personnel, demonstrations and pilot projects, regional journals, newsletters and other publications. Support will be given for the participation of specialists from developing countries for research related activities and for conferences.

Coming to the South East Asian region, UNESCO in 1973 assisted in the formation of an Association for Engineering Education in South-east Asia. Every two years the association has a conference. In the 1979 conference held in Perth, it was coupled with a workshop on Alternative Energy Sources. In 1978 UNESCO assisted in the formation of the Federation of Engineering Institutions of South East Asia and the Pacific. This Federation has already established five working groups, one of which is involved in the development of alternative energy sources, especially for rural areas. The first meeting of this working group was held in 1978 and the second meeting is scheduled to be held in Manila in December this year. UNESCO looks forward to the outcome of this Seminar. In fact we hope that some of the ideas generated will serve as input for the UN Conference on New and Renewable Sources of Energy which is due to be held in mid-1981.

Distinguished scientists and engineers, I wish you all success in your deliberations. Thank you.

SPEECH BY MR CHAI CHONG YII, SENIOR MINISTER OF STATE FOR EDUCATION

The sun is known to be the original source of energy that sustains all forms of life on our planet. Without solar energy, there can be no life on earth. Even human life is nothing more than a form of stored energy in action under the control of the soul. But human knowledge on how to harness solar energy commercially for the betterment of life is still quite limited.

Modern society today derives its comforts mainly through the use of energy produced by oil. With the multi-fold increase in oil prices since 1973, the whole world is suddenly realizing that it has depended too much on a single source of energy for too long a period. To rectify this error, without lowering level of comfort, one or both of the following measures must be taken: existing oil energy must be used more efficiently or reliance on oil must be lessened, to be replaced partly by a greater reliance on other forms of energy such as coal or solar energy. Replacing oil with other energy sources is easier said than done, but conservation of oil energy can definitely be put into effect more easily and much more quickly. Conservation does not mean lowering efficiency in production or sacrificing living standards. It just means realizing the true importance of energy and having a lifestyle that prevents wastage.

Following the energy conservation conference held in 1976, the Building Control Division of the Public Works Department embarked upon a research programme. This has led to the implementation of new building regulations to minimise electricity consumption in lighting and air-conditioning. To increase public awareness of the importance of energy conservation, two national campaigns were held in Singapore, one in 1973 and the other in 1979.

As Singapore depends entirely on imported oil to generate electricity, every drop must not be wasted. To drive home this point to big consumers of electrical energy, a 20% surcharge on electricity bill will be imposed on existing commercial buildings, if they exceed the Overall Thermal Transfer Values of 45 watts per square metre. This punitive move is balanced by the introduction of tax incentives to owners who spend money on improving the fabric of buildings to conserve energy.

Singapore is in the tropics where energy directly from the sun and wind is abundant. These renewable energy sources are now given more attention throughout the world. Research work is being carried out to explore their application in water heating, in cooling, refrigeration and generation of electricity. While it is still uncertain to what

Bill B. P. Lim (ed.), Solar Energy Applications in the Tropics, 7–8.

extent they can replace oil energy, the mere fact that more attention is being directed towards the search for and application of new energy sources, as in the Seminar you are holding today, points to the urgency of the problem.

It is important that our scholars and researchers keep abreast of the technological development in this field. With up-to-date knowledge of the practical usage of alternative energy sources, some electricity generation may be diversified to lessen our total dependence on oil. Oil, as we all know, is an exhaustible source of energy. At the rate it is now being used by the whole world, it will not last for many more decades. Even with the addition of new discoveries through exploration or the improvement of extraction techniques, the day that most developing countries will find themselves unable to pay for the high cost of oil is not far off. Until the techniques for practical application of renewable energy sources are perfected and their costs lowered to manageable levels, the only sensible way to deal with the problem of energy today is still conservation.

Programmes to conserve energy and to develop renewable energy sources would not only optimise the use of available resources, but also provide a platform for various countries to work together. The sharing of experience among nations would also foster international understanding and co-operation. Your gathering here today is indicative of your common interest in energy matters. It will definitely lead to further dialogue and the exchange of ideas among experts in this field.

Finally, I have much pleasure in declaring this Semimar open and wish you every success in your discussions.

SOLAR DECISION MAKING FOR THE TROPICS

Jeffrey Cook,
Professor of Architecture,
Department of Planning,
College of Architecture,
Arizona State University,
Tempe, AZ USA 85281.

ABSTRACT

Solar applications are viewed as being a particularly efficient and effective energy source for urban tropical locations. The goal of encouraging the inevitable transition to the permanency of solar energy will require many human commitments. Capital formation requires education, hardware and software investments in addition to a business atmosphere that encourages innovation and entrepreneurship. Commercialization requires initial experimental and demonstration stages. Solar decision incentives and patterns of early solar adoption in the United States are characterized. In conclusion, the rate of popularization appears to be the key to measuring solar commercialization and early small scale solar applications appear to precede larger urban and industrial solar installations.

1. SOLAR ASSETS AND LIABILITIES

It may be curious to invite a northern solar enthusiast to address the solar potentials of Southeast Asia. Superficially, the climatic patterns, energy needs, industrial and commercial potential and socio-political contexts of tropical urban locations would seem to have little in common with North American practice.

In particular, the conditions in the United States where a solar revolution has been initiated in less than 10 years are conditions that are not readily transferrable to other locations. For instance, the United States has a wide variety of seasonal climates and of accessibility to solar energy. Only in the Southwest deserts where I live is the sun available in large quantities and in highly dependable patterns. But in the desert one does not need the sun, at least for space heating. In those cold northern regions of the U.S. where winter conditions come close to an Arctic climate the sun may not be seen for weeks at a time when it is needed, and then it is low on the horizon. Thus, in the U.S. where solar applications continue to accelerate,

Bill B. P. Lim (ed.), Solar Energy Applications in the Tropics, 9–24.

there is a diverse mix and match of solar accessibility and thermal needs. This solar potential is further confused by the uneven accessibility to other energy sources at a regional level.

In contrast, the solar potential in the urban tropics appears, at least from a distance, to be much more homogeneous. You have lots of sun in dependable patterns and always high in the sky. Because of your stable climates your end use of energy sources is much more uniform. And to the degree that the interest of this conference is an urban context, you have identified a ready market with access to materials and a skilled technical force - commercial assets - that are lacking in rural locations throughout much of the world.

Nevertheless, solar potentials in the tropics share some of the same assets and liabilities as any other location:

(a) The sun is an <u>irregular</u> energy source. It does not shine at night.

(b) Thus, energy <u>storage</u>, whether of solar energy or some other source, is necessary to provide continuity of energy supply.

(c) The sun, even when strong and dependable, is a <u>diffuse energy source</u>. It is available virtually everywhere man wants to live on earth. But its universal distribution is not in a concentrated format with the high quality of density associated with other energy sources. Thus, the density of energy requirements within a city must be matched with the solar potential. Theoretically, the tropics may offer the best opportunity for urban solar applications because needs are not so high but solar access is dependable.

(d) But even in the tropics solar energy is diffuse. Thus, <u>collection surfaces must be large</u> - architectural in size.

(e) And conversion processes may be more complex and, therefore, require <u>larger spaces</u> and more careful integration, both within buildings and within the urban fabric; and spaces for mechanical equipment as well as the equipment itself must be paid for.

(f) Thus, while the sun is free, its collection and conversion can require expensive equipment and <u>high investment costs</u>. In contrast to the economics of most industrial investments in the world where low capital initiation and high running costs are the standard, solar requires high front end investment, but requires small running costs. <u>Solar is a capital intensive investment</u>.

(g) Integration with backup or alternative energy sources may require duplication of systems, or redundancy of equipment which will add to the initial high costs of solar investment. Even when end needs can be programmed by technical designers to run exclusively on solar, other decision makers may insist on such duplication. Thus, for instance, in the U.S.A many solar installations have been expensive because bankers or mortgage lenders have insisted on duplicated backup systems.

(h) The success of any solar application does not depend just on the expertise of the systems designer. It is also dependent on the confidence and trust of bankers and building inspectors as well as on the technical competence of construction contractors and equipment installers. Solar applications then are very much an inter-disciplinary activity involving dependence on several occupations and professions and many levels of competence and understanding. All of these assets and liabilities must be recognized and orchestrated within the context of local conditions and regional resources.

2. GENERAL GOALS AND STRATEGIES

In general, we may summarise two solar strategies that are appropriate to applications regardless of location:

(a) The sun is a universal energy source. It is available everywhere. And since solar technology responds to the same laws of physics throughout the world the sun is a kind of world common denominator of energy. It cannot be a monopoly of a particular political entity. It cannot be exclusively exploited by select groups within a country. In comparison, all of the other energy sources are, in fact, secondary solar sources - coal and petroleum are, in reality, time-delayed concentrations of ancient sunlight. And nuclear power is an imperfect, even crude miniature imitation of the sun. The advantage is that the solar nuclear power station is a safe 93 million miles away (or 140 kilometers, or 8.3 minutes at the speed of sunlight). All of these concentrated secondary solar energy forms have finite sources. Various experts and authorities may provide conflicting estimates of how many years supplies are remaining - but all agree that sources are limited - in some cases within our lifetime.

Thus the sun is not only universal in the sense that it is available in varying degrees throughout the world, but it is also universal because it is the only guaranteed long term energy source. It is our earth's common and almost infinite denominator of energy. In fact, solar energy should be regarded as the standard and almost infinite source; and other energy sources, such as coal, petroleum or nuclear, should be regarded as alternative and uncertain sources. So the transition to permanent and stable economies may be largely dependent on the success of transfer from present erratic energy sources to the permanent source of the sun.

In the sense that solar is the only assured long term energy source it would be a prudent goal for individuals, companies and countries to begin that inevitable transition. Transformation of economies or technologies does not happen overnight, particularly when economies are highly energy dependent. For them, the goal of a solar future is so obvious that the real question is not whether, but simply when? Dependable energy is critical, both for the maintenance and enhancement of our present standards of living, and certainly for the retention of an urban lifestyle. Without permanent economical energy sources, cities as we know them today will rapidly become obsolete.

(b) A solar energy economy requires a whole infrastructure of support that differs from our present fossil fuel and nuclear fuel based economy. Both the hard and soft investments necessary for the commercialization and popularization of an innovative technology require many human commitments. Those commitments are similar to the shifts necessary in any historical innovation transfer. While the particular technology of solar applications may be unique, the processes of technology adoption are similar throughout history. Thus, the experiences with solar adoption, such as in the U.S., can reveal how it may happen elsewhere. The development of other technical adoptions such as automobiles, refrigerated air conditioning, computers or even bows and arrows can instruct us of the patterns by which new thinking and new products become standard. Thus, human interactive behaviour is another common denominator which is universal, regardless of geographic location, and which can apply to any innovation adoption.

3. CAPITAL FORMATION

An innovative technology such as solar energy requires the formation of a support base which will generate, extend and direct its development. In the broadest sense this is capital formation - both of physical resources and financial leverage; but also of human intelligence and resources. I would like to concentrate on the latter. If there are compelling human reasons for action, the physical resources can usually be found - witness the U.S. space program. "Where there is a will, there is a way." However, shortages of human intelligence abound in every land and every age.

3.1. Education

As an educator, it is not surprising to identify education as a high priority. At the simple level of information transfer, obviously there needs to be agreement on a problem and its proportions before one proposes solutions. The size and consequences of the energy problem are already common knowledge and it is an elementary informational goal to be informed as to the solar solution.

When a professional educator talks about education, he is usually talking about education for the professional. When a researcher talks about education, he is usually referring to Ph.D.'s doing basic research work in a particular field. And when a plumber refers to education, it is typically the acquisition of abilities to earn his daily bread! Each of these levels of education is critical to the comprehensive success of solar applications. Interest in solar can never mature without all levels of education being involved.

Fortunately, solar is not a rare new discipline. Rather, it is a new blend of old skills and perceptions. The traditional expertise of architects, engineers and physicists are among the professional and academic disciplines that are involved. But don't forget the plumber. I have personally inspected the record of success and failure of solar applications during the 1950's. Typically, failures were plumbing. It was not the sun that failed, nor an understanding of the physics of heat transfer.

An evaluation of solar applications today may imply that the technical solutions are all well within reach. Indeed, the long term problem for solar may be aesthetic acceptability - once there are no leaks, it will be a question of human acceptability.

Creating popular demand tends to be missed by most solar enthusiasts who tend to get absorbed in the intricacies of technical issues. It is interesting that in the U.S. children appear to play important roles in solar decision making. In the promotion of solar domestic hot water heaters, salesmen often display their wares and pitch their sales stories at shopping centers or exhibitions. Family involvement in solar equipment enquiries is typical - and often the children understand the concepts first and proceed to explain them to their parents.

Children, as the decision makers of the next generation, have a particular educational opportunity through their schools. Not only are special courses, or new educational materials in existing instruction important vehicles, but the environment itself can be highly instructive. One of the early valuable U.S. government sponsored programmes in the winter of 1973-74 was the so-called "little red school house solar program". Here five schools in different locations of the country were fitted within two months with carefully monitored solar space heating systems. Science and industry were well served by these 1/2 million dollar crash programs. But the school systems and communities were also immediately involved in the issues of the new technology. Thus, the success measures of such a program must be evaluated on a much broader base than BTU/hr.

More recently the interpretation of passive solar space heating features in school designs is producing learning environments that make us all want to go back to school.

One of the elementary observations about the educational quality of solar issues is that the sun is an integrative subject. The sun and solar applications are so easily a component part of every aspect of human life that its emphasis serves to enrich human perceptions, regardless of the level or direction of instruction.

3.2. Hardware

The development of appropriate hardware is an early and obvious necessity in the emergence of a solar economy. Typically, early recognition and identification of solar applications is associated with the solar hardware that is exposed. In some cases the understanding and appreciation of solar potential never gets beyond solar collectors - sometimes blatantly displayed. Flat plate solar collectors are the visible hallmark of a solar installation. Indeed, the social recognition of solar commitment has caused at least one instance in California of false solar collectors - plastic dummies, so you can let your neighbours believe you have the real thing. There is a certain parallel with the larger Russian cities in the 1950's where TV antennae were always prominently displayed on the houses-but seldom were there TV sets inside.

Typical solar hardware designs require construction materials and fabrication techniques which are typical to industrialized locations. Thus, they are part of a world economy of copper, steel, glass and plastics, each of which has local and regional market characteristics. Glass or various plastic glazings are commonly necessary to intensify the solar greenhouse effect. In tropical locations and for low temperature locations, glazings may not be so necessary as in colder locations. Sheet metal is necessary to form the plates, boxes and ducts of flat plate collectors and to form the reflective troughs of concentrating collectors. Indeed, the general high availability of direct solar radiation in the tropics favours economical collector designs from a materials and hardware point of view. Thus, unglazed collectors and trough collectors may have special potential in the regions near the equator.

Solar thermal storage materials and processes tend to be less specialized from a technical point of view than are thermal conversion processes, such as the production of cooling from solar heat, which require substantial technical innovation. Control systems typically utilize electronic devices and, thus, are part of an advanced worldwide electronic industry.

In summary, it can be seen that while solar applications may involve some minor dislocations within the existing construction and supply industries, the general trend will be to provide more business and to encourage certain specializations. As a more capital intensive way of building, solar is a growth mode for construction materials suppliers. Since it requires more fabrication, it promotes and thrives on craftsmanship and technical ability. And since the processes of solar applications are more complex, they will require increase involvement from professional designers - thus stimulating a new and large cadre of well-informed specialists,

Economic studies of the impact of solar industry verify that it increases employment and increases investment return better than any other energy application. Because the characteristics of solar collectors are especially sensitive to local solar conditions, there is a certain finely tuned match between hardware and solar availability that favours local and regional manufacture. In addition, the weight and bulk of solar collector hardware would also discourage shipping and enhance the potential for local industry.

3.3. Software

Beyond the specially trained skills of the new solar specialists, who include researchers, designers, consultants and even inventors, are a band of professional decision makers often ignored. These include the lawyers who now have a new property right to guard--now solar access should be included in the legal descriptions of property privileges. Obviously, no prospective owner will invest in a solar application when there is the possibility that the development of a neighbour or even a growing tree could shade the sun. Even more uncertain might be the design of a delicately balanced cooling system in which the performance could be rendered ineffective by solar reflections from a neighbour.

Financial opportunity is another critical area of access to solar applications. If mortgage bankers or government tax officials make no allowance for the new investment strategies of solar energy, there will be little change from the old energy economy. Particularly in urban situations the price of money, life cycle costing and fiscal rates of return are much more likely to influence solar decisions than in rural or private situations where installations may be small in size and, typically, might be owner-financed.

3.4. Entrepreneurship

Lastly is the importance of a general business climate in which change is not only possible but might even be encouraged. Innovators or entrepreneurs need an atmosphere that not only permits but even stimulates the formation of new businesses and involvement in new areas of activity. Highly regulated economies and highly structured societies are not likely to stimulate new technical or social solutions such as solar.

Within a highly mixed economy, such as the U.S., it has been possible for new entrepreneurs to become established relatively easily. Indeed, it has been the large corporations who have had most difficulty in establishing themselves as solar leaders. Their lack of resilience has made their progress in the new solar field difficult. Typically, large businesses have succeeded in the solar field best by purchasing successful smaller businesses.

4. COMMERCIALIZATION CHARACTERISTICS

Typically the introduction of an innovative technology occurs not at the grass roots, and not in the middle of a society but at dis-separate fringes. Thus, researchers who are regarded as eggheads, inventors who are regarded as crazy and independent innovators who are also regarded as way off centre, are responsible for the first installations. Often these installations don't work; sometimes because the construction is not completed. The purpose of the installation is to satisfy intellectual curiosity or to respond to the need for adventure. Typically, the goal of efficiency or investment return has little interest compared with the objective of scientific, technical or personal accomplishment.

At this level of installation the results are always curious and must be regarded as experimental. At the next level of installation where funding is more adequate and research goals are more precise, the results may be more presentable, but the application must still be regarded as experimental.

An intermediate level of application is the demonstration. Here an installation with predictable results and, hopefully, sound engineering, is erected to display or exhibit commercial potential. A series of hybrid cooling techniques on tract house models for which I was consultant is an example of a neat, ready technology which is not yet ready for sale. Potentially, after further experience, the builder may be confident enough to consider selected sales.

The earliest commercial solar applications are ones where the prospective owner has goals of commercial or economic gain. Typically, he has paid someone else for part if not all of the installation - and, often there are other strong incentives to encourage the decision to be made. The experience in the U.S. has been that the earliest commercial solar applications have typically resulted from decisions with economic justifications although personal and associated satisfactions have also been strongly present. After installation, owners normally self-justify and reinforce the basis for their solar decision. In fact, a careful examination of the economic facts of early solar adopters reveals that the economic arguments were not always very sound. Often the belief in economic justification is in belief only - and, often that rationalization covers a series of other less measurable but equally strong decision parameters that relate to the quality of the experience.

It should be noted that early installations are usually personally financed, are small in size and are initiated by decision makers who have surplus resources and thus can afford some risk. They are also people with a higher than average education, and, particularly, have technical backgrounds. Middle-aged doctors are one professional group that could characterize many solar adopters of early solar installations.

Once a solar decision maker has made a positive solar decision and has lived with his solar installation, he is, typically, highly likely to want another. He is a predictable second customer. Thus, solar decision making is a chain effect within the same person in addition to the networking and leadership/imitation modes that influence others. In general, early solar installations in the U.S. had an assumed payback period of between 8 and 15 years. Generally, those payback periods for the same installation should now be much shorter. The general goal now appears to be to keep payback periods to under 8 years. However, typical industrial and corporate investment in such mechanical advantages as solar energy are often built only if the investment is under a 4 year payback.

5. DECISION INCENTIVES

It is difficult to isolate the precise bases on which solar decisions are made. Typically, there is a matrix of incentives each of which carries varying weights in various decision processes. Preservation of present standards, capital enhancement, lack of alternatives are practical incentives. However, personal values, such as patriotic and environmental concerns as well as personal and associational satisfactions, are also strong motivations for positive solar commitments.

The report of a survey which this author conducted for the State of Arizona in the spring of 1977 reveals the bases of solar decisions within metropolitan Phoenix:

5.1. Justification of Solar Installation

"Each respondent was asked to list the major considerations which resulted in his decision to install solar equipment. Responses were subsequently reduced to five general categories:

(i) Cost Effectiveness - Respondent emphasized the reduction of costs for provision of basic mechancial services (i.e. hot air, hot water);

(ii) Personal Gratification - Respondent emphasized personal satisfaction through involvement with solar technology (i.e. challenge of skills, desire to be a pioneer, experimentation with the unknown, etc);

(iii) Financial Return - Respondent emphasized expected monetary benefits through association with solar technology, but not as a result of savings from operation of their units (i.e. future market potential);

(iv) Social-Environmental Impact - Respondent emphasized improvement of social or environmental conditions through use of solar energy.

(v) Quality of Services - Respondent emphasized utilization of solar hot air or hot water for a particular purpose or stressed the quality of these media as provided by the solar equipment (i.e. year round swimming).

"Cost effectiveness was reported to be the most common justification for a solar purchase; 46% of the decision makers indicated that it was their primary consideration. An additional 27% reported cost effectiveness as their second most important consideration. A summary of the major considerations is tabulated below:

Ranked Reasons for Solar Installation

	First Consideration	Second Consideration	Third Consideration
Cost Effectiveness	46%	27%	27%
Personal Gratification	19%	31%	-
Financial Return	15%	8%	-
Social-Environmental Impact	12%	19%	4%
Quality of Services	8%	4%	-

"

Since tax refunds and other government financial incentives were not yet in place, this cross section of early adopters represents a solar market situation without massive outside influences.

6. INFORMATION SOURCES FOR SOLAR DECISION MAKERS

"In the Arizona survey on the sources of information that influenced positive solar decision tended to be popular, not professional. This may be disheartening to professional solar designers who ranked very low. The following are the information sources ranked by order of influence:

Sources ranked by order of influence on decision to adopt solar technology

(1) Public libraries
(2) Owners of other installations
(3) Donated professional advice
(4) Relatives or friends
(5) Sales personnel
(6) Educational institutions
(7) Government agencies or representatives
(8) Paid professionals (such as architects, engineers)
(9) Special interest groups (such as Sierra Club, etc.)"

7. PATTERNS OF SOLAR UTILIZATION

Within a large country such as the U.S. there are a series of patterns that have emerged as part of the popularization and commercialization of solar applications. When compared with the socio-economic circumstances and regional characteristics of other locations, they may provide some anticipation of how early solar applications are initiated.

7.1. Personal Characteristics of Decision Makers

Within Arizona of 1977 the major findings that characterized early solar consumers were thus described:

Solar Consumers. "The stereotype of solar users as young, liberal environmentalists living on minimal budgets, appears to be inaccurate. Most solar users are middle-aged, conservative professionals with relatively high incomes. Their adoption of solar technology was prompted by a combination of personal characteristics: (a) a practical outlook in life; (b) the ability to make independent judgments; (c) monetary resources sufficient to underwrite the purchase; and (d) high achievement motivation. The only significant variation within this pattern was in association with individuals who built their own equipment. They tend to be more concerned with social and ecological issues, to have more extensive technical training and to work with slightly smaller budgets. However, they are by no means the antithesis of those who purchased manufactured systems."

Although the types of solar applications in other regions have been different, there has been no evidence that early solar adopters themselves were very much different elsewhere. It must be noted that large corporate installations are not part of early solar adoption. These tend to grow out of the small and personal and prior commitments made by their directors as individuals, usually as initiated by the few mavericks and adventurous leaders present in most successful businesses.

7.2. Regional Climate and Type of Applications

Although the U.S. is usually assumed to be a homogeneous economy and culture, the inception of solar applications has emphasized divergent local and regional characteristics. Thus, in the desert regions of Arizona solar heating of swimming pools to extend their use throughout the spring and fall and into the winter has been by far the most popular first solar application. Thereafter, domestic hot water has been the next solar application in popularity. Space heating has not been of interest because little winter heating is required. Summer space cooling is a significant problem but solar cooling is still considered experimental and, thus, there are almost no applications. In no other part of the U.S. are solar swimming pool heaters a significant part of the early commercialization of solar.

In contrast, on the high arid plateaus of northern New Mexico, with cold winters but bright skies, the popularity of passive solar space heating has initiated a whole international consideration of space conditioning. Solar greenhouses that provide both food and heat, together with other innovative passive applications, have been a particular regional contribution to national solar usage.

In New England both domestic hot water and space heating have been popular - but not because it is easy. But, because whatever contribution the sun can make will be welcome: cold, cloudy winters develop a high winter heating demand and alternative sources are available only at high cost.

7.3. Motivations of Early Solar Adopters

A recent U.S. survey just completed in the state of New Mexico of solar greenhouses by Scott Morris and Clair Reiniger reveals that, at least in this instance, a highly successful regional solar activity has not benefited significantly from massive government programmes, nor from organized voluntary solar energy groups or associations. Again, access to professionals for design information was of no importance-popular literature and friends were the primary information sources.

"How did you find out about Solar Greenhouses?"

Local solar association	25%
Popular literature	48%
Friends	47%
Lectures, courses	23%

"How was the project actually funded?"

Cash	73%
Bank (home-improvement loan)	5%
Bank (rolled into mortgage)	5%
Government grants	3%
Private grants	1%
Workshop assistance	4%

"How involved are you with solar energy groups or programmes?"

Very involved	12%
Involved	29%
Not involved	59%

"In general, how satisfied are you with your greenhouse?"

Very pleased	73%
Pleased	18%
Satisfied	7%
Not satisfied	1%

"Which of the following factors inspired you to build your greenhouse?" (Check all applicable)

Saving money on heat	81%
Saving money on food	55%
Quality of food	54%
Horticultural interests	59%
Having a nice sunspace	53%
Self-reliance	43%
Survival-type self-sufficiency	31%
Closer relationship to nature	33%

Again, the Arizona survey of 1977 confirms the same tendencies:

Testimonials. "Solar consumers are a constant source of information about solar equipment for potential customers. In the last month, sixty-five percent of the sample has been contacted by one or more persons seeking information about the solar industry. A smaller, but more potent group talked with at least twenty people. As owners of solar installations are ranked second in the listing of information sources used by potential solar consumers, this exposure will undoubtedly be significant. At this point, the existing solar consumers would appear to be satisfied with their purchases and encouraging to others. However, they are frequently not up-to-date in their information or have forgotten particulars of their own decision making process.

Information Demand. "Potential solar consumers who have talked with solar users are interested in four very practical concerns: (a) monthly savings on utility bills; (b) equipment reliability; (c) hardware costs; and (d) equipment outlets. Answers to their questions are most frequently found in public libraries or through interaction with solar users, relatives, friends, and helpful professionals in the field. Advertisements and contacts with sales personnel were generally characterized as having marginal influence on the decision-making process. It is worth noting that, as information sources, governmental agencies, special interest groups, and paid professional services were of limited value to the solar consumers."

7.4. Influences of Government

The encouragement of solar applications has many supporters and few enemies. Although in the U.S. solar appears to be very strongly supported by federal government budgets, it has miniscule funding compared with other energy budgets.

Everyone agrees that government support of basic research is necessary and desirable, especially in a field without immediate commercial return. It is generally believed that tax incentives for solar adoptions is a fair way of balancing solar against the subsidies that other energy sources have received. However, government funding of commercialization programmes has had a mixed reception in the United States. One area of criticism can be levelled at government support of demonstration projects. Why should governments pay for systems that are not economically viable? Even more troublesome has been the government support in constructing solar applications which are, at present, uneconomical, while competitive solar application systems which are already economical get no support.

7.5. Influences of Electric Utilities

Within the U.S. there are many electric utilities - of all sizes and including all dispositions. Typically, utilities have the large capital and technical resources that could make a major solar impact on an urban location. In general, the response of electric utilities to solar potentials has been guarded if not, in some cases, totally discouraging. Generally speaking, few electric utilities have regarded solar as an opportunity to expand their social or business responsibility. A more typical response has been to penalize solar houses that use electricity for a backup energy source in the belief that this would add to their peak load.

One exception has been the activity of the largest electricity producer - the Tennessee Valley Authority. This organization has sponsored massive programs of solar domestic hot water installations in which the costs are time-financed at low interest rates and included in the monthly bill. TVA has also designed and built a number of solar buildings of all sizes - including a programme of select passive solar house designs for the builder market. However, the solar leaderships of TVA has come not without internal problems or external setbacks, and it is too early to evaluate the impact of their various activities on the cities and regions they serve.

8. CONCLUSION

Solar decision making for any region is a complex subject whose interpretation requires some match between various incentives and the local conditions which affect energy applications and use. Solar technology itself is a broad and diverse field with many parameters. Some aspects of solar application, such as cooling and electrical production, are still at relatively immature stages of development and thus are not ready for commercialization.

The term 'commercialization' implies that a new technology has already gone through research, experimental and demonstration stages and is at a point of readiness for widespread application. However, technical preparedness is not the only prerequisite to extensive commercial application. There are a whole range of human factors pertinent to diffusion of a new technology. Thus, popularization is a critically important aspect of solar decision making - and the rate of popularization appears to be the best measure of positive decision-making.

Everywhere early solar applications have been associated with small scale systems and domestic contexts. Thus, solar has been assumed to be most appropriate to rural or suburban uses. However, the experience in the U.S. has indicated that these early adoptions are, in effect, the forward edge of larger, more complex, and more comprehensive applications that do indeed match urban and industrial needs. And although some early solar applications have been made simply as a substitute for existing energy sources, there has quickly emerged a solar ethic in which the particular qualities of solar have been increasingly sought.

From both a professional and commercial point of view, the current world-wide energy realignments represent a changed and uncertain time of energy wealth. However, the solar challenge may best be understood in the Chinese world for "crises". There are two characters: one for "danger", the other for "opportunity".

危 機

SOLAR RADIATION, MEASUREMENTS AND AVAILABILITY IN THE COUNTRIES OF SOUTH EAST ASIAN REGION

K. R. Rao,
M.Sc., Ph.D., F.Inst.P.(U.K.), F.I.E.(India)
Professor and Head, Department of Building Science,
National University of Singapore.

1. INTRODUCTION

Solar energy being a renewable energy resource, increased efforts are being made in developed as well as developing countries towards its utilisation. The number of seminars and conferences held in recent years on solar energy is a clear indication of the growing research and development activities in many countries, including the region of South East Asia. Several applications of solar energy, ranging from simple solar water heating to complex megawatt power generation systems, are under extensive investigation. We now know that even the fossil fuels on which the world is so dependent are nothing but stored solar energy over a long long period (millions of years). The sole objective of all our present day efforts is to find out how far we can use today's solar energy for today's needs.

At the outset it is necessary to know all about this energy source. A detailed knowledge of solar radiation (its nature and characteristics, variability over geographical regions and availability at a given time and place) is essential for its successful application. Solar radiation data is required not only for the design of solar devices but also for the design of energy efficient buildings. Therefore, it is worthwhile to review the present statue of art in radiation measurements, estimation methods, and types of radiation data required for different purposes.

2. EXTRA TERRESTRIAL SOLAR RADIATION

Extra terrestrial solar radiation and its spectral energy distribution is of considerable importance for all solar studies on the earth's surface. These form the basis for all computational methods of estimating solar radiation intensities at any place on the earth's surface. The energy emitted by the sun is in the form of electromagnetic radiation and the range of wave-lengths extends from a fraction of a millimicrometer ($< 10^{-9}$m) to hundreds of meters.

Bill B. P. Lim (ed.), Solar Energy Applications in the Tropics, 25–56.

Thekaekara(1) has provided reliable data on the solar spectral irradiance above the earth's atmosphere. The peak spectral irradiance is in the blue-green region for a wave-length of 0.48μm. The intensity of radiation falls off from its maximum on either side i.e. towards shorter (blue) and longer (red) wave-lengths. However, the energy diminishes rapidly towards the blue and more gradually towards the red and near infra-red. The region between the wave-lengths of 0.28μm and 3.0μm carries about 98% of the total energy. The best estimates indicate broadly a spectral distribution of energy as 7-8% in the ultraviolet region, 41-42% in the visible region and 51% in the infra-red region.

The integrated area under the spectral energy distribution curve provides the total solar energy received. An important finding is that the energy received on a surface normal to the sun's rays outside the earth's atmosphere does not depend on the solar altitude. The solar energy received at mean solar distance (149.6 x 10^6Km) is called "Solar Constant" and its currently accepted value is 1353 W.m^{-2}, with an estimated error of $\pm$ 1.5%. This is subject to slight annual cyclic variations from 3.42% to -3.27% due to variations in actual earth-sun distance owing to the earth's orbital eccentricity.

3. CHARACTERISTICS OF SOLAR RADIATION REACHING THE EARTH'S SURFACE

The intensity and spectral energy distributions of solar radiation undergo significant modifications while passing through the layers of the earth's atmosphere. The extent to which the solar energy is depleted will depend upon the attenuation characteristics of the atmospheric constituents and the path length the sun's rays have to traverse through the atmosphere before reaching the earth's surface. The main constituents of the atmosphere such as air molecules, water-vapour molecules and dust particles scatter the incoming radiation in all directions. Part of the radiation is absorbed, mainly by ozone and oxygen in the upper atmosphere and by water-vapour and carbon dioxide in the lower layers of the atmosphere. The scattering and absorption coefficients of each atmospheric constituent further depend on the wave-length of the solar energy spectrum. The scattering effects depend on the particle size with respect to the wave-length. In a dense layer of atmosphere the incident radiation may be scattered many times. The downward scattered radiation is of a diffuse nature and comes from the entire hemisphere dome of the sky.

Whereas atmospheric scattering is a continuous function of the wave-length, atmospheric absorption takes place in selected spectral bands. Oxygen molecules in the uppermost layers cut off the extreme ultraviolet in the spectral range of 0.12 to 0.18μm. Ozone absorbs in the spectral range of 0.20 and 0.33μm and also to a small extent in the visible region of 0.44 and 0.76μm. Owing to the oxygen and ozone absorptions, the energy in the spectral region up to 0.29μm does not, in practice, reach the earth's surface. Water vapour has a number of absorption bands in the near infra-red region, viz. 0.93, 1.13, 1.42, 1.47, 1.9, 2.7, 3.2 and 6.3μ m. Carbon dioxide also absorbs in the near infra-red region at spectral bands of 1.6, 2.0, 2.7 and 4.3μm. The total precipitable water content of the atmosphere will vary widely from place to place, with seasons and from day to day at any given place.

The path length the sun's rays have to traverse through the atmosphere is expressed in terms of air mass. Typical solar spectral energy distribution curves on clear days at sea level for air masses 1 and 5 for an assumed atmospheric condition of precipitable water 30mm; dust particles of 400/cm^3 and ozone content of 2.5 mm are shown in Fig. 1.

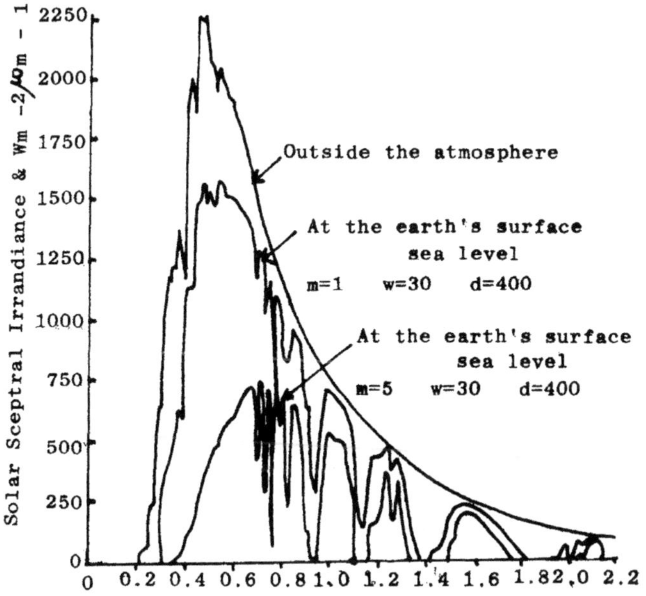

Fig. 1 - Solar spectral energy distribution curves above the earth's surface (sea level) for air masses 1 and 5 (Adapted from ASHRAE Trans., 64, 50)

The air mass is a function of the solar altitude. Thus, the attenuation of solar radiation intensity will be a function of the solar altitude which, in turn, is determined by the latitude of the place, day of the year and time of the day. It is obvious that even on clear days the solar radiation intensity at the earth's surface is expected to be highly variable. The diurnal and seasonal cycles in solar energy flux due to the rotation of the Earth and to its revolution around the Sun are well-known and predictable. From the foregoing it can be seen that the solar radiation received at the earth's surface has two components, namely, direct (beam) radiation and diffuse sky radiation. On clear days the beam radiation may be as much as 90% of the total radiation which may be of the order of 1000 Wm^{-2} at noon-time. Usually, clear sky conditions do not prevail every day and the presence of clouds further reduces the solar radiation received at the earth's surface. The ratio of the beam to diffuse radiation will also be altered considerably by the clouds. Under part cloudy conditions, diffuse radiation may vary from 300 to 500 Wm^{-2}.

The effect of clouds on solar radiation is highly complex. The cloud formation varies on a geographical, seasonal and diurnal basis and makes the solar radiation received at the earth's surface highly variable.

4. SOLAR RADIATION DATA REQUIREMENTS

Solar radiation data is required by scientists and engineers working in a variety of disciplines. Their needs on the type of information and the accuracy of the data differ. In order to design and evaluate solar collectors and other solar devices, detailed information on the availability of solar radiation and its variability is required. Insolation data is needed for selecting optimum sites, i.e. sites with high irradiance, small cloud cover and with very few days of overcast conditions, for locating large scale solar conversion systems. For many urban and rural applications like heating and cooling of buildings, domestic water heating, crop drying, water pumping and so on, actual solar radiation at the location is required. Hence data on insolation should be available for subregional to microclimate scales.

The types of solar data required will also depend on the type of solar device or system employed for energy conversion. These requirements have been summarised by Machta(2) and include duration of sunshine, normal beam, global and diffuse sky and reflected radiation on planes of different tilt angles, the spectral shortwave composition of solar radiation and atmospheric longwave radiation. The time scale requirements of these data vary from instantaneous to averages over long periods of hours, days, months and years.

For designing solar systems that use flat plate collectors, the amount of solar energy received by a tilted surface oriented towards the equator is needed. Data on normal-beam radiation is required for the designing and evaluation of the concentrating types of collectors. The relationship between direct and diffuse radiation is important for both flat-plate and concentrating types of collectors. For solar cell applications and photosynthetic processes of the growth of plants etc. data on not only the global radiation but also its broad band spectral energy distribution is of considerable importance. Spectral values of the solar radiation received at the earth's surface are required for the evaluation of spectrally selective surfaces, filters, reflectors, heat absorbing and heat reflective glasses used in building and for the development of solar simulation sources. Information on the ultraviolet radiant component of the solar spectrum reaching the earth's surface is required as it affects the durability of materials. Complete solar spectral radiation data may not be required on a continuous day to day basis for the evaluation of solar energy systems. But it is necessary to know spectral energy distribution in wave-length bands and its variation with atmospheric conditions, solar altitude and azimuth angles during each season.

Solar radiation data for vertical and sloping surfaces of different orientations is required for the calculation of cooling loads, computer evaluation of a building's energy requirements and for the design of energy efficient buildings and passive solar energy applications. Atmospheric physicists and meteorologists need solar radiation data in their studies on the physics of the atmosphere and for long-term forecasting of radiation trends.

Monthly mean values of hourly and daily global, diffuse, normal beam radiation and percentage sunshine hours over several years are commonly required for most applications. Data on sunshine hours is often made use of for the estimation of solar radiation where the measured data are not available. Statistical treatment of the available data and determination of the frequency of occurrence of specified radiation levels are also of practical interest. Further, data on other meteorological factors such as cloud amounts, atmospheric radiation, ambient temperatures and wind velocities is also required for estimating the efficiency of performance of solar energy systems at a given place.

The accuracy requirements also vary depending on the application in mind. For example, for most of the engineering needs an accuracy of 10% is considered adequate, especially in view of the variability of weather conditions. For the evaluation of collector efficiency, accuracies of the order of 5-10% are required. On the other hand, for atmospheric physics studies and for accurate forecasts of trends, accuracies better than 5% are required. It may be noted that the monthly average values of sunshine and solar radiation can differ from one year to another and this uncertainty is much greater than the errors in measurements or estimates of solar radiation.

5. INSTRUMENTATION FOR SOLAR RADIATION MEASUREMENTS

The direct approach to obtain the required solar radiation data is, of course, by direct measurements at the site itself. For many decades measurements of duration of sunshine, global radiation on horizontal surfaces have been regularly made and recorded in many countries by National Meteorological Services or Weather Bureaus. This data is collected mainly for the purpose of weather prediction, agricultural and hydrology applications. Long-term measurements of diffuse radiation on horizontal surfaces and beam radiation on a surface normal to the sun's rays are made at relatively few places. Spectral energy distribution measurements are much rarer. Great progress has been made in the last two decades, especially since 1956 which was declared The International Geophysical Year (IGY), during which intensified efforts were made by all participating national meteorological services of the world and co-ordinated by the World Meteorological Organisation (WMO). In the past, solar energy conversion has not been a major application area for insolation data and it is not surprising that the data collected for other purposes is not that those most needed in solar technology. In recent years several groups, organizations and individuals have embarked upon regular solar radiation measurements. The network of stations maintained by national meteorological services in many countries are also being expanded. In view of these developments there is a greater need now to employ standard instruments and procedures for a meaningful comparison of data collected. The types of instruments available and their suitability for specific measurements of solar radiation are discussed below. WMO Guide to Meteorological Instrument and Observing Practice(3) listed and classified the radiation measuring instruments according to accuracy.

5.1. Sunshine Duration

The duration of bright sunshine is an important parameter in solar measurements and easy to measure. Cambell-Stokes sunshine recorder which is shown in Fig. 2 is extensively used all over the world. The measurement is simple and less expensive. This works on the principle that a clear solid glass globe concentrates the sun's rays to a focus point on a special card with time markings. The card is burnt whenever bright sunshine conditions exist. At any moment if the sun's disc is covered by a cloud cutting off the direct sun rays, the card will be left unburnt for that duration. The portion of the burnt area will provide the duration of sunshine at each hour and for the day. It is observed that even for clear sky conditions, this type of sunshine recorder fails to respond around the dawn and dusk times as the intensities of the sun's rays are too weak until the sun is 5° above the horizon. Errors of more than 10% can occur in routine measurements. The errors can be reduced by the use of standard cards and standard procedures for the evaluation of the data.

Fig. 2 - Cambell-Stokes Sunshine Recorder

In recent years a variety of electronic types of instrument, with photo sensors arranged in such a way that the net output sensed is only of direct rays, have been developed for recording duration of sunshine.

5.2. Global Radiation

The instruments that measure global solar radiation on a horizontal surface are called pyranometers. These have a view angle of 2 steradians. Three types of pyranometers are in common use.

(a) The first type is the instrument that operates on the principle of thermopile whose electrical output voltage (in millivolt range), generated by the temperature difference between the black and white surfaces or exposed and unexposed surfaces indicates the intensity of solar radiation. The output can be recorded on a strip chart or digital printout of integrated values over a preselected duration (quarter, half or one hourly) can be obtained. The most likely used makes of this type are, Eppley (U.S.A.), Kipp and Zonen (Netherlands) and EKO (Japan). The response time is of the order of several seconds. This is a fairly precise instrument and is employed by most of the first class meteorological observatories. A typical instrument of this type is shown in Fig. 3.

Fig. 3 - Eppley Black and White Pyranometer

(b) The second type of pyranometer uses silicon solar cells as sensors. These are less expensive compared to the thermopile types. The response time is fast, less than a millisecond.

However, it has a limited spectral sensitivity and is not recommended where precise measurements are required.

(c) The third type uses bimetallic strips as sensors. The receiving surface consists of a bimetallic strip painted matt black mounted between two similar strips of bimetal shielded from solar radiation. The bimetallic strips are connected to a liver mechanism with a pen arm and the movements of the pen which are proportional to the temperature differentials between the exposed and shielded bimetallic strips are recorded on a rotating drum fixed with a chart. The magnitude of the graph traced on the chart is thus proportional to the solar radiation intensity. The response time of the instrument is several minutes. The accuracy of measurements is also limited (of the order of 10%). Fig. 4 shows a typical recorder of this kind.

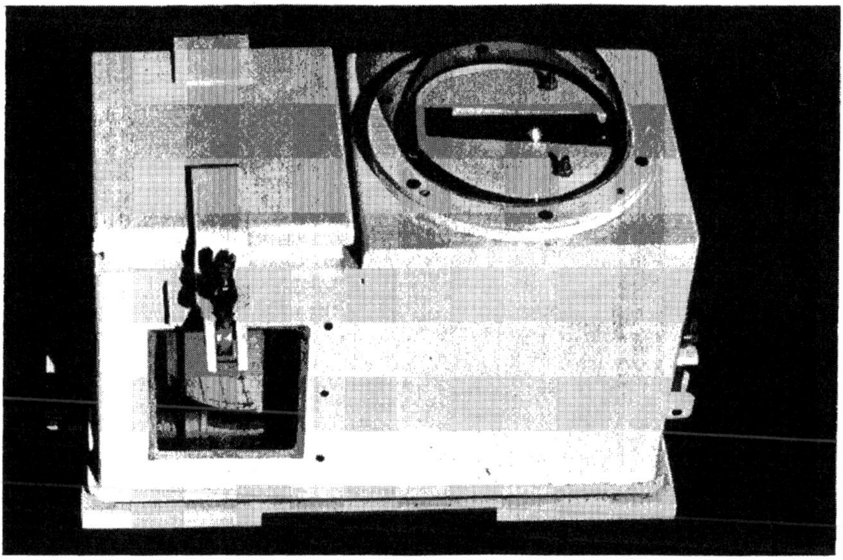

Fig. 4 - Cassella Bimetallic Actinograph

5.3. Diffuse Sky Radiation

For the measurement of diffuse sky radiation, pyranometers are used with a shading ring or disc to shade the direct sun. The amount of diffuse radiation received is usually small compared to global radiation and hence larger errors in diffuse radiation measurements may be expected. In view of this, it is preferable to use the thermopile type of pyranometer for better accuracy. Electronic chart recorders or integrated digital printout systems can be used for obtaining continuous records. It may be noted that the shading ring screens out a certain amount of diffuse sky radiation in addition to the direct radiation. Thus, a correction factor has to be applied to the measured value obtained by this method, as recommended by Drummond(4). Fig. 5 shows a typical set up for the measurement of diffuse sky radiation.

Fig. 5 - Kipp & Zonen Solarimeter with Shading ring to measure diffuse sky radiation

5.4. Normal Beam Radiation

To measure the direct beam radiation on a surface normal to the sun's rays, a pyrheliometer is used. The sensor is a thermopile at the end of a cylindrical tube which limits the view angle to about 5.43°. The instrument is mounted on a motor driven heliostat. The axis of the heliostat is to be adjusted frequently, say 2 or 3 times a week, for the changes in the sun's declination. A commonly used Normal incidence pyrheliometer (NIP) is shown in Fig. 6. Since the sun's disk represents an angle of about 0.53°, the pyrheliometer measurements include the circum solar radiation. Sometimes, it is necessary to make measurements of circum solar radiation from the centre of the sun to 2° to 5° angles at 0.1 degree intervals. Scanning spot pyrheliometers provide data on intensity versus angle. The pyrheliometer measurements provide data on atmospheric turbidity and a clearness index of the atmosphere.

Fig. 6 - Eppley Normal incident pyrheliometer with motor-driven Heliostat

5.5. Spectral Energy Distribution

The instrumentation for the measurement of solar spectral energy distribution as a function of wave-length is more complex and varied than total irradiance measurements. Monochromators with a wave-length range of 0.29 to 3.0μm with integrating sphere attachments are employed where high wave-length resolution is required. The instruments should be calibrated for wave-length and spectral sensitivity at frequent intervals. For most purposes broad band spectral energy distribution data is sufficient. Interference filters which transmit the energy over a known spectral band or glass cut off filters which transmit all wavelengths beyond a known limit are used in combination with a total irradiance detector like a pyranometer. The wave-length resolution is 10-100 times less than for a monochromator. The Eppley spectral pyranometer with 6 filter hemispheres as outer domes is often used for these broad band spectral energy mesurements. The commonly used cut off filters are:

Filter	Wave-length range
WG 7	0.290 - 2.8 μm
CG 400	0.400 - 2.8 μm
CG 495	0.495 - 2.8 μm
CG 610	0.610 - 2.8 μm
RG 715	0.715 - 2.8 μm
RG 805	0.805 - 2.8 μm

By subtraction of the energy in the above broad bands the spectral irradiance in 0.100 wm bands or in any of the other bands such as ultraviolet, visible and near infra-red regions of the spectrum can be obtained.

5.6. Atmospheric Radiation

In the past there was no direct method of measuring atmospheric (longwave) radiation. Thermopile sensors with polyethylene cover which is transparent to long wave radiation would provide total short and long wave (solar and atmospheric) radiation. By substracting the short wave solar radiation as measured by a global pyranometer from this total radiation and by applying a correction for the longwave radiation emitted by the instrument, the required atmospheric radiation can be obtained. Recently, Eppley Laboratories have introduced a precision infra-red radiometer (pyrgeometer) for the direct measurement of incoming long wave atmospheric radiation. The elimination of short wave solar radiation is achieved by using a specially coated KRS-5 hemisphere dome, which limits the transmission to about 4-50μm. A temperature sensing thermistor circuit is incorporated to compensate for the instrument emitted radiation.

5.7. Possible Errors in Solar Radiation Measurements

There are several factors that would contribute to the errors in routine solar radiation measurements. Firstly, all the types of instruments used for radiation measurements are not of the same accuracy. Depending on the type, the instrument accuracies vary from 3 to 10% or more. Knowing the instrument accuracy limits alone is not sufficient to evaluate the data quality of the measurements. Anson(5) made the following observations based on his visits to several solar radiation measuring stations in the state of California.

(a) With time, sensors tend to degrade by the change of colour from the original flat black;

(b) In some cases the parson's black paint tend to flake off over long periods of exposure;

(c) The neglect of clearing the glass dome at regular intervals would affect the transmission.

(d) The recommended yearly calibration of the instrument with reference standards is not often followed.

(e) The instruments are not properly levelled at the time of installation and are not completely free from obstructions.

(f) Keeping the dessicant dry is essential to prevent moisture from entering inside the dome. Condensation of moisture inside the dome of the instrument may reduce measured insolation values and may also degrade the sensor coatings.

(g) In general, poor instrument maintenance and the lack of interaction between measuring stations is responsible for wide variation in the type, availability and accuracy of solar data.

These observations appear to be universally true for all regions. Hence an efficient system of verification and quality control needs to be evolved for the data collected by various individuals and organisations, especially in view of the expected growth of radiation measuring stations in the region, for any meaningful comparisons to be made.

5.8 Standards for Radiation Measurements

For any measurements, standardization of instruments is an important prerequisite. Primary and secondary reference standards are to be established. The instruments used in field measurements are required to be checked against standards at periodic intervals. The 'primary standards' are assumed to maintain an accurate scale of radiometry. In the past there were two radiation scales, the Angstrom Scale of 1905 (AS 05) and the Smithsonian Scale of 1913 (SS 13) based on the Angstrom Compensation Pyrheliometer and the Abbot silverdisc Pyrheliometer as reference standards, respectively. In 1956, for the International Geophysical Year (IGY) a new scale called International Pyrheliometric scale (IPS 56) which was a compromise between AS 05 and SS 13 scales which differed by as much as 5%, was approved by the World Meteorological Organization (WMO). In recent comparisons it was observed that three primary standards that are supposed to maintain the IPD 56 scale differ among themselves by 22 or 3%. The need for more accurate radiation measurements has led to the demand for an absolute radiometric reference standard of total irradiance in SI units. Recently developed self calibrating absolute Cavity radiometers have been proved to give the highest acuracy better than $\pm$ 0.5 %. The EMO Commission on Climatology and Instruments and Methods of Observation (CIMO) has recommended that a new World Radiometric Reference (WRR) based on absolute Cavity radiometers be brought into international use from 1 July 1980.

The World Meteorological Organization has designated World, Regional and National Radiation Centres and a hierarchy of standards for the calibration of instruments, to ensure regional and world-wide comparability of radiation data. According to WMO recommendations, all the field pyrheliometers and pyranometers are to be calibrated once a year against the national standards. The national and regional standards in turn are to be calibrated against the regional and world standards, respectively, once in five years.

6. ESTIMATION OF SOLAR RADIATION

For various reasons the solar radiation measurements are very limited compared to the observations of other meteorological elements. The increasing demand for data on the availability of solar radiation for solar energy utilization and solar architectural needs has highlighted the inadequacy of the current observational networks. This inadequacy occurs in three basic ways:

(a) limited spacial coverage

(b) limited length of record, and

(c) the available data is mostly restricted to global radiation on a horizontal surface, while inclined and vertical surface data is more appropriate for solar energy work. This has led to the development of several computational methods, for the estimation of solar radiation on any surface, to supplement the observational data base.

6.1. Estimation of Direct (Beam) Radiation on Clear days

Theoretical procedures are well established for the Computation of normal beam radiation (I_N) at any place on the surface of the earth, as a function of solar altitude (Θ). These computational methods are based on the precisely measured data of (i) extraterrestrial solar radiation and its spectral distribution and (ii) the attenuation characteristics of the atmospheric constituents as a function of wave-length. Trigonometric equations relating solar altitude and other parameters such as latitude, solar declination and hour angle are well known. Declination varies with the day of the year and the hour angle with the time of the day. Rao and Seshadri(6) have produced a set of solar insolation curves of normal beam radiation as a function of solar altitude following the theoretical calculation procedures, for easy application. Spencer(7) has fitted this data in an equation form of a higher degree polynomial of Sin θ, as

$$I_N = a_1 \text{ Sin } \theta + a_2 \text{ Sin2 } \theta + a_3 \text{ Sin}^3 \theta + a_4 \text{ Sin}^4 \theta + a_5 \text{ Sin}^5 \theta \quad - \quad (1)$$

Once the normal beam radiation (I_N) is obtained, the direct (beam) radiation on any tilted surface (I_s) can easily be calculated by the formula

$$I_s = I_N \text{ Cos } i_s \quad - \quad (2)$$

where i_s is the angle of incidence of the sun's rays with respect to the surface and is given by

$$\text{Cos } i_s = \text{Sin } \theta \text{ Cos } \phi + \text{Cos } \theta \text{ Cos } (\beta - \alpha) \text{ Sin } \phi \quad - \quad (3)$$

where θ is solar altitude
α is solar azimuth
β is azimuth of the surface (orientation)
ϕ is the tilt angle with respect to horizontal

The accuracy of the estimated direct solar radiation depends upon how precisely the date on precipitable water and dust content of the atmosphere are known.

6.2. Estimation of Global Radiation from Bright Sunshine hour Data

For the estimation of long term averages of global radiation on a horizontal surface from the data of bright sunshine hours, empirical formulae were developed by many investigators. The basic procedure is to develop correlations between the ratio of the monthly mean daily global radiation ($\bar{H}$) to extra terrestrial radiation ($\bar{H}_o$) on a horizontal surface and the ratio of the monthly mean bright sunshine hours ($\bar{n}$) to maximum possible sunshine hours ($\bar{N}$). Such correlations are applied to other locations where the bright sunshine hours data is only available to estimate the global radiation. The correlations are obtained by regression analysis. The most commonly obtained form of linear empirical equation is of the type

$$(\bar{H}/\bar{H}_o) = \bar{H}/\bar{H}_o = a + b\ (\bar{n}/\bar{N}) \qquad - (4)$$

where $\bar{H}$ is the monthly mean daily global radiation on a horizontal surface

$\bar{H}$ is the monthly mean daily extraterrestrial radiation on a horizontal surface

$\bar{n}$ is the monthly mean daily bright sunshine hours

$\bar{N}$ is the monthly mean daily maximum possible sunshine hours

$$\bar{H}_o = \frac{24}{\pi} I_{sc} \left(1 + 0.033 \cos \frac{360n}{365}\right) \left[\cos\lambda \cos\delta \sin W_s + W_s \left(\frac{\pi}{180}\right) \sin\lambda \sin\delta\right] \qquad - (5)$$

$$\text{and } \cos W_s = (-\tan\lambda \tan\delta) \qquad - (6)$$

λ is the latitude of the place
δ is the solar declination
I_{sc} is the solar constant ($1353 Wm^{-2}$)
W_s is the sunset hour angle
n is the day of the year that represent the mean daily radiation of the month

a, b are correlation coefficients.

Iqbal(8) has developed empirical equations on similar lines correlating monthly mean daily diffuse radiation on a horizontal surface (H_d) and the bright sunshine hours (n). This equation is of the form

$$\bar{H}_d/\bar{H}_o = c + d\ (\bar{n}/\bar{N}) + e\ (\bar{N}/\bar{N})^2 \qquad - (7)$$

where c, d and e are correlation co-efficients. Co-efficient is negative in sign.

Once the diffuse radiation is determined the direct (beam) radiation component ($\bar{H}_D$) can be obtained by subtracting $\bar{H}_d$ from $\bar{H}$

i.e. $\bar{H}_D = (\bar{H} - \bar{H}_d)$ - (8)

The accuracy of these estimation methods is limited. However, as greater spatial coverage and longer periods of recording bright sunshine observations are available in comparison to solar radiation measurements, the solar radiation data base can be expanded considerably.

6.3. Estimation of Diffuse Radiation from Global and Extraterrestrial Radiation on a horizontal surface

For many applications it is necessary to separate diffuse and direct components from the measured global radiation. As the measured data on diffuse radiation is very much limited to a few places compared to global radiation, computational methods have been developed for its estimation. The estimation methods are based on empirical correlations as in the previous case. The quantities which are generally correlated are:

(a) Correlation between monthly mean, daily global radiation ($\bar{H}$), diffuse radiation ($\bar{H}_d$) and extraterrestrial radiation ($\bar{H}_o$) on a horizontal surface.

(b) Correlation between monthly mean hourly global radiation ($\bar{I}$), diffuse radiation ($\bar{I}_d$) and extraterrestrial radiation (I_o) on a horizontal surface.

The monthly mean daily and hourly extraterrestrial radiation on a horizontal surface can easily be calculated from the data on solar constant, and latitude of the place. Liu and Jordan(9) employing statistical methods have developed a correlation between the ratio ($\bar{H}_d/\bar{H}$) and the clearness index $\bar{K}_T$. The clearness index is defined as the ratio of ($\bar{H}/\bar{H}_o$). Klein(10) expressed their results as a non-linear equation as below:

$$(\bar{H}_d/\bar{H}) = 1.390 - 4.027\,\bar{K}_T + 5.531\,\bar{K}_T^2 - 3.108\,\bar{K}_T^3 \qquad - (9)$$

Page(11) employing a regression analysis approach, presented a linear relationship between ($\bar{H}_d/\bar{H}$) and $\bar{K}_T$ for U.K and other latitudes as

$$(\bar{H}_d/\bar{H}) = 1.00 - 1.13\,\bar{K}_T \qquad - (10)$$

The above two relationships are extensively used by designers of solar devices all over the world. Long term averages of monthly mean hourly values of diffuse and direct radiation are equally important for many design problems. The interrelationship between hourly and daily values of global and diffuse radiation has also been investigated by Liu and Jordan. The ratios of hourly and daily global 'r_T' and hourly and daily diffuse 'r_d' were presented graphically.

$$r_T = (\bar{I}/\bar{H}) \quad - (11)$$

$$\text{and } r_d = (\bar{I}_d/\bar{H}_d) \quad - (12)$$

Iqbal(12) reviewed various correlations and has recommended the use of Liu and Jordan's assumption of

$$(\bar{I}_d/\bar{H}_d) = (\bar{I}_o/\bar{H}_o) \quad - (13)$$

for the estimation monthly mean hourly diffuse (I_d) radiation. The effects of solar altitude and type of cloud on these correlations are to be further investigated for precise estimates of hourly and daily diffuse radiation on a horizontal surface at any place. For clear days, based on the analysis of measured hourly diffuse radiation (I_d) data at Melbourne, Spencer(7) expressed it as a fifth order power series of Sin as

$$I_d = C_1 \sin\theta + C_2 \sin^2\theta + C_3 \sin^3\theta + C_4 \sin^4\theta + C_5 \sin^5\theta \quad - (14)$$

where C_1, C_2 ....C_5 are regression coefficients which are to be determined for the locality

Liu and Jordan obtained an empirical relation of the form

$$(I_d/I_o) = (0.2710 - 0.2939\,(I_D/I_o) \quad - (15)$$

for clear sky conditions.

It may be pointed out that the simple empirical models presented above suffer from the drawback that the empirical coefficients show substantial spacial and temporal variations. This poses the problem of applying correlations of one place to another place. Perhaps similarities in meteorological factors and altitudes above the sea level of the places of interest can be taken as a guideline.

6.4 Estimation of Global Radiation on a Tilted or Vertical Surface

A number of mathematical models have been developed to compute hourly and daily global radiation on a tilted surface from the insolation data on a horizontal surface. At any instant the global radiation on a tilted surface consists of three components, namely, direct beam, diffuse sky and reflected round radiation. Hence

$$I(\phi) = I_D(\phi) + I_d(\phi) + I_g(\phi) \quad - (16)$$

The direct Component $I_D(\phi)$ can be calculated by simple geometric relationships and is given as

$$I_D(\phi) = (I - id)r_b \quad - (17)$$

$$\text{where } r_b = (\text{Cos}_{is}/\text{Sin}\,\theta) \quad - (18)$$

ϕ is the tilt angle with respect to horizontal
i_s is the angle of incidence of the sun's rays on the tilted surface
θ is the solar altitude angle

Because of the variety and complexity of sky and ground reflected radiation, it is commonly assumed that they are of isotropic distribution and expressed as

$$I_d(\phi) = I_d\,(1 + \text{Cos}\,\phi)/2 \quad - (19)$$

$$I_g(\phi) = \rho I\,(1 - \text{Cos}\,\phi)/2 \quad - (20)$$

where the ground reflectance is ρ.

Thus the final expression for global radiation on a tilted surface would then be:

$$I(\phi) = (I - I_d)\frac{\text{Cos}_{is}}{\text{Sin}\,\theta} + I_d\frac{(1 + \text{Cos}\quad)}{2} + I\frac{(1 - \text{Cos}\quad)}{2} \quad - (21)$$

The non-isotropic character of diffuse fields includes higher intensities in the region near the sun and the horizon, and minimum intensities opposite the sun. The isotropic assumption results in either under- or over-estimation of diffuse radiation depending upon the tilt angle of the surface concerned, position of the sun, cloud amount and distribution.

Temps and Coulson(13) suggested that for clear days two correction factors be applied to account for the anisotropy in the diffuse radiation field and one for reflected ground radiation. The corrected expression for global radiation on a tilted surface for clear days is given as

$$I(\phi) = (I - I_d)r_b + I_d \left(\frac{1 + \cos\phi}{2}\right) \left[1 + \sin^3\left(\frac{\phi}{2}\right)\right]$$
$$\left[1 + \cos^2 i s \ \sin^3 (90 - \theta)\right] + \rho I \left(\frac{1 - \cos\phi}{2}\right)$$
$$\left[1 + \sin^2 \left(\frac{Z}{2}\right)\right] [\cos \alpha] \qquad - (22)$$

where Z is the zenith angle of the sun
and α is the solar azimuth angle

It was observed that the isotropic assumption of diffuse sky radiation worked well for overcast days and the Temps and Coulson model worked well for clear days. For all sky conditions including patially cloudy conditions a further correction factor is to be applied. The following expression is found to be satisfactory for this general case.

$$I_{(\phi)} = (I - I_d)r_b + I_d \left(\frac{1 + \cos\phi}{2}\right) \left[1 + F \sin^3\left(\frac{\phi}{2}\right)\right]$$

$$\left[1 + F \cos^2 \sin^3 (90-\theta)\right] + \rho I \left[\left(\frac{1 - \cos\phi}{2}\right)\right] \left[1 + \sin^2 \left(\frac{Z}{2}\right) [\cos\alpha]\right]$$

where $F = 1 - (I_d/I)^2$ - (23)

The global radiation on the tilted surface, $H(\phi)$ is obtained by the summation of the hourly values, i.e.

$$H(\phi) = \sum_{day} I(\phi) \qquad - (24)$$

The monthly mean values of hourly and daily global radiation on a tilted surface are obtained by the following expressions, respectively using isotropic assumption.

$$\bar{I}(\phi) = (\bar{I} - \bar{I}_d)\ \bar{r}_b + \bar{I}_d \left(\frac{1 + \cos\phi}{2}\right) + \left(\frac{1 - \cos\phi}{2}\right) \qquad - (25)$$

where $\bar{r}_b$ is the ratio of the monthly mean hourly beam radiation on a tilted surface to that on a horizontal plane which is given by

$$\bar{r}_b = \frac{\cos(\lambda-\phi)\cos\bar{\delta}\ \cos\bar{w} + \sin(\lambda-\phi)\sin\bar{\delta}}{\cos\lambda\ \cos\bar{\delta}\ \ \cos\bar{w} + \sin\lambda\ \sin\bar{\delta}} \qquad - (26)$$

$\bar{r}_b$ can be evaluated at the mid-point of the monthly average of hour angle ($\bar{w}$) and using the monthly average declination value ($\bar{\delta}$). Monthly mean daily values ($\bar{H}(\phi)$) can be calculated by the equation:

$$\bar{H}(\phi) = (\bar{H} - \bar{H}_d)\,\bar{R}_b + H_d\left(\frac{1 - \cos\phi}{2}\right) + \bar{H}\left(\frac{1 - \cos\phi}{2}\right) \qquad - (27)$$

Where $\bar{R}_b$ is the ratio of monthly mean daily beam radiation on a tilted surface to that on a horzontal surface.

An approximate value of $\bar{R}_b$ can be obtained by the ratio of extraterrestrial radiation on a tilted surface and a horizontal surface. i.e.

$$\bar{R}_b \simeq \frac{\bar{H}_o(\phi)}{\bar{H}_o} \qquad - (28)$$

It may be noted that vertical surface is a special case of tilted surface for the tilt angle (ϕ) equals 90°.

The daily method implicitly assumes symmetries of diffuse and total radiation around solar noon, whereas the hourly method takes account of the asymmetries of actual data. However in the case of hourly method the computational effort is considerable. Comparison of the hourly and daily methods of computing insolation on inclined surfaces have been made by Iqbal(14) for Canadian Cities and found to be in close agreement.

6.5. Estimation of Atmospheric Long wave Radiation

Data on atmospheric long wave radiation is required for the study of nocturnal radiant cooling possibilities of buildings, and of heat balance analysis of solar collectors. The measured data is pratically non-existent in this region. However estimational methods are being developed by several investigators to calculate its intensity from the meteorological data to a reasonable accuracy. Burnt(15) had developed the following empirical equation for the estimation of atmospheric radiation for clear skies as far back as 1932.

$$I_{AT} = \sigma\, T_A^4 \quad (a + b\sqrt{Pv}) \qquad - (29)$$

where I_{AT} is the atmospheric longwave radiation.

σ is Stefan-Boltzman Constant (5.67×10^{-8} Wm^{-2} K^{-4})
T_A is the absolute temperature of the ambient air
a and b are correlation coefficients
Pv is the water vapour pressure

The effective emissivity of the clear sky '$\mathcal{E}_{cs}$' is then equal to the term $(a + b\sqrt{Pv})$ in the above equation. The constants 'a' and 'b' obtained by Burnt are a = 0.55 and b = 0.33 for a horizontal surface. These coefficients vary with atmospheric conditions and hence with time and place. The problem of low temperature radiation incident on building surfaces was examined by Holden(16). For an overcast sky it is reasonable to assume that the effective emmissivity '$\mathcal{E}_{os}$' as equal to unity. For partially cloudy conditions an intrapolation between clear and overcast values may be used.

$$\text{Thus } \mathcal{E}_{av} = [\, n.\mathcal{E}_{os} + (1 - n)\, \mathcal{E}_{cs}] \qquad - (30)$$

where 'n' is the fractional cloud amount, n = o for clear sky and n = 1 for overcast sky.

This does not take account of the type of cloud layer or layers and their heights, and it requires corrections to be applied.

In recent years considerable interest has been taken in this problem by several investigators(17, 18). A comprehensive treatment of the subject was made by Exell(19) and also by Cole(20). It was observed that the variability and selectivity of the emissivity of clouds and underlying surfaces is considerable. Idso and Jackson(21) have developed correlations for clear skies which appear to be valid at all latitudes and seasons. It has the form

$$R_o = \sigma\, T^4 \left\{ 1 - C \exp\, [\, - d\, (273 - T)^2] \right\} \qquad - (31)$$

where R_o is the atmospheric downward radiation on a horizontal surface for clear skies.

$$c = 0.261 \text{ and } d = 7.77 \times 10^{-4}\ k^{-2}$$

To estimate the atmospheric radiation under more general conditions two corrections are to be made to Idso and Jackson's equation. The first is a diurnal correction caused by variations in the air temperature lapse rate near the surface. The second is as a result of the effect of thermal radiation from clouds. The corrected equation is given as

$$R_A = R_o + (\sigma T^4 - R_o)\, F \qquad - (32)$$

where R_A is the atmospheric longwave radiation under partially cloudy conditions.
F is the combined correction factor for all clouds and is given as

$$F = \mathcal{E}_c\, \dot{\gamma}\, n^{1.4} \qquad - (33)$$

$$\gamma = (1 - a_1 Z + a_2 Z^2 - a_3 Z^3) \qquad - (34)$$

Where ε_c is the emissivity of the cloud
γ is a factor depending on the height of the cloud base
n is the fraction of the sky covered by the cloud
z is the height of the cloud base in Km.

If there are two cloud layers simultaneously present the corresponding factors F_1 and F_2 are combined to get the combined correction factor as

$$F = (F_1 + F_2 - F_1 F_2) \qquad - (35)$$

It was observed that the typical annual mean values of atmospheric radiation lie around 400 Wm^{-2} with diurnal and seasonal changes. The emissivities of clouds depend on the type of cloud and their suggested mean values are as follows:

Type of Cloud		ε_c
(a) Thin high Cloud	-	0.4
(b) Thick high cloud	-	0.6
(c) Translucent medium cloud	-	0.7
(d) Opaque medium cloud	-	0.9
(e) Low Cloud	-	1.0

The longwave radiation incident upon a surface inclined at an angle (ϕ) to the horizontal is a combination of atmospheric and ground radiation, i.e.

$$R(\phi) = R_A(\phi) + R_G(\phi) \qquad - (36)$$

The atmospheric radiation incident upon a tilted surface is given by

$$R_A(\phi) = R_A \cdot K_1 + (K_2 - 0.5\, K_1)\, b^1 . \sigma T^4 \qquad - (37)$$

The values of K_1 and K_2 are a function of the slope angle ϕ and are given in the table below(22):

ϕ	K_1	K_2
0	1.0000	0.5000
10	0.9924	0.5184
20	0.9698	0.5523
30	0.9330	0.5890
40	0.8830	0.6214
50	0.8214	0.6447
60	0.7500	0.6554

ϕ	k_1	k_2
70	0.6710	0.6514
80	0.5868	0.6315
90	0.5000	0.5957

The ground Component R_G (ϕ) is given by

$$R_G(\phi) = \mathcal{E}_g \, \sigma T_g^4 \quad \sin^2(\phi/2) \qquad - \ (38)$$

Where ground temperature is not available, it is often approximated as $\mathcal{E}_g T_g^4 = T^4$ with an expected error of 10-15% and the modified equation becomes - (39)

$$R_G \ (\phi) \simeq T^4 \ \sin^2 \ (\phi/2) \qquad - \ (40)$$

7. SOLAR ENERGY AVAILABILITY IN THE REGION OF SOUTH EAST ASIA

In almost all the countries of the region, Meteorological Services or Weather Bureaus are operating a number of weather stations. Most of them make routine measurements of various weather elements which include cloud amount and sunshine hours. Continuous solar radiation measurements are made only in a few places and these too are limited to global radiation. In the last few years, meteorological services in most of the countries have been in the process of expanding their network of solar radiation measurements. In addition, a number of universities and research institutes are engaged in research and development of solar energy projects for various applications. They have also been recording solar data over varied periods. In some places the data has been analysed and is available in published form. However, so far no concerted efforts have been made to collate and present the data on a regional basis. The author made an effort in this direction and issued a questionnaire (see appendix) to a number of organizations and individuals in the countries of this region. Only some have responded to it and provided the published and/or unpublished information they possess. So the data collected does not cover all the countries. A summary of the available information is presented below.

7.1. Singapore

Global solar radiation on a horizontal surface has been measured by the Meteorological Department at the airport since 1961. Data on mean daily, highest monthly and lowest monthly values have been published in climatological summaries of Singapore(23). Ten years data was analysed by Rao and Lim(24) and the frequency distribution and mean hourly values were presented. The annual mean daily global radiation in Singapore forms about 45% of that extraterrestrial radiation on a horizontal surface. Whereas on clear days it will be about 67%, the annual mean daily value of 16.06 MJM^{-2} occurs for 66.5% of the days.

Sunshine hour data has been measured for more than 40 years. The average daily sunshine hours amount to 5.61 hours, i.e. 46.4% of the maximum possible sunshine.

7.2. Malaysia

Data on solar radiation in Peninsular Malaysia was presented in a paper by Chua and Lee(25). Measured data on global radiation is available for three major towns, viz. Kuala Lumpur, Penang and Kota Bahru over a period of four years (1975-1978). The mean monthly daily values vary from 12.2 to 21.6 MJM^{-2} D^{-1}. In the last two to three years, the Meteorological Department has increased the number of solar radiation measuring stations to 14. Records of duration of sunshine hours are available for 12 towns over a period of 15 years. The annual average daily sunshine hours vary from 5.07 to 6.85 hours. Chua and Lee have presented the estimated global solar radiation data based on the correlations worked out between the sunshine hours and the mean monthly daily radiation for another nine towns of Peninsular Malaysia. They have also made rough estimates of mean daily diffuse radiation for three towns. According to their estimates, the diffuse component seems to form 45 to 60% of the global radiation value. This appears to be on the high side and needs confirmation by future measurements.

7.3. Indonesia

The National Centre for Meteorology and Geophysics of the Ministry of Communications is the official body that makes regular solar radiation measurements in Indonesia. These measurements have been made in four places(26), viz., Djakarta, Bandung, Medan and Kupang since 1975. Solar radiation is also monitored at the Institute of Technology of Bandung (ITB), Gaja Mada University at Yogyakarta, and at the National Aeronautics and Space Institute. It is reported(27) that the average value of global radiation in Yogyakarta is 15.60 $MJM^{-2}D^{-1}$. In Djakarta it is 13.38 $MJM^{-2}D^{-1}$and at Bandung it is 16.8 $MJM^{-2}D^{-1}$. It is about 45% of the extraterrestrial radiation on a horizontal surface for a corresponding latitude.

7.4. Thailand

In Thailand it is reported that the data on global solar radiation measurements is available for two places, namely, Bangkok and Chiang Mai, from 1968 onwards. The sunshine hour data is available for 118 stations, and cloudiness data is available for 44 places spread over the various parts of the country. Exell and Saricalli(28) have estimated global and diffuse radiation from sunshine and cloudiness data, using empirical correlations, for all other places and have presented radiation maps for Thailand for eight 1½ monthly intervals. They have also made a detailed analysis of the percentage frequencies of days with global radiation divided into radiation classes of the width of 50 Cal cm^{-2} D^{-1} (2.09 MJM^{-2} D^{-1}) and the frequencies of occurence of these classes during each 1½ month period. The main observations are that

(a) the highest mean values are above 20.39 MJM^{-2} D^{-1} and are widespread in spring;

(b) the lowest values are below 15.32 MJM^{-2} D^{-1} in localities with heavy rainfall in autumn - the estimated diffuse radiation averages about 8.37 MJM^{-2} D^{-1};

(c) the mean mid-day hourly global radiation intensities range from 2.09 MJM^{-2} Hr^{-1} in autumn to 2.93 MJM^{-2} Hr^{-1} in spring;

(d) the seasonal variations are easily noticeable.

It was reported(29) that the meteorological department has a phased programme to extend the number of solar radiation measuring stations to 26 places between 1978 and 1980.

7.5. Philippines

At present, the data on solar radiation in Philippines is limited to a few localities. The data on global and diffuse radiation on a horizontal surface for Metro Manila have been available since 1960. The annual daily mean value is 16.39 MJM^{-2} D^{-1} and it varies from 13.64 MJM^{-2} D^{-1} in December to 20.28 MJM^{-2} D^{-1} in April. Monthly mean hourly data on global and diffuse radiation for Metro Manila was supplied by Mahilum(30) of the Centre for Non-conventional Energy Department.

The annual mean daily percentage of diffuse Component is 39%. It varies from 29% in May to 48% in July. The mid-day hourly global radiation varies from 1.93 MJM^{-2} Hr^{-1} in January to 2.55 MJM^{-2} Hr^{-1} in April.

It is reported that the Philippine Atmospheric Geophysical and Astronomical Services Administration (PAGASA) has a programme of establishing a network for monitoring stations, in addition to the usual meteorological data now being monitored by the National Weather Bureau.

7.6. Hong Kong

Hong Kong Royal Observatory measures regularly the global solar radiation on a horizontal surface and the data has been available in published form since 1969. Based on 10 years measured data (1969-78) Leung and Cheung(31) made a statistical study on the frequency distribution of daily total global solar radiation. The expected number of days with radiation level greater than 12.57 $MJM^{-2}\ D^{-1}$ is 230 a year - between 12.57 and 16.76 $MJM^{-2}\ D^{-1}$ on 82 days, between 16.76 and 20.95 $MJM^{-2}\ D^{-1}$ on 78.7 days and above 20.95 $MJM^{-2}\ D^{-1}$ on over 71 days. Leung(32) had made comparisons of the diffuse radiation estimated by three different methods using the Correlation Coefficients found elsewhere. The annual daily mean diffuse radiation is of the order of 6.00 $MJM^{-2}\ D^{-1}$. The ratios of diffuse to global radiation obtained by the three methods are found to differ by as much as 30%, as shown below:

		Variation of the ratio of $\bar{H}_d/\bar{H}$
Liu and Jordan's method	-	0.354 to 0.521
Page's method	-	0.411 to 0.601
Iqbal's method	-	0.425 to 0.726

Iqbal's method estimates from sunshine hour data and appears to give higher values compared to the other two methods. There is a need to make diffuse measurements locally to establish realistic correlation coefficients applicable to local atmospheric and cloud conditions.

The available monthly mean daily global radiation in different places in the countries of this region is presented in Table I for a comparison.

India has at present a network of 30 radiation stations distributed all over the country. Diffuse radiation is measured at 13 stations only. The sunshine hour data is measured at over 10 stations. The geographical, monthly, seasonal and annual features of global and diffuse radiation and sunshine hour data were presented by Rao and Ganesan(33, 34) in two monographs of the Indian Meteorological department. A detailed discussion of the data is beyond the scope of this paper. Those interested in Indian radiation data are referred to the above publication.

8. CONCLUSIONS

From the survey on the extent of availability of solar energy it becomes apparent that the documentation on measured data in the countries of the

Table I - Measured Monthly Mean Daily Global Solar Radiation For Different Cities in the Region

PLACE	Monthly mean daily Global Radiation - $MJM^{-2}D^{-1}$ JAN	FEB	MAR	APR	MAY	JUN	JUL	AUG	SEPT	OCT	NOV	DEC
A. SINGAPORE (24) 1° - 22'N	16.09	17.48	17.86	16.85	15.74	15.63	16.22	16.07	16.32	16.13	14.70	13.72
B. MALAYSIA (25)												
1. MALACCA 2° - 16'N	18.36	16.49	19.98	18.79	17.71	16.27	15.41	15.48	17.32	16.52	14.69	16.24
2. KUALA KUMPUR 3° - 07'N	17.1	17.4	18.32	17.82	16.74	16.02	16.38	16.34	16.45	16.34	14.58	15.41
3. IPOH 4° - 31'N	19.44	18.86	20.23	20.77	18.29	17.53	18.00	15.84	16.74	17.24	17.17	18.32
4. PENANG 5° - 18'N	21.10	20.74	21.10	20.91	18.43	17.89	18.32	17.24	16.56	17.21	17.68	18.04
5. KUALA TRENGGANU 5° - 20'N	18.47	14.51	19.84	21.60	19.12	16.96	17.78	16.06	16.78	16.38	10.66	13.32
6. KOTA BAHRU 6° - 10'N	16.56	18.29	20.52	21.35	18.83	17.42	17.64	17.93	18.36	16.85	12.24	12.38
C. PHILIPPINES												
1. MANILA (30) 14° - 40'N	14.56	17.34	18.29	20.28	19.15	16.44	14.07	16.72	17.67	14.84	13.66	13.64
D. HONG KONG (31) 22° - 11'N	11.48	11.76	11.95	13.87	16.13	17.06	18.84	18.16	17.02	15.33	13.87	12.51

PLACE	JAN 14 to FEB 26	FEB 27 to APR 12	APR 13 to MAY 28	MAY 29 to JUL 15	JUL 16 to AUG 31	SEPT 1 to OCT 15	OCT 16 to NOV 29	NOV 30 to JAN 13
E. THAILAND (28)								
1. BANGKOK 13° - 15'N	16.71	18.88	18.19	15.87	15.70	15.32	16.46	16.58
2. CHIANG MAI 18° - 55'N	18.04	18.76	20.39	17.54	15.70	18.04	16.96	15.95

South East Asian region needs greater attention. The estimated diffuse radiation based on sunshine hour correlations appears to give higher values. To establish reliable correlations, coefficients for the empirical equations are to be determined based on locally measured data on diffuse radiation. Research efforts should be made in each country in this direction. Reference data tables on solar radiation on vertical and tilted surfaces on clear and average days are to be produced for the use of designers of buildings and solar devices. As mentioned in the text, standardization of the equipment used, as recommended by WMO, and necessary precautions with regard to its maintenance to minimise measurement errors are quite important for reliable comparisons of the data collected. If a set of accepted procedures of analysis and presentation of the data and common units for radiation quantities are followed, this will further facilitate such comparisons. It is suggested that a solar energy information exchange centre be established in each country which can serve as the contact point for similar centres in other countries of the region.

9. ACKNOWLEDGEMENTS

The author gratefully acknowledges the information and data supplied by various organizations and individuals from Malaysia, Hong Kong and the Philippines.

REFERENCES

1. M.P. Thekaekara - "Proposed standard values of the Solar Constant and the Solar Spectrum". J. Environmental Sciences, 13(4), pp.6-9, (1970)

2. L. Machta - "Monitoring Solar Radiation for Solar Energy". UNESCO/WMO Solar Energy Symposium, Geneva, 1976

3. WMO Guide to Meteorological Instrument and Observing Practice. WHMO Secretariat, Geneva (1971)

4. A.J. Drummond, et al - "The Measurement of the Components of Solar short-wave and Terrestrial Long-wave Radiation". Solar Energy 9, pp.127-135 (1965)

5. Dean Anson - "Errors in Solar Energy Measurements". Solar World, No. 4, pp.19-21 (1977)

6. K.R. Rao and T.N. Seshadri - "Solar insolation Curves". Indian J. Met and Geophysics, 12(2), pp.267-72 (1961)

7. J.W. Spencer - "Computer estimation of Solar Radiation on clear days". Solar Energy 13, pp.437-8 (1972)

8. M. Iqbal - "Correlation of average diffuse and beam radiation with hours of bright sunshine". Solar Energy 23(2), pp.169-173 (1979)

9. B.Y. Liu and R.C. Jordan - "The interrelationship and characteristic distribution of direct, diffuse and total solar radiation". Solar Energy, Vol. 4(3), pp.1-19 (1960)

10. S.A. Klein - "Calculation of monthly average insolation on tilted surfaces". Solar Energy 19, pp.325-329 (1977).

11. J.K. Page - "The estimation of monthly mean values of daily total short wave radiation on vertical and inclined surfaces from sunshine records for latitudes 40N - 40S". Proc. UN. Conf. New Sources of Energy (1961)

12. M. Iqbal - "Prediction of Hourly Diffuse Solar Radiation from Measured hourly global radiation on a horizontal surface". Solar Energy Vol 24(5), pp.491-503 (1980)

13. R.C. Temps and K.L. Coulson - "Solar radiation incident upon slopes of different orientations". Solar Energy 19, p.179 (1979)

14. M. Iqbal - "Hourly vs daily method of computing insolation on inclined surfaces." Solar Energy 22(1), pp.81-86 (1979)

15. D. Burnt - "Notes on radiation in the atmosphere". Quart. J. Roy. Met. Soc. 58, pp.389-420 (1932)

16. T.S. Holden - "Calculation of incident low temperature radiation upon building surfaces". ASHRAE Journal 3(4), pp.51-54 (1961)

17. W.C. Swinbank - "Longwave Radiation from clear skies". Quart. J. Roy. Met. Soc. Vol 89, p. 339 (1963)

18. M.J. Unsworth and J.L. Monteith - "Longwave radiation at the ground (1) Angular distribution of incoming radiation". Quart. J. Roy. Met. Soc. 101, pp.1-13 (1975)

19. R.H.B. Exell - "Atmospheric Radiation in a tropical climate". Research Report No. 71 Asian Institute of Technology, Bangkok (1978)

20. R.J. Cole - "The longwave radiation environment around buildings". Build. Environ. 11(1), pp.3-13 (1976)

21. S.B. Idso and R.D. Jackson - "Thermal Radiation from the Atmosphere". J. Geopysics Vol 74, pp.5397-5403 (1969)

22. R.J. Cole - "The longwave radiation incident upon inclined surfaces". Solar Energy Vol 22, pp.459-462 (1979)

23. "Climatological Summaries - Singapore". Meteorological Services Singapore. (1973)

24. K.R. Rao and B.P. Lim - "Solar Energy Availability in Singapore"

25. G.S.D. Chuah & S.L. Lee - "Solar Radiation in Peninsular Malaysia". National Physics Symposium, 28-30 June. University Kebangsaan, Malaysia, Kuala Lumpur - 19.

26. Filino Harahap - "Solar Energy R & D in Indonesia". A Report presented at the meeting of The ASEAN Working Group on Solar Energy, New Delhi (1978)

27. M.S.A. Sastroamidjojo - "A multipurpose solar air dryer for Research and Development". Solar Energy Research Centre - Gajah Mada University, Yogjakarta, Indonesia

28. R.H.B. Exell and K. Saricali - "The Availability of Solar Energy in Thailand" - Research Report No. 63. Asian Institute of Technology, Bangkok (1976)

29. Sompongse Chantavorapap - "Solar Energy Programmes of Thailand". A report presented at the meeting of ASEAN Working Group on Solar Energy, New Delhi (1978)

30. "Solar Radiation Measurements at Metro Manila Station" (1960-1974). (Personal Communication - by David M. Mahilum - Center for Non-conventional Energy Department. Ministry of Energy. Metro-Manila, Philippines).

31. C.T. Leung and Y.L. Chueng - "A statistical study of Solar Radiation in Hong Kong". Hong Kong Engineer, September 1979, pp.27-30.

32. C.T. Leung - "Estimation of the monthly average diffuse solar radiation in Hong Kong". Hong Kong Engineer, April 1980, pp.27-30

33. K.N. Rao and H.R. Ganesan - "Global Solar and Diffuse Solar Sky Radiation Over India". Meteorological Monograph. Climatology/No.2/1972. Indian Meteorological Department. Poona, India.

34. K.N. Rao and H.R. Ganesan - "Sunshine Over India". Meteorological Monograph. Climatology/No. 3/1972. India Meteorological Department (IMD), Poona, India.

Appendix

Questionnaire on Solar Radiation Measurements and Data

Please provide the following information for your country.

1. SOLAR RADIATION MEASUREMENTS:

   (a) Number of solar radiation measuring stations in operation, their location and period of operation (years).

   (b) Type of measurements made at each place.
   (i) Global (ii) Diffuse (iii) Direct Normal Beam and (iv) Sunshine hours

   (c) Type of equipment used for these measurements.
   (i) Thermoelectric type pyranometers and pyreheliometers
   (ii) Bi-metallic type actinographs
   (iii) Silicon solar cell type pyranometers
   (iv) Sunshine recorders

2. SOLAR RADIATION DATA AVAILABLE FOR EACH PLACE:

   (a) Monthly mean hourly global radiation on a horizontal surface
   (b) Monthly mean hourly diffuse sky radiation on a horizontal surface
   (c) Monthly mean hourly direct normal beam radiation
   (d) Monthly mean daily totals of global radiation
   (e) Monthly mean daily totals of diffuse radiation
   (f) Mean maximum and mean minimum daily totals of global and diffuse radiation for each month
   (g) Monthly mean hourly percentages of sunshine
   (h) Monthly mean daily percentages of possible sunshine hours
   (i) Percentage frequencies of days with global radiation between specified limits
   (j) Percentage frequencies of days with diffuse radiation between specified limits

3. PUBLICATIONS:

   Meteorological Services, Research and Higher Educational Institutions engaged in solar energy work might have produced some publications on this important topic. Please provide the list of publications and reprints from your country on this topic.

SOLAR AIR CONDITIONING

Norman R. Sheridan
Ph.D., F.I.E.Aust.,
Reader in Mechanical Engineering, University of Queensland

ABSTRACT

Passive and active solar air conditioners are briefly described. Evaluation of passive systems in terms of comparative values of the Predicted Percentage Dissatisfied is advocated and the method applied to a building with a roof pool. Performance of active systems and the possible avenues for performance improvements are considered. The solar fraction is suggested as the objective function. After making recommendations for system design, the paper continues by analysing cost data for several Australian installations. Finally, system costs are estimated for the end of this decade when these systems could be economic for remote areas.

1. INTRODUCTION

One of the main arguments for the concept of air conditioning by solar energy has been the phase relationship between the supply of solar energy and the demand for cooling. In other words, since the radiant energy that drives the cooler also raises the environmental temperature, there is close agreement between the input and required output of the air conditioning system. Thus energy storage is not as important as for a space heating system, where the supply and demand are almost out of phase.

In the tropics, average daily temperatures can exceed the comfort temperature for long periods with the result that air conditioning has a great deal of appeal. If solar energy could reduce the overall cost of the air conditioning, a demand for solar air conditioning could be developed in urban areas where there are large multi-storied buildings and the homes of affluent citizens.

In rural areas and the suburbs with low cost housing, air conditioning would be of low priority whatever the power source. However, there would be a case for the use of passive solar systems which would ameliorate the climate at a small cost.

Bill B. P. Lim (ed.), Solar Energy Applications in the Tropics, 57–89.

2. SYSTEMS

In accordance with recent trends, two major approaches to solar systems will be identified:

(a) Passive
(b) Active

Passive systems are those that function with natural circulation of fluids, either the ambient air or a heat transfer liquid. Generally, a passive cooling system will seek to optimise the variables associated with the design of the building envelope so that the average ambient indoor temperature is a minimum. Design variables include orientation, wall and roof insulation, shading, glass fraction, thermal capacity, induced ventilation and floor properties.

Active systems require a pump/or pumps to circulate the heat transfer fluid and/or the room and ventilating air. A typical active solar air conditioner would consist of collectors, energy converter, energy storage, auxiliary energy source and air conditioner (Fig. 1).

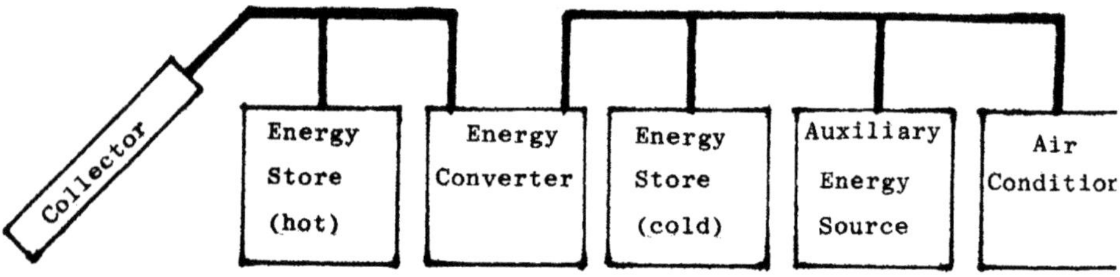

Fig. 1 - Schematic circuit showing the components of an active solar air conditioning system

2.1. Passive Systems

Generally, passive solar energy systems are involved with building heating, either for the interior space or domestic water. However, cooling can be assisted by high reflectance building surfaces, shading of outside walls and evaporative cooling.

For centuries, the cooling effect of evaporating water has been recognised and used in buildings by trickling water down screen walls exposed to the prevailing breeze, or by forming a pool of water on the roof. Recently an improvement to the roof pool was made by Hay(1), who introduced a movable insulating cover over the pool. For summer, the pool is covered in the day-time to avoid solar heating and to allow the water to absorb heat from the room below. At night, the uncovered pool loses heat by evaporation to prepare it for the next day. The pool tends to increase the thermal capacity of the building and is best constructed of metal with the floor of the pool forming the ceiling of the space. Then the resistance to heat transfer between the pool and the space is small. In winter, the pool can be uncovered in the day time to absorb solar heat, and covered at night while the energy is transferred to the space. To reduce evaporation and the consequent cooling in winter, a transparent cover over the pool will be an advantage. While roofs may be more suited to the drier climates, they are useful in the humid tropics.

These ideas are consistent with climatic design and are applicable to most buildings. A good design approach would use those passive components which are cost effective in their own right or as a means of reducing the cost of active air conditioning systems.

2.2 Active Systems

Several categories of active system should be mentioned:

(a) Thermal system
(b) Electrical system
(c) Total energy system

(a) Thermal System. Radiant energy is collected as thermal energy which is conveyed by a liquid circulating fluid to the liquid store or to the energy converter, a heat operated refrigerator. Generally, this latter will produce chilled water which can be stored or sent to a fan-coil air conditioner. Collectors need to operate at temperatures in excess of 85-90°C, values that can be achieved by flat-plate, evacuated tube, and concentrating collectors. Several heat operated refrigerators have been developed. Basically, these machines receive heat from the source and pump heat from the room, while rejecting an equivalent amount of heat to the environment by way of the condenser.

(i) Rankine refrigerator - This consists of a Rankine cycle engine driving a vapour compression refrigerator. Usually, the Rankine cycle will have an organic working fluid to achieve better efficiency at the relatively low temperature available from solar thermal energy collectors and in the small size plants needed for individual buildings. If the same fluid is used in the engine and refrigerator, the turbine and compressor can be close-coupled and the system sealed.

(ii) Absorption refrigerator - The basis of absorption refrigeration is that the gaseous refrigerant from the evaporator is dissolved in an absorbent fluid before pumping the solution to the high pressure of the cycle, where the refrigerant is released by the addition of heat. Since a liquid is pumped, the work is much less than the work involved in pumping the refrigerant as a gas. The temperature required to release the refrigerant is comparable with the operating temperature of solar thermal energy collectors.

(iii) Ejector refrigerator - A motive gas is expanded to a high velocity in a nozzle located in the evaporator of the refrigerator. Then it enters a diffuser where recompression occurs. But on the way to the diffuser, the expanding motive gas entrains the refrigerant and both gases are compressed together. Generally, the same material will need to be used for the motive fluid and the refrigerant since separation of the compressed gas into components is not practicable. Steam and organic refrigerants are suitable working fluids.

(iv) Open cycle - An open cycle cooler uses ambient air as the working fluid. In one arrangement, ambient air is chemically dried and cooled back to ambient temperature. Then water is evaporated into the air to reduce its temperature and to allow it to be used for supplying air. Room air leaves through a heat recovery wheel. In a solar collector, ambient air is heated to allow it to be used for regenerating the chemical drier.

Conventionally, thermal energy is stored as sensible heat in water or rocks. Storage may occur either at the collector temperature (hot-side storage) or at the chiller temperature (cold-side storage). Alternatively, refrigerant storage, involving latent heat of evaporation and providing a high storage capacity per unit volume, may be practicable.

Auxiliary energy may be provided to supplement the solar collectors at times of insufficient sunshine. It may be gas, which is supplied to a boiler and produces hot water for the absorption chiller, or electricity supplied to a standby vapour compression chiller.

(b) Electrical System. If solar electricity became cheap enough, it may be feasible to use it to drive vapour compression refrigerators as in conventional air conditioning. Such a system could have a photovoltaic generator with battery storage. The refrigerator would operate only when required to meet the load.

(c) Total energy system. Solar cells range in efficiency from 10-15% - i.e they convert 10-15 % of the incoming energy into electricity while the remaining 85-90% is dissipated as heat. It would be possible to divert a large proportion of this reject energy for useful purposes as low grade heat.

Indeed, the demand for low grade heat in domestic residences could exceed the electricity demand, especially if some of the energy were used for heat operated air conditioners.

The production of both electrical energy and low grade heat from the same solar aperture will increase the overall efficiency and the collection or, from another viewpoint, would increase the specific output of useful energy. The result should be an effective cost benefit.

Flat-plate or concentrating photovoltaic arrays can be designed to operate at temperatures up to 100°C and provided with water cooling for the extraction of the low grade energy. The photovoltaic cell efficiency is reduced at the elevated temperatures but the loss is more than compensated for by the gain in thermal energy.

3. PERFORMANCE

The objective of an air conditioning system is to make people comfortable in their thermal environment. Conditions that are found to satisfy 95% of the population have been specified on the American Society of Heating, Refrigerating and Air Conditioning Engineers' Comfort Chart and cover a range of temperatures from 24°C to 26.5°C and a range of relative humidities from 20% to 60%. Where energy is available by choice, a mechanical air conditioner can provide the desired conditions all the time.

In the case of an active solar air conditioner with auxiliary energy, it could be expected that the desired conditions can be produced at will. Then the effectiveness of the solar source could be measured by the solar fraction, the fraction of the energy input supplied by the solar source. And the objective of the solar system could be to provide the most cost effective system which will usually result from a solar fraction in the range 0.7 - 0.9, depending on the relative costs of solar equipment and auxiliary fuel.

But for passive solar systems without auxiliary energy, the desired conditions will not be available all the time. A method of comparing the performance of competing passive systems could be on the basis of the percentage of time that the comfort zone is produced. However, this approach does not take account of conditions outside the comfort zone which satisfy some people and which are likely to occur frequently with passive systems. Based on responses of subjects over a wide range of conditions, the percentage of persons likely to be dissatisfied by the given thermal environment can be calculated. This is called the Predicted Percentage Dissatisfied, PPD(2). In an hourly analysis, the PPD based on the average hourly conditions can be found. An average of these hourly values $\overline{PPD}$ can be used as an index of system performance for long term occupancy.

3.1. Passive system

A study of a roof pool with movable insulation illustrates the application of PPD(3). One calculation for summer performance over two days is shown (Fig. 2). The pool was uncovered in the night and covered during the day. Windows were shaded to prevent direct sun penetration and ventilation provided at two air changes per hour. It was assumed that the occupant would adjust his clothing in an attempt to improve his comfort and that the minimum clothing would have a resistance of 0.2 clo. The maximum clothing resistance chosen by the occupant is seen to be about 1.0 clo. For the climatic data, a typical week of summer weather in Brisbane was used. Only the outside air temperature is shown for comparison with the calculated pool and inside air temperature. Although the average indoor temperature is only about 2.5°C less than the average outdoor temperature, the indoor temperature at the time of the outdoor maximum has been reduced about 7°C below the outdoor temperature. PPD is seen to rise to about 12% compared to an assumed minimum level of 5%. For evaluation, this calculated level would need to be compared to some other passive cooling ystem to decide which had the best performance.

3.2. Active System

System performance can be developed from a knowledge of the performance of the components - the collector, the heat operated refrigerator, the storage and the fan coil.

(a) Collectors. For the collector, the efficiency is the best measure of the performance and is defined by:

$$\eta = \frac{Q_u}{AG_T}$$

η = efficiency

Q_u = useful energy collected, W

A = collector area, m^2

G_T = radiant energy on collector aperture, Wm^2

Efficiency is determined by a field test or with a solar simulator and is usually expressed as a function of the variable t/G_T where

$$\Delta t = \frac{t_i + t_c}{2} - t_a$$

t_i = inlet fluid temperature

t_o = outlet fluid temperature

t_a = ambient temperature

Some results of collector efficiency tests are shown (Fig. 3)

Curve A is for a flat-plate collector with two panes of low-iron glass, an anti-reflective coating on the glass and a good selective absorber. The performance of a similar collector with a single pane of glass is also shown (Curve B). At $\Delta t/G_T = 0$, the efficiency of the single pane collector is higher than the double pane collector, due to the lower transmittance loss when the radiation passes only one pane. However, the convective and radiative losses of the single pane are higher than the two panel alternatives as shown by the greater slope of Curve B compared to Curve A.

The flat-plates (Curves A and B) show higher efficiency at small values of $\Delta t/G_T$ and greater losses than the other collectors shown.

Curve C is for an evacuated tube collector. The efficiency at $\Delta t/G_T$ is lower than that of the flat-plate due to the effect of tube spacing but the losses are lower. Thus at higher values of $\Delta t/G_T$,

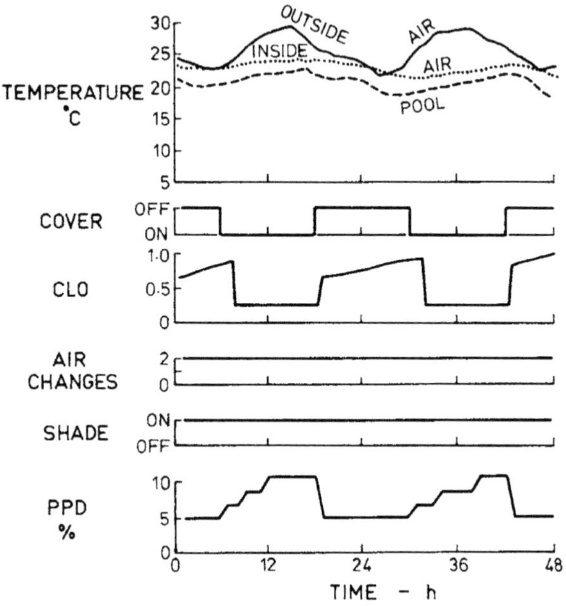

Fig. 2 - Calculated Performance of a Passive cooling system

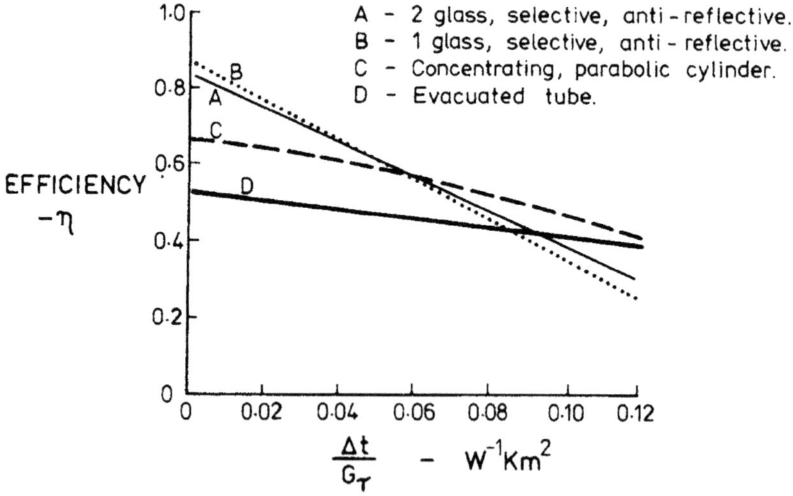

Fig. 3 - Efficiency of several solar collectors

such as occur at higher temperature, the efficiency is higher than that of the flat-plate.

Curve D is for a parabolic cylinder concentrating collector which exhibits an increasing loss coefficient as the temperature increases due to the bare absorber.

Any of the above collectors could be suitable for air conditioning systems - the choice will depend on a detailed study of the system performance with these units together with an economic study.

(b) Refrigerators. For the refrigerator, the performance is usally shown as the coefficient of performance (COP) for different values of the heat supply temperature, which will be close to the collector outlet temperature. Several different machines are compared (Fig. 4) on the basis of a condenser and/or absorber temperature of 40°C, an evaporator temperature of 5°C and a boiler or generator temperature 6°C less than the supply temperature.

The COP of the Carnot machine (Fig. 4) is the maximum efficiency possible due to limitations of the thermodynamic laws and represents the COP of a reversible 2-T refrigerator. Any practical engine must have a COP less than the Carnot machine.

Two Rankine refrigerators are shown (Fig. 4) - a machine with water in the engine and Refrigerant 12 ($C\ Cl_2F_2$) in the refrigerator (H_2O - R12), and a machine with R12 in both the engine and refrigerator (R12 - R12). Turbine efficiency $_t$ and compressor efficiency n_c are indicated. The Rankine cycle is assumed to be a saturated cycle with a pressure corresponding to the boiler temperature, until the maximum operating pressure is reached. With further increases in temperature, the Rankin fluid is superheated at the maximum cycle pressure.

The performance of the Rankine machine (R12 - R12) has been calculated with a turbine efficiency of 0.8, a compressor efficiency of 0.75 and a maximum pressure of 3 MPa, which is just below the critical (Fig. 4A). The rate of increase of the COP decreases once the gas must be super heated i.e. at the supply temperature of 100°C corresponding to boiler conditions of 3 MPa, 94°C.

For the H_2O - R12 machine, a lower turbine efficiency $\eta_t = 0.6$ shows the effect of higher friction losses with water, a low molecular weight fluid. The COP is lower than that of the R12 - R12 machine until a temperature of 110°C when the advantage of the saturated cycle becomes of greater significance.

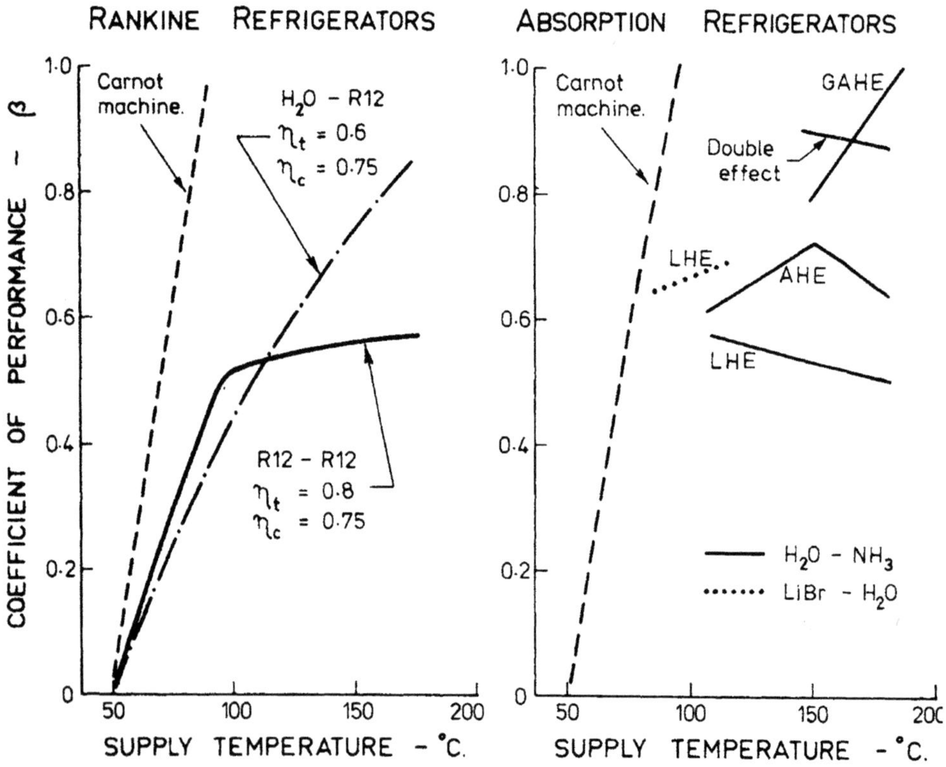

H_2O - R12	-	Water-Refrigerant 12 Rankine Refrigerator
R12 - R12	-	Refrigerant 12 Refrigerant 12 Rankine Refrigerator
H_2O - NH_3	-	Water-Ammonia Absorption Refrigerator
LiBr - H_2O	-	Lithium Bromide-Water Absorption Refrigerator
LHE	-	Liquid heat exchanger
AHE	-	Absorber heat exchanger
GAHE	-	Generator absorber heat exchanger

Fig. 4 - Performance of heat operated refrigerators

It is clear that the working fluid significantly affects the performance of Rankine machines. If a satisfactory high molecular weight fluid, giving good turbine performance and allowing saturated cycles to be used up to the desired supply temperature, can be found, there could be some advantage in operating at the higher temperatures.

The common absorption refrigerator, lithium bromide-water ($LiBr-H_2O$) has a limited range of performance 70-100°C due to crystallisation and other problems (Fig. 4). Over the range, the COP increases only slightly. Water-ammonia machines with liquid heat exchanger (LHE) can operate over a wide temperature range but the COP does not increase with an increase of temperature. Without any increase in performance, it will be better to operate at as low a temperature as possible.

But performance improvements are possible with the addition of internal heat exchange, which is somewhat similar to regenerative heating in the Rankine cycle, e.g. absorber heat exchanger (AHE) or generator-absorber heat exchange (GAHE) does show a gain of efficiency with increase in supply temperature (Fig. 4). Double effect generation also helps(4).

(c) Collector-refrigerator combination. When the refrigerator and collector are operated together, the COP of the combined unit will be the product of the component COPs at the operating temperature. The COP will show a maximum at a temperature considerably below the maximum collector operating temperature (Fig. 5). It will be seen that the flat-plate system with the R12 - R12 cooler has an anticipated COP of 0.27 at 110°C, while the concentrator unit peaks at 0.30 at 130°C.

(d) Piping. Heat will be lost from the piping used to interconnect the various components of the system. A factor of 0.7 to 0.9 could be used to allow for these losses.

(e) Storage. In heat operated systems, thermal energy can be stored as sensible energy in water or rocks. On the high temperature for input side a temperature change of 20 - 25°C may be possible and would produce a storage capacity of 100 MJm^{-3} for water and 35 MJm^{-2} for rocks with a void fraction of 0.33. Losses from such storage would occur over a temperature difference of 50 - 60°C from the mean storage temperature to the ambient temperature and can be susbstantial. On the output side, chilled water could be stored, but an effective temperature change of 7 - 10°C would limit the storage capacity to 40 MJm^{-3}. Still ,losses could be low since the temperature difference of 15 - 20°C between the average water temperature and the ambient is less than half that for storage on the input side. In addition, since converted energy is being stored, less storage is required per unit cooling capacity than for the high temperature case.

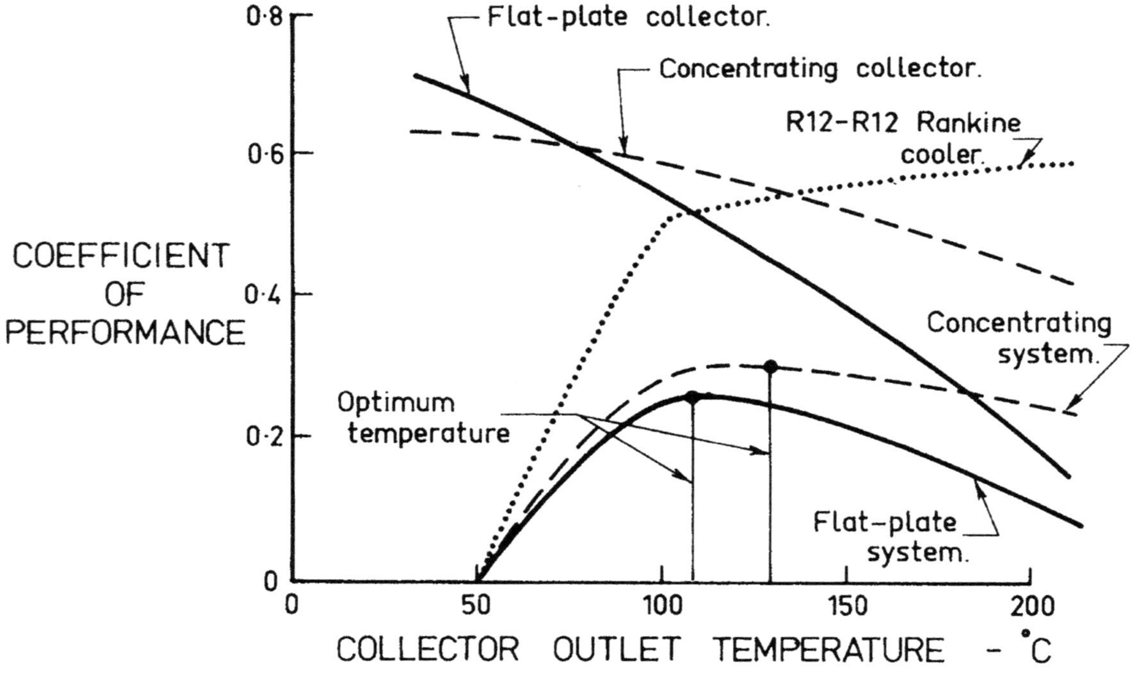

Fig. 5 - Peformance of a combined collector refrigerator system

As the store size is increased relative to other system components, each new increment of thermal capacity is less effective than previous increments in improving the solar fraction achieved by the system. A modelling study based on water heating systems determined the solar fraction for different storage capacities and found a relation with a typical "knee" (Fig. 6). Cost studies indicated that the optimum energy cost would occur at storage capacities in the neighbourhood of the knee, with conventional systems(5).

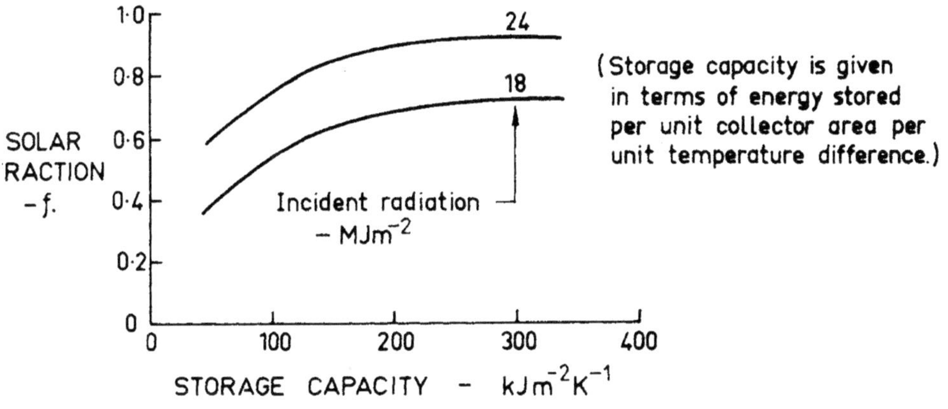

Fig. 6 - Fraction of load supplied for various sizes of store

Refrigerant storage, by utilising latent heat of evaporation, offers a thermal capacity up to 2500 MJm^{-3} and a minimum loss since the store temperature is close to the ambient. A study of lithium bromide - water machine(6) showed that 800 MJm^{-3} of total storage volume (including storage of additional absorbent) could be achieved. The machine stores refrigerant at the condenser pressure and additional absorbent at the absorber pressure. When such a machine is operated by solar energy, it will store refrigerant during periods of sunshine so that cooling can be provided at other times. Energy flows for the system studied are shown (Fig. 7) while the relevant temperatures are given (Fig. 8). It will be noticed, from the condenser heat transfer Q_C (Fig. 7), that the production of refrigerant commences soon after effective solar collection commences as is indicated by generator heat transfer, Q_G. At all times the possible heat transfer from the evaporator Q_E exceeds the building load Q_B, though with careful control, the actual evaporator load will equal the building load. About 8.00 h, the possible load equals the building load.

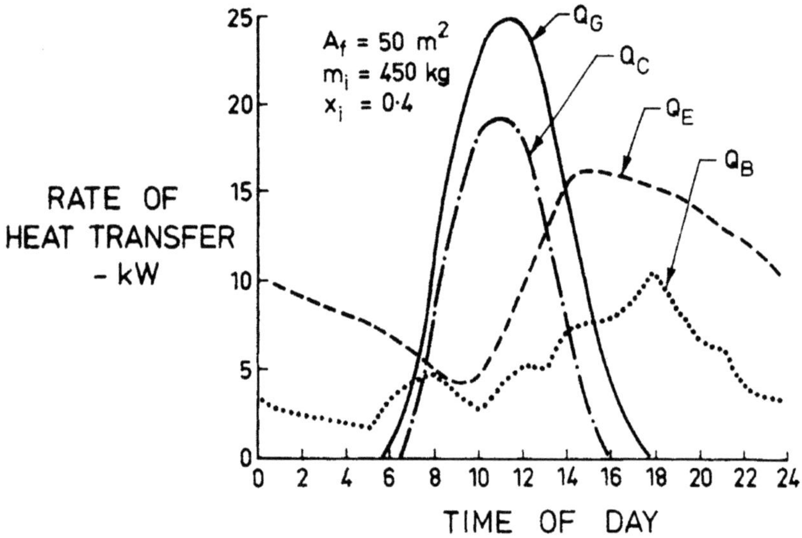

Fig. 7 - Rates of heat transfer with given building load

Generator temperature t^G increases to a maximum of 100°C about 14.00 at which time the minimum evaporator temperature t_E occurs together with the maximum absorber temperature of 48°C (Fig. 8).

The quantity of refrigerant in the store varies from a minimum at 7.30 h to a maximum at 14.30 h. Under the conditions, the machine has some excess capacity as can be seen from the fact that there is always some refrigerant in storage (Fig. 9).

(f) Design day performance. System performance for a design day can be calculated by quasi-steady state methods and component energy balances.

Insolation and temperature data are assumed to be available and allow calculation or building load and collector input. For an optimum system, component sizes can be determined by computer modelling.

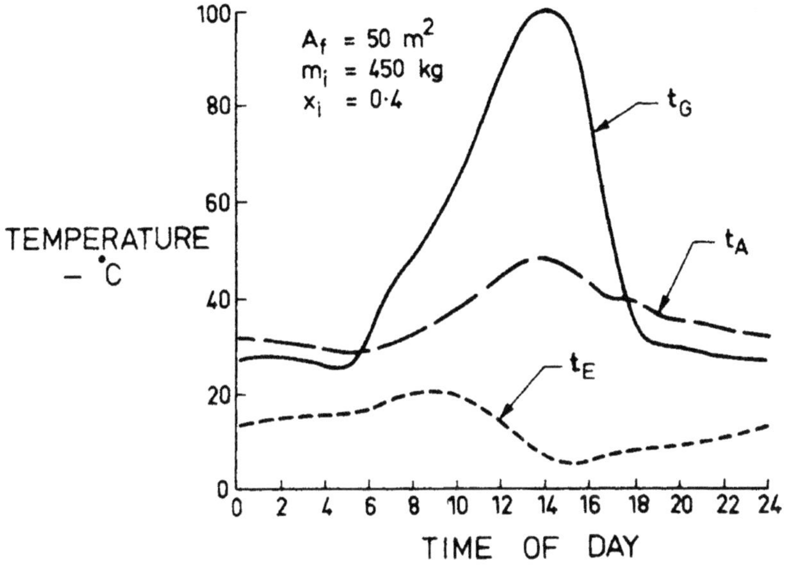

Fig. 8 - System temperatures with given building load

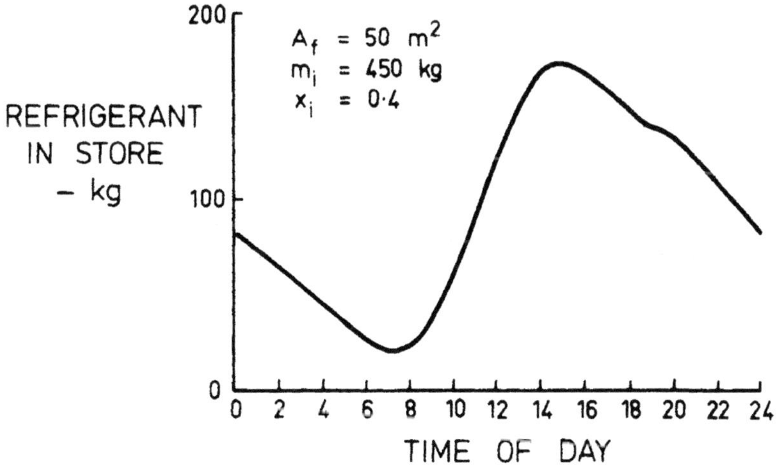

Fig. 9 - Refrigerant storage with given building load

The radiation falling on the collector aperture can be found from flux density and collector area. Energy output is determined from collector efficiency which varies throughout the day. The efficiency at peak radiant input is approximately 50% but the resulting daily efficiency, based on collector energy output to energy input is only about 30%. The chiller starts at 8.15 h and reaches full capacity at 9.00 h at which time its COP is 0.67 (Fig. 10). Excess collector energy from 9.00 h to 14.50 h is placed in the Thermal Energy Store (TES). Recovery from the TES commences at 14.50 h and continues until 19.15 h, though the refrigerator output reduces once the storage temperature starts decreasing. During the operating time of the refrigerator, chilled water is stored since the refrigerator output exceeds the building load. During the night and early morning, 19.15 h to 8.15 h, the stored chilled water supplies the building load.

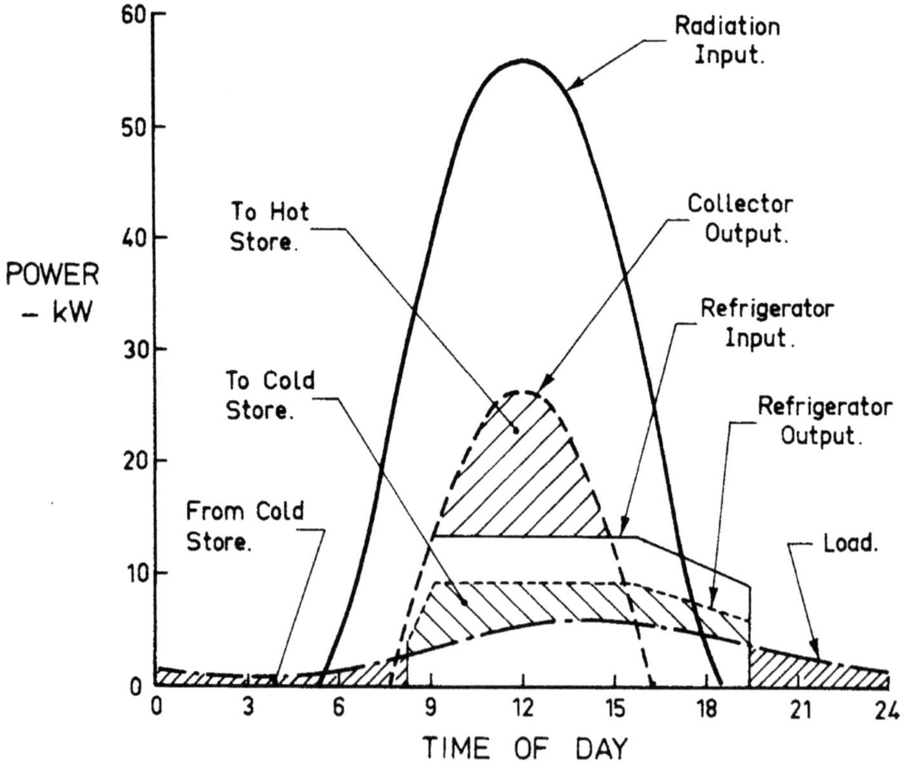

Fig. 10 - Design day performance

(g) Solar fraction. The fraction of energy demand supplied by solar can be found on an instantaneous basis by calculations similar to those previously quoted (Fig. 10), Here the solar fraction is unity at all times since the solar produced cooling always exceeds the demand. On less sunny days, the cooling demand may exceed the supply and auxiliary energy would be needed to make up the deficit. Such calculations for every day of the year would enable an annual value for the solar fraction to be estimated.

For simple design procedures, calculations on a monthly average basis may be adequate, and have been applied to the preliminary design of a solar heated and cooled building in Peoria, Illinois (Fig. 11).

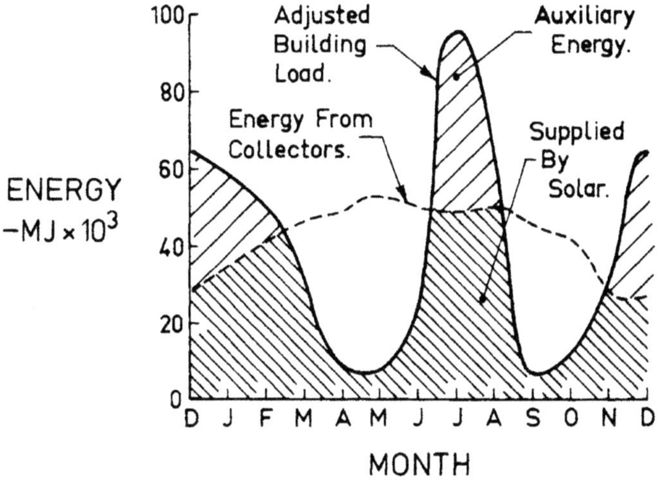

Fig. 11 - Computed performance of a solar heating and cooling system at Peoria, Illinois

The energy from the collectors is shown in comparison to the adjusted building load, i.e. the load adjusted to allow for the conversion efficiency of the absorption refrigerator. Auxiliary energy is needed whenever the adjusted load exceeds the energy available from the collectors. For this system, the solar fraction varies from a low value of 0.427 in December to unity. It has an annual average value of 0.737 (Table I).

Table I - Solar fraction for a solar heating and cooling system at Peoria, Illinois

Period	Jan	Feb	Mar	Apr	May	Jun	Jul	Aug	Sep	Oct	Nov	Dec	Year
Solar Fract-ion	.609	.860	1.0	1.0	1.0	1.0	.636	.885	1.0	1.0	.986	.427	.737

4. SYSTEMS MODELLING

The performance of solar enegy systems is very dependent on the correct matching of components and on the control strategies used for operating the systems. Also, since the energy input and the energy demand are dependent on the climate which varies from hour to hour and day to day, the best design can only be determined by a dynamic modelling of the system.

Computer modelling is a practical approach in which each component of the system is described by a function which gives the component output for a given input in terms of the component parameters. As an example, the collector may be described by the efficiency equation

$$\eta = F_R (\tau\alpha) - F_R U_L (\Delta t/G)$$

in which $F_R(\tau\alpha)$ and $F_R U_L$ are sufficient design parameters of the collector to characterise it and $\Delta t/G$ defines the climatic variables. The energy output will be given by η x input. Connections between components must also be specified so that the output of one component becomes the input to another component. The boundary values of the system, radiation and ambient temperature, are defined by time series. Operation of the model begins at an initial time when the corresponding system inputs are assumed to remain constant for the time interval which may be 0.25 h to 1 h. Resulting outputs are calculated as the model steps through time with equal intervals.

Typical results have been shown for a solar air conditioner with refrigerant storage (Fig. 7, 8 and 9).

The energy results can be integrated over given periods to determine the solar fraction for the period, whether this be a day, a month or a year.

Modelling is expensive and can only be justified for prototype design. But it can be used to develop simpler design methods for more general application.

While programmes can be written for specific models, generalised modelling programmes for solar energy system study are available. TRNSYS(7) is a research oriented programme in modular form with about twenty sub-routines describing components. A program must be written to describe the interconnection of the components for the particular system under study. Input is climatic data and output can include the temperatures of the various components as well as the energy flows.

5. SYSTEM DESIGN

5.1. Building

For solar air conditioning design, the building should be considered as an integral part of the system, since only then can an optimum solution result. Hopefully, energy conscious design applied to the building envelope will save the installation of excess capacity in the solar collectors and air conditioner.

Selection of a site to allow the desired building orientation without shading from surroundings is the first essential. Generally, the longer side of a rectangular building should be oriented North or South to ensure best control of solar penetration.

Insulation of the envelope is needed to reduce thermal exchange. Choice of the mount of insulation is an economic problem for an air conditioning situation; insulation costs are balanced against a saving in air conditioning costs. It is preferable that the insulation resistance be variable such as can be organised with openable windows and/or shutters since the ambient air can be utilised for cooling when its temperature is suitable.

Another factor of importance is the thermal capacity, which could be particularly useful for solar systems, since it provides energy storage. Thermal capacity is best located on the inside of the building such as in slab floors or in masonry partitions. Where insulation of a masonry building is used, it may be preferable to place the insulation on the outside since this will usually be of lower thermal capacity than the masonry.

Little quantitative work has been recorded on the desirable amount of insulation or thermal capacity for tropical air conditioned buildings. However, it would appear that an overall thermal transmittance U_o better than 1.0 $Wm^{-2}K^{-1}$ should be used for the envelope with higher resistance in the roof and a maximum glass area of 15% of wall area (glass forms a low thermal resistance path). Where the average found temperature is within or below the comfort level, slab-on-ground construction is desirable for domestic buildings to obtain access to the thermal capacity of the ground.

Generally, the roof will be required to carry the solar collectors and should be adapted to the type of collector chosen. Flat-plate collectors should face either N or S and be inclined at an angle $s = \phi - 15°$, where ϕ is the latitude angle. In the tropics, a minimum inclination of 10° may be specified to ensure that any air in the water circuit will be moved through the collectors. Generally, collectors in the southern hemisphere will face N, since the objective is to intercept the maximum possible flux in summer when high air conditioning loads occur. Evacuated tube and concentrating collectors will have specific requirements which should be accommodated by the roof shape.

5.2. Equipment

Only solar absorption air conditioners will be discussed in this section since they are the only commercially available alternative. However, some reference will be made to photovoltaic generator - vapour compression systems when cost estimates are made (Section 7).

Collectors should be chosen on performance and cost. A relatively good efficiency will be needed to ensure that the operating temperature of the chiller will be reached. A minimum efficiency of 50% at

$$\left[\frac{t_{fi} + t_{fo}}{2} - t_a\right] = 0.07\ Km^2W^{-1}$$

should be specified as this can be attained by most quality collectors. Reliability and corrosion resistance are other important criteria that are hard to quantify. Collector types that could satisfy the requirements include flat-plates, evacuated tubes, parabolic troughs and linear fresnel concentrators.

At this time, the only viable commercial chiller appears to be the lithium bromide absorption machine which has been well developed and is known to be reliable. Thus its performance should be used as a minimum specification by which other machines that may become available will be judged. A coefficient of performance of 0.6 should be required with collector water at 90°C, cooling water at 30°C and chilled water at 8°C.

It would appear that sensible heat storage in water would be the logical choice for thermal energy stores.

The hot water store can operate in the range of 100°C to 80°C and store 80 $kJkg^{-1}$. Alternatively, the chilled water store, which can operate in the range of 6-14°C, can store only 34 $kJkg^{-1}$. However this latter storage, being of converted energy, is equivalent to a hot store capacity of $(\frac{10}{6} \times 34) = 67$ $kJkg^{-1}$. While the sizing of the plant would preferably be confirmed by a modelling study, some approximate criteria can be stated as a guide (Table II).

Table II - Approximate component sizes for absorption air-conditioning

Collector area/unit air-conditioned floor area	0.2 - 0.45 m^2m^{-2}
Collector area/unit chiller capacity	3 - 6 m^2kW^{-1}
Capacity of hot-side TES/ unit collector area	0.05 - 0.075 m^3m^{-2}; 4 - 6 MJm^{-2}
Capacity of cold-side TES/ unit chiller capacity	0.2 m^3kW^{-1}; 5 - 8 MJ kW^{-1}
Solar fraction	0.65 - 0.8
Parasitic power/unit chiller capacity	0.1 - 0.2 kW kW^{-1}

It will be important to keep the piping as simple and compact as possible; otherwise, the losses will be significant. Also the cost of piping can be a major factor in system cost and should be reduced to a minimum. Associated with the piping is the parasitic power, i.e., the energy required to circulate the various fluids. This power will need to be supplied by electricity. It should be kept as low as possible.

5.3. Control

Absorption chillers usually require that the collector temperature be considerably above the minimum operating temperature before they will start. To achieve the starting temperature takes considerable time due to the chiller thermal capacity. If the chiller shuts down because of low radiation or because the cooling demand is satisfied, there will be a time delay before it can start up again. Generally, it is preferable for the chiller to run continuously once it starts and to store chilled water whenever the room load is satisfied.

A fluid circuit providing for both "hot" and "cold" storage is shown (Fig. 12). Both the hot and cold circuits are provided with a single pump which achieves three modes of operation by opening or closing the valves. The modes of operation are: input to output; input to store; and store to output.

Thermostats and differential thermostats will be used as detectors for the controller which will start the pumps and fan, set the valves for the various modes, bring in auxiliary energy, and protect the system against adverse conditions. With so many levels of control, the circuits can be quite complex.

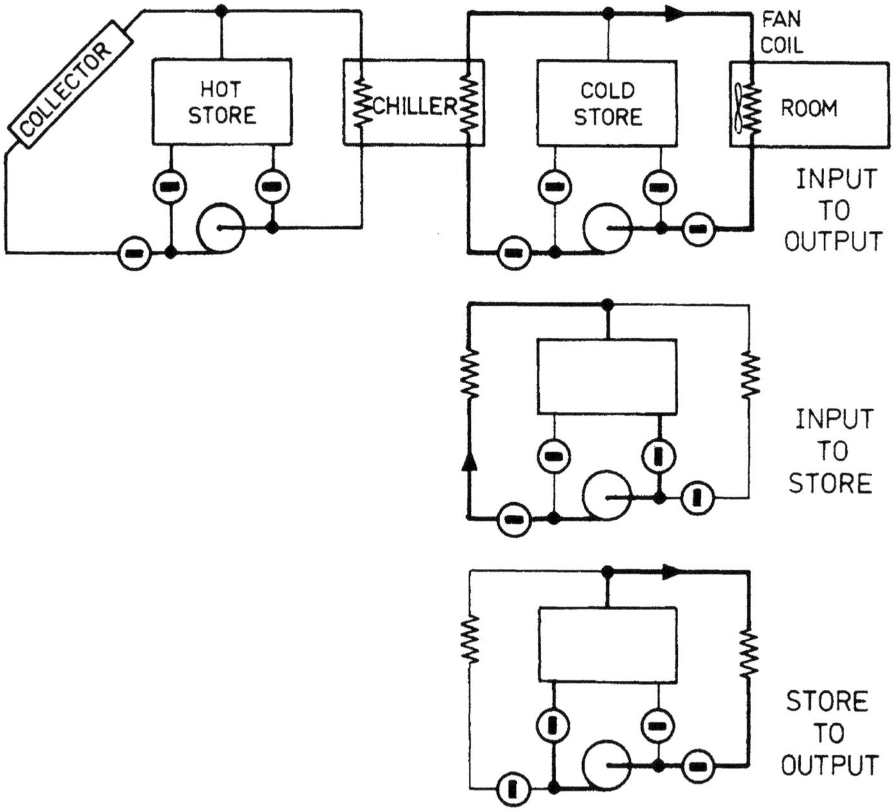

Fig. 12 - Control of fluid circuit to achieve various operating modes

6. DEMONSTRATION PROJECTS

An ambitious programme of demonstrating solar air conditioning has been undertaken by the Solar Energy Research Institute, Western Australia. Seven projects were reported at a recent symposium(8). All used absorption chillers but a variety of collectors were installed. The buildings concerned were used for a variety of purposes (Table III and Table IV). Also reports on several other projects by different authorities are included.

Air conditioned spaces ranged from 30 - 1130 m^2 in area, and were cooled by plant ranging from 7 - 105 kW cooling capacity. To enable better comparisons to be made several indices were calculated.

The collector area per unit area of air conditioned space varies from 0.18 to 0.93. Low values, about 0.2, apply to buildings of good thermal design that are used for offices and occupied for 10 - 11 h per day. Values around 0.4 are common and would seem suitable for domestic installations. High values are not thought to be optimised.

The area of air-conditioned space per unit cooling capacity (m^2kW^{-1}) ranges from 2.8 - 15.4, while values in the range 8 - 10 are common. Since unspecified factors such as occupancy and overall thermal transmittance of the building affect this index, the correctness of the design cannot be evaluated.

Collector area per unit cooling capacity gives values from 1.4 to 6.1. The value chosen will depend on collector efficiency, storage and chiller coefficient of performance. Considerable variation can be expected.

The ratio of hot storage to cold storage is quoted on a volume basis. Since the cold store is needed mainly to stop chiller cycling, it might be expected that the cold store would be small and result in values of this ratio greater than unity. The values of 1.5 - 21. appear reasonable whereas the value of 0.49 would be low.

Values of hot storage volume per unit collector area show a wide variation from 0 - 0.104. It is thought that values should err on the low side of the recommended values (Table II) at around 0.05. A value of 0.104 is almost certainly too big. No storage is hard to justify.

Where cold storage is provided, the storage capacity per unit cooling capacity is fairly consistent.

A lack of optimum design for most of these projects is the main criticism. One of the biggest faults concerns the building design since

Table III - Description of components for solar air conditioned demonstration projects

No.	Project Name	Collector*	Chiller	Stand-by	Application
1	National Gas Depot, Ballajura, W.A.	FP, selective	Yasaki		Office
2	National Iron and Steel Co., Perth, W.A.	concentrating parabolic trough	Yasaki	Boiler	Office
3	Yanchep Sun City, W.A.	Evacuated tube, Sanyo	Sanyo		Domestic
4	Kounis Metal Industries, Perth, W.A.	FP, selective	Yasaki	Boiler	
5	West Australian Institute of Technology, Perth, W.A.	FP, selective	Yasaki	Boiler	
6	Paraburdoo, W.A.	FP, selective "boiling tube"	Yasaki	Boiler, LP Gas	Domestic
7	Port Hedland, W.A.	FP, selective some "boiling tube"	Yasaki	Boiler LP Gas	Domestic
8	Government Building, Garbutt, Q.	FP, selective Beasley and Yasaki	Yasaki	Electric Boiler plus vapour compression plant	Office
9	Gas and Fuel Corp., Dandenong, Vic.		Yasaki		
10	Yasaki-Pacific Pty. Ltd., Melbourne, Vic.	FP, selective Yasaki	Yasaki		
11	University of Sydney, N.S.W.		Yasaki		
12	Merchant Builders Vermont, Vic.		Yasaki		
13	Energy Information Centre, Melbourne, Vic.		Yasaki		

* FP - flat plate; selective - selective absorbing surface; "boiling tube" - vaporising fluid in collector

Table IV - Comparison of design parameters for solar air conditioned demonstration projects

No.	Project Name	Air Conditioned Area - m^2	Cooling Capacity - kW	Solar Collector Area - m^2	Hot Store Capacity - m^3	Cold Store Capacity - m^3	Collector Area / Floor Area - $m^2\ m^{-2}$	Floor Area / Cooling Capacity - $m^3\ kW^{-1}$	Collector Area / Cooling Capacity - $m^2\ kW^{-1}$	Hot Store Capacity / Cold Store Capacity - $m^3\ m^{-2}$	Hot Store Capacity / Collector Area - $m^3\ m^{-2}$	Cold Store Capacity / Cooling Capacity - $m^3\ kW^{-1}$
1	National Gas Depot	345	35	144	8	8	.42	9.8	4.1	1.0	.056	.23
2	National Iron & Steel	300	35	148	0	0	.49	8.6	4.2	-	-	-
3	Yanchep Sun City	160	17.5	53	3	0	.33	9.1	3.0	-	.057	-
4	Kounis Metal Industries	600	94	260	9	0	.43	6.4	2.8	-	.035	-
5	W.A.I.T.	81	7	43	0	0	.53	11.6	6.1	-	-	-
6	Paraburdoo	120	14	50	1.1	2.25	.41	8.6	3.5	.49	.022	.16
7	Port Hedland	120	14	37	1.1	2.25	.31	8.6	2.6	.49	.030	.16
8	Garbutt	1076	69.8	202	21	14	.19	15.4	2.9	1.5	.104	.20
9	Gas and Fuel Corp.	1130	140	200	40	-	.18	8.1	1.4	-	.20	-
10	Yasaki-Pacific	1013	105	508	20	-	.47	9.6	4.8	-	.04	-
11	University of Sydney	52	7	34	3	1.4	.65	7.4	4.9	2.1	.09	.20
12	Merchant Builders	30	10.6	28	1.4	-	.93	2.8	2.6	-	.05	-
13	Energy Information Centre	30	10.6	28	103	-	.93	2.8	2.6	-	3.7	-

in only a few projects has the building been given a reasonable thermal design. Computer modelling of the systems should also have been used and would prove especially valuable in analysing the results.

7. COSTS

An objective of demonstration projects should be to provide realistic cost estimates. Since the solar energy equipment being demonstrated is little more advanced than the prototype (and all too frequently may be the prototype), it cannot be expected that the capital costs will be representative of future market prices once normal commercial production has commenced. Costs will be high due to limited scale production but they may also be loaded with development costs. As a result, demonstration capital costs should be subject to some interpretation before being used as a basis for estimates of future costs. However, operational cost data should give a fair indication of the likely future costs, though maintenance costs of prototype grade equipment may not be representative of future costs.

7.1. Capital Costs

For the Solar Workshop in Western Australia, a questionnaire was sent to project supervisors, requesting capital and installation costs of each component of the system. The resulting costs have been divided into eight sections (Table V). To determine a specific cost for each section, a size is quoted which is representative of the section e.g. area of collector, volume of storage and cooling capacity of chiller. The sectional cost as a percentage of system cost is also given.

There is some lack of uniformity in the data, especially with regard to work that can be done by contract or within the organisation concerned. Some projects quote contract prices while others quote only direct labour and material charges. The result is a factor of two or more difference between the costs for the same item.

Comments on the different sectional costs are as follows:

(a) Solar collectors. The cost ranges from \$87-477 m^{-2}. The low value \$87 is a factory cost for concentrating collectors and not a true market cost. Evacuated tube collectors account for the high value \$477. Flat-plate collectors range from \$168-260, but none could be said to be in quantity production.

The potential for cost reduction in flat-plate collectors is not good since the labour content is small and a relatively large amount of material is used in their manufacture. It is considered unlikely that

Table V - Cost data on several demonstration projects

No.	Name		Solar Collectors	Storage	Chillers	Aux Boiler	Cooling Tower	Fan Coil	Water Circuit	Elec Control	System
1	Na:ional Gas Depot, Ballajura	Cost - $	27468	11840	7868	4157	3530	13980	38237	9674	116746
		Size	144 m^2	16m3	35 kW		35 kW	35 kW	35 kW		35 kW
		Sp.cost-$	191 m^{-2}	740 m^{-3}	225kW^{-1}		101 kW^{-1}	399 kW^{-1}	1092 kW^{-1}		3336 kW^{-1}
		%	24	10	7	4	3	12	33	8	101
2	National Iron ard Steel	Cost - $	12909	-	5873	1300	2151	6036	8992	2000	39261
		Size	148 m^2	-	35 kW		35 kW	35kW	35 kW		35 kW
		Sp. Cost-$	87 m^{-2}	-	168 kW^{-1}		61 kW^{-1}	172kW^{-1}	257 kW^{-1}		1122 kW_{-1}
		%	33	0	15	3	5	15	23	5	99
3	Yanchep Sun City	Cost - $	25300	2700	12300	-	2500	5300	14500	8600	71200
		Size	53 m^2	3 m^3	17.5 kW		17.5 kW	17.5 kW	17.5 kW		17.5 kW
		Sp. Cost-$	477 m^{-2}	900 m^{-3}	703 kW^{-1}		143 kW^{-1}	303 kW^{-1}	829 kW^{-1}		4069 kW^{-1}
		%	36	4	17	-	4	7	20	12	100
4	Kounis	Cost - $	43800	2700	20700	900	2600	4600	13150	3500	91950
		Size	260 m^2	9 m^3	94 kW		94 kW	94 kW	94 kW		94 kW
		Sp. Cost-$	168 m^{-2}	300 m^{-3}	220 kW^{-1}		28 kW^{-1}	49 kw^{-1}	140 kW^{-1}		978 kW^{-1}
		%	48	3	23	1	3	5	14	4	101
5	Gas and Fuel Corp.	Cost - $	52000	18800	35000	10000	6500	75600	53100	31000	282000
		Size	200 m^2	40 m^3	140 kW		140 kW	140 kW	140 kW		140 kW
		Sp. Cost-$	260 m^{-2}	470 m^{-3}	250 kW^{-1}		46 kW^{-1}	540 kW^{-1}	379 kW^{-1}		2014 kW^{-1}
		%	18	7	12	4	2	27	19	11	100

$ = Australian currency

costs can be reduced much below $\$130 m^{-2}$. The collector is a major cost component of these systems, as can be seen from the percentage cost, and is likely to remain as such.

(b) Storage. Costs range from $\$300$-$900\ m^{-3}$. If the lowest cost store is disregarded, the specific cost of the others shows the effect of scale, i.e. the specific cost decreases as the size increases. It cannot be expected that the cost of water storage could be reduced much below $\$250\ m^{-3}$, which price is frequently quoted for American systems.

(c) Chillers. Three projects have a chiller cost of $\$220$-$250\ kW^{-1}$ which is independent of system size. The high price $703 is a typical. The chiller is the most sophisticated part of the solar air conditioner but it only contributes 7-23% of the cost. Factory production is responsible for the low cost which is not expected to decrease greatly with further increases in production rate.

(d) Auxiliary Boiler. Few details are available for the boilers, but they do not affect the overall cost significantly.

(e) Cooling Tower. Cooling tower prices appear to be somewhat random and not related to size, These are already standard items so little cost reduction can be anticipated for the future.

(f) Fan Coil. The considerable range from $\$49$-$540 kW^{-1}$ for the fan coil and ducting installation must be brought about by different costing methods and/or different quality standards. It must be expected that the cost of this item should be a significant fraction of the system cost since it is difficult to avoid some on-site labour for duct and register installations. The technology is known so significant cost reductions cannot be anticipated.

(g) Water Circuit. On site labour and design costs must account for the relatively high cost of the water circuits but the wide range of costs, $\$140$-$1092\ kW^{-1}$,, must be influenced by different costing approaches. It is imperative that the fractional cost of this item be reduced much below the high value of 33% to achieve reasonable system costs.

High piping costs are a feature of distributed collector systems as was found in the Department of Energy study for the comparative evaluation of distributed collector and centralised receiver solar thermal systems. As a result of this study, the centralised receiver system was chosen for further development partly due to the saving in piping costs.

(h) Electrical and Control Circuit. The costs of this item are not of great concern. Switchboards and sub-circuits can be pre-wired so that on-site labour for wiring and testing is reduced.

(i) System. The costs range from \$978-4069 kW^{-1}. System size has an effect on the specific cost but the major difference occurs due to cost methods that are not comparable.

From all these data, an estimate can be made of what solar air conditioners might cost in today's dollars when their production rate is reasonable and their technology is well known at the tradesman level. These values are compared to the costs of equivalent vapour compression systems (Table VI). The cost per unit output of each section was estimated and a total found for the system cost. Then the percentage of the system cost for each section was found.

Table VI - Sectionalised costs of air conditioning sytems - Size 10-20 kW

	Solar Absorption Air Conditioner		Vapour Compression Air Conditioner	
Cost of section (sub-assembly or component)	Cost per kW	Percentage of System Cost	Cost per kW	Percentage of System Cost
Solar collectors	640	39	-	
Energy Storage	120	7	-	
Chillers	180	11	180	26
Auxiliary Boiler	60	4	-	
Cooling Tower	50	3	50	7
Fan coil and air ducts	320	19	320	46
Water Circuit	200	12	100	14
Electrical and Control Circuit	80	5	50	7
System	1650	100	700	100

With an area of 5 m^2kW^{-1} of cooling, the collectors were estimated on the basis of a unit cost of \$128 m^{-2}, which could be considered a target figure for present flat-plate collectors. At 39% of the system cost, the collectors are by far the largest item of cost.

Energy storage is based on storage equivalent to 0.2 m^3kW^{-1} cold storage and 0.06 m^3m^{-2} of collector hot storage (which gives 0.3 m^3kW^{-1}

of cooling capacity after allowing for 5 m^2kW^{-1} for the collector area). This total of 0.5 m^3kW^{-1} was costed at \$240 m^{-3}. This value is considerably below costs quoted (Table V).

Absorption machines and vapour compression machines are assumed to cost the same at a value that might be reached in a competitive market.

Obviously the cost of the fan coil and ducts will depend on the quality of the interior installaton, the difficulty of duct installation, etc. The value quoted is within the range of \$49-540 (Table V).

The water circuit cost is a major problem. To reduce this cost, a much greater percentage of the plumbing work must be performed in the factory. The present estimate allows for 12% of the system cost for plumbing.

It is clear that the system cost of \$1650 kW^{-1} represents a minimum cost for solar air conditioners that could be built with existing techniques. A significant point is that they must be expected to cost about 2.4 times the cost of vapour compression units for the same cooling output. (Possibly, the peak output of solar air conditioners would be somewhat less than vapour compression machines in the same installation due to the energy storage. If the solar air conditioner is only 0.8 the size of the vapour compression machine, the effective cost ratio would be reduced to 1.9).

7.2. Energy Cost

To determine energy cost, a hypothetical plant is considered for a domestic situation. Assumed values are as follows:

Air conditioned space	-	100 m^2
Annual cooling load	-	94600 MJ
COP - vapour compression	-	2.0
COP - absorption	-	0.6
Efficiency - gas boiler	-	0.75
Solar fraction	-	0.85
Parasitic energy to solar system	-	0.33 x electrical load for vapour compression plant.
Vapour compression cooling capacity	-	10 kW
Solar absorption cooling capacity	-	8 kW
Cost of vapour compression	-	\$700 kW^{-1}
Cost of absorption chiller	-	\$1650 kW^{-1}

Interest and depreciation rates of 11% give a discount rate of 14% for a 15 year life.
Maintenance costs ae allowed at 3% of capital cost

Gas cost	- $0.006 MJ^{-1}
Electrical input to vapour compression	$= 94600 \times 0.5 \times \frac{1}{3.6}$ kWh
	= 13140 kWh
Gas cost to solar	$= (1 - .85) \times 94600 \times \frac{1}{6} \times \frac{1}{.75} \times .006$
	= $189
Annual cost of vapour compression	= .17 x 10 x 700 + 13140 x C
Annual cost of solar system	= .17 x 8 x 1650 + 189 + .33 x 13140 x C
For equal cost	C = $0.14 kWh^{-1}

This cost of $0.14 kW^{-1} is comparable with the cost of diesel generated electricity for which fuel alone costs in excess of $0.04 kWh^{-1}. (A study of 220 kWh remote area power supply in Australia indicated an average cost of $0.12 kWh^{-1}). This cost has been achieved with a solar system cost of $1650 kW^{-1}, which must be considered as the target cost to make the systems competitive. It is suggested that the cost break-down has indicated the maximum costs allowable for each section and has shown which sections need to be developed to reach this target, viz. storage and water piping. Storage probably requires some new development while reduction of water piping costs is essentially a matter of reducing on-site erection costs of collectors and chilled water piping.

7.3. Preliminary costs of photovoltaic air conditioner

A detailed study on photovoltaic systems for remote area power supply has indicated that these systems should become competitive with diesel generated electricity by the end of this decade(9). It seems appropriate to consider such generators as the power supply for air conditioning.

Two systems will be considered:

(i) Passive flat-plate array producing direct current, a small battery storage and a direct current motor driving the vapour compression air conditioner

(ii) Concentrating total energy system supplying electricity and domestic hot water as well as sufficient energy to run an absorption air conditioner.

The system capacities are the same as previously (Section 7.2).

(i) This unit will have an 8 kW photovoltaic generator, 10 kWh storage, 5 kW motor and kW chiller. The cost is estimated as

8 kW generator with storage at $900 kw	= $ 7200
10 kW air conditioner at $700	= $ 7000
	$14200

(ii) The required energy flows for the various energy components are:

Electricity	: daily energy 50 MJ (14 kWh)
Domestic Water	: daily energy 50 MJ
Air conditioner	: daily cooling energy 260 MJ: daily energy input 432 MJ

To meet the total heat output of 482 MJ, the unit will be able to produce 240 MJ (67 kWh) of electricity. It will include adequate electrical storage and power conditioning i.e. conversion to A.C.

The rating of the unit would be:

Electricity	13kW
Air conditioner	8 kW cooling

Cost is estimated as:

13 kW total energy generator with storage at $2400 kW^{-1}	= 31200
8 kW absorption unit at $940 kW^{-1}	= 7520
	$38720

It is as well to note that the absorption unit with flat-plate collectors will cost $13200 (8 kW at $1650 kW^{-1}). On these figures the flat-plate absorption system is cheapest but the passive flat-plate array system is not greatly different. Although the total energy system does much more than the others, it is clear that it would not be cost effective.

8. CONCLUSION

Solar air conditioning is relatively expensive but will be able to compete in remote areas where the alternative energy source is diesel power by the end of this decade. At that time, solar absorption air conditioners could be challenged by solar photovoltaic vapour-compression systems.

The building and the solar equipment should be designed as a system and optimised with respect to minimum cost over the building life. Factors of importance include the building envelope resistance and capacity, the collector area, storage capacity, converter capacity and air conditioner capacity.

The thermal performance of low cost tropical housing designs should be analysed by computer modelling and the best design selected on the basis of the closest approach to thermal comfort.

Note: Costs quoted are in Australian currency.

REFERENCES

1. Hay, H.R, and Yellott, J.L. (1969) International Aspects of air conditioning with movable insulation. Solar Energy Journal, Vol. 12, pp 427-438

2. Fanger, P.O. (1970) Thermal comfort. McGraw-Hill, New York.

3. Ahmad, I. (1976) The roof pool and its influence on the thermal environment, Ph.D. Thesis, University of Queensland.

4. Phillips, B.A. (1976). Absorption cycles for air cooled solar air conditioning. Paper 4/15, Dallas Conference, ASHRAE, New York.

5. Sheridan, N.R., Bullock, K.J. and Duffie, J.A. (1967) Study of solar processes by analog computer, Solar Energy Journal, Vo. XI, pp 69-77.

6. Grassie, S.L. and Sheridan, N.R. (1977) Modelling of a solar-operated absorption air conditioner with refrigerant storage. Solar Energy, Vol. 19, pp 691-700

7. Solar Energy Laboratory (1975) TRNSYS, a transient simulation program, University of Wisconsin - Madison, Wisconsin.

8. Solar Energy Research Institute of Western Australia (1980) Proceedings of Solar Cooling Workshop, SERIWA, Perth.

9. Sheridan, N.R., Yeow, Y.T. and Hall, C.T. (1980) Solar photovoltaic power systems for remote area power supply, Report, National Energy Research, Development and Demonstration Council, Canberra

DISCUSSION

Mr. D. Hampton:

Prof Cook, have there been any landmark decisions on "sun rights" in USA? I realise that there are sun rights declared by various States. Have any of these ever been challenged in court or has anybody or any individual actually gone through the process of going through courts to ensure these sun rights?

Prof. J. Cook:

To my knowledge there have not been any landmark court decisions. Probably the most interesting piece of legislation in the US related to solar rights is the one that the State of New Mexico has put on the books which is a statement of prior right. In essence, whoever got there first is entitled to the solar right. The Solar Energy Research Institute, a government-funded organisation, publishes a quarterly periodical called Solar Law Reporter. But, to-date, legal questions seem not to have been the problem that people thought they might be. One should understand that the higher the latitude the lower the sun angle and the more critical sun access becomes, particularly in urban locations. Access to the sun can become a major planning problem, and buildings need to be protected for the life of the application. However, these may not be such critical questions in the tropics.

Question:

Is there very much difficulty in keeping collectors clean, and if so what is the best method?

Prof. K. R. Rao:

In Singapore, the required inclination for solar collectors is only a few degrees and hence, there is the problem of dust collection. The only way to prevent it is by frequent cleaning. Here we get rain frequently and probably to some extent it washes away the dirt. However, there will still be dirt marks on coverglass unless a regular cleaning programme is undertaken.

Prof. J. Cook:

Further away from the tropics, dirt and other pollutants have not been a problem. They seem to have very little effect on the collectors, the reduction of efficiency is less than 5%. Even if the collector looks filthy, the radiation does get through. There is quite a different problem however with photovoltaics. I have been involved in an installation where we have a complete roof covered by solar cells. Our experience there is that even something as small as a bird dropping had a major impact on the production of electricty. So the performance of that type of system is rather delicately related to keeping it clean.

Dr. N. Sheridan:

I agree with you that if there is a reasonable inclination and a fair amount of rain, collectors do not seem to have any problems in keeping themselves clean. I think we have a little problem in trying to use concentrating collectors because these are going to be more subject to dust problems.

I will make one comment about the "sun right" problem in Australia. We have to remember that if we provide "sun rights" to one property, we are taking away from another property the right to use and to enjoy the benefits of their property. This is not a simple legal problem and in trying to provide a reasonable approach to the access to sunshine, the New Mexico approach certainly is one way out, and this is really going back to an old English approach to recognise the prior rights of people that were there first.

Question:

The capital cost of solar air-conditioning is about 2 - 2 times more than that of conventional methods. This appears to be exorbitant compared to the present cost of energy. Is there any research being done at all to improve efficiency and bring down the cost rather than waiting for the present energy cost to go up to match it?

Dr. N. Sheridan:

What I was hoping to indicate was the present installation cost of the system. I analysed in some detail and the cost was A$3,000 per kilowatt whereas the hypothetical system which I was postulating that we might be able to build if we had mass production, was about half the cost. So, if we are trying to get solar air-conditioning to be all competitive, we have to reduce the cost to about half of what any actual installation cost at this time, and that is a fairly big programme.

We have done some studying on how flat plate collectors might be produced, and we think that the problem might be that it uses too much material. About 75% or more of the production material cost, and so we do not have much opportunity of reducing cost by mass production, because there is not too much labour content. Hopefully, one might be able to keep this present price at the same level with some future developments of different materials used, but the cost may still increase anyway because the life span of the system might be reduced. And so, I do not see any simple way to reduce the cost of this system.

Question:

Why would you suggest that the government should support projects that were already economical in preference to those that were not already economical? I would have thought that the preference would be the other way around.

Prof. J. Cook:

We have the experience in the United States of systems which were known to have some economic value which were not supported, because they already appeared to be viable. That puts these systems at a disadvantage. The experience had been that the government was supporting uneconomic systems. So there is a question whether commercialization is really helped if one is constantly supporting unviable solutions. Do you understand the kind of dilemma? There is no simple answer. I think it is a very controversial kind of issue.

There may be some question whether governments should be involved in supporting certain applications of solar energy at all. Perhaps, government intervention should only be in terms of research and development, in the dissemination of knowledge and perhaps in tax incentives, but not in the commercial sector at all. What is the experience then in Australia in terms of promoting commercialisation of solar applications?

Dr. N. Sheridan:

The government in Australia has not been very sympathetic towards solar energy. In Western Australia, the government has supported a number of demonstration projects. Personally, I believe that the government should not have to single out solar energy as a support measure. I think solar energy is good enough and that it can eventually find its own level in the market. It really ought to be encouraged to stand on its own feet and not ask for government handouts because, in the long run, if it can establish itself it will become a much stronger industry. Certainly, if governments are to support solar energy to any extent, it should be in a limited amount at the start of the industry, just to get it over the hump, rather than guaranteeing it long-term support. Generally, any tax rebate on solar energy ought to be short term just to get the industry going rather than extending it in the long run, allowing the industry to go strong on its own initiative, and thereby becoming a good competitor of other forms of energy.

REFRIGERATION AND AIR COOLING

R. H. B. Exell,
M.A., D. Phil. (Oxon),
Associate Professor of Applied Physics,
Asian Institute of Technology.

ABSTRACT

The scientific principles involved in the use of solar energy for refrigeration and air cooling are reviewed together with mention of some of the early experimental work and discussions of more recent developments. The topics covered include passive systems, compression refrigeration, open cycle dehumidification, ammonia-water absorption refrigeration, water-lithium bromide absorption refrigeration, and solid absorption refrigeration.

1. INTRODUCTION

It has always been recognised that refrigeration and air conditioning are potentially attractive users of solar energy because of the demand for cold is greatest when the sun is hottest. Yet research and development efforts throughout the world over the years have been slow to produce results. The reason for this are both economic and technical. The economic reasons is that conventional electrically operated refrigeration equipment is still cheaper and more convenient than its solar counterpart. The technical reason is that in order to obtain solar operated refrigeration at the rate of consumption to which we are now accustomed, and continuously throughout the day and night in both clear and cloudy weather, relatively large solar radiation collecting systems and solar energy storage systems have to be used. Nevertheless, the rising demand for cold coupled with limits in the supply of power, is certain to stimulate its widespread development in the future. This is particularly true for the tropics.

This paper discusses the principles upon which are based various methods of using solar energy to produce refrigeration and air cooling, namely:

(a) compression refrigeration systems using solar heated Rankine cycle engines, or photovoltaic solar cells and electric motors;

Bill B. P. Lim (ed.), Solar Energy Applications in the Tropics, 95–127.

(b) systems using dehumidified air and evaporative cooling with solar desorption of the desiccant;

(c) solar heated absorption refrigeration systems.

2. PASSIVE SYSTEMS

The methods listed aove are all described as active systems in that they require an energy input for their operation. Before examining them in detail we shall consider the possibility of using passive cooling which, instead of using energy to extract heat from the space to be cooled, contrives to lose heat from a storage medium to the sky at night and then use this cooled medium to lower the temperature of the required space during the day.

Fig. 1 shows the basic design by Hay and Yellott known as the Skytherm House(1). It obtains air conditioning by using the atmosphere as a heat sink. A movable insulated roof permits water ponds in thermal contract with a metal ceiling to dissipate heat to the sky at night, and keep the interior of the building cool during the day.

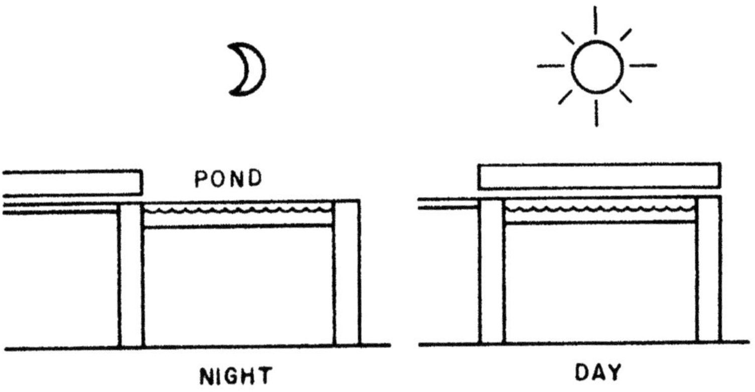

Fig. 1 - Skytherm House

A full size house with this system in California"2" used 20 cm of water sealed in plastic bags resting on the metal ceiling covering the $102m^2$ living space. The movement of the 5 cm thick polyurethane insulating panels was automatically controlled to optimize performance. During August the maximum and minimum outdoor temperatures averaged 31°C and 12°C respectively. The room temperature remained almost constant throughout the day at about 21°C.

Another example from Perth, Western Australia(3) is shown in Fig. 2. Air cooled by passing it through a solar collector at night is drawn by means of fans through "thermal banks" of brick, and is distributed to the interior of the house during the day.

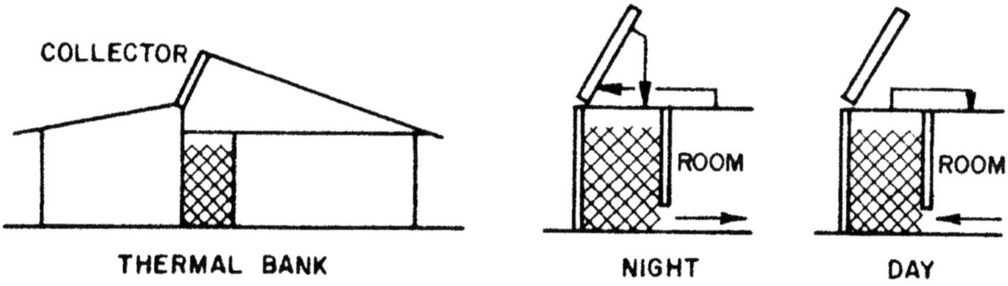

Fig. 2 - Solar House, Perth

The success of this method depends on the extent to which the nocturnal cooling can be effected, which, in turn, depends on ambient conditions, which are unfavourable in the tropics. Meteorological data for Songkhla in the south of Thailand show, for example, that the effective sky temperature at night is fairly constant throughout the year at about 16°C, compared with daytime air temperatures averaging 32°C(4). Therefore, some comfort cooling might be obtainable with difficulty by the efficient radiation of heat to the night sky. This would require a radiating surface completely isolated from the wind, to prevent convective warming by the ambient air, with the protective cover transparent to longwave radiation. A possible arrangement is shown in Fig. 3(5) in which the air to be cooled is blown under the radiating surface. Ideally, the radiating surface should be spectrally selective, having a high emissivity in the wavelength range 8 m to 13 m of the "atmospheric window", and a high reflectivity outside this range. The cover must be transparent at least in the atmospheric window range. If the cover can be made transparent to infra-red radiation and opaque to solar radiation, cooling will occur during the daytime also.

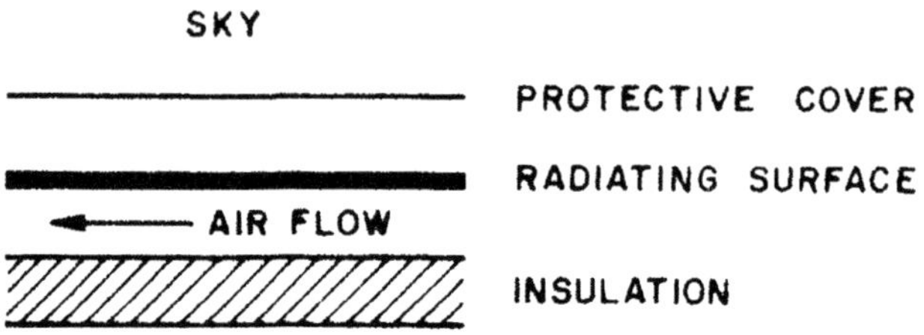

Fig. 3 - Nocturnal Radiation

The achievement of this result has recently been reported in the literature(6)(7) using aluminized Tedlar sheet as the selective radiator and a specially treated double polyethylene film as the cover. This system might be useful in the tropics for the partial cooling of large spaces, such as warehouses and factories, where full air conditioning is not needed.

All these arrangements need a roof area equal in area to that of the cooled floor space. This prevents them from being used in tall buildings in cities. The same conclusion holds for all solar powered active cooling systems. An alternative idea has been proposed and studied by Rao(8). In this system, air heated in solar collectors near the ground rises through ventilation stacks in the centre of the building, see Fig. 4. In a building 60 m high with a collector area ten times the stack cross section area, and solar radiation flux 840 W/m^2, the updraught would be 1.5 m/s. This draught could then be used to draw a flow of air from the windows of the building through the rooms. Although no air cooling is produced, comfort is increased by the solar-induced air movement.

3. COMPRESSION REFRIGERATION

The commonest refrigeration system in commercial equipment employs the liquid-vapour compression system illustratd in Fig. 5. Typical refrigerants are R12 ($C\ Cl_2\ F_2$) and ammonia. The refrigerant near ambient temperature passes from a high pressure at A through the expansion value to a low pressure at B. There it evaporates absorbing heat to produce the refrigeration effect. The low pressure vapour at C then passes through the compressor, where its temperature and pressure are adiabatically increased. To complete the cycle the vapour at D releases latent heat through the condenser and returns to A in the liquid state. The compressor must be driven by mechanical power.

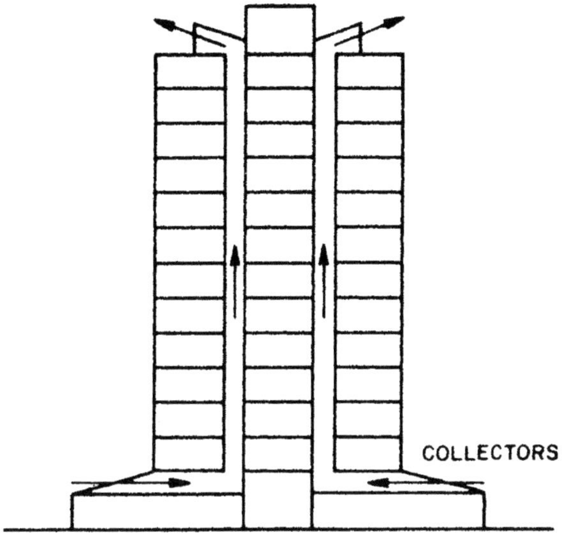

Fig. 4 - Solar Ventilated High Rise Building

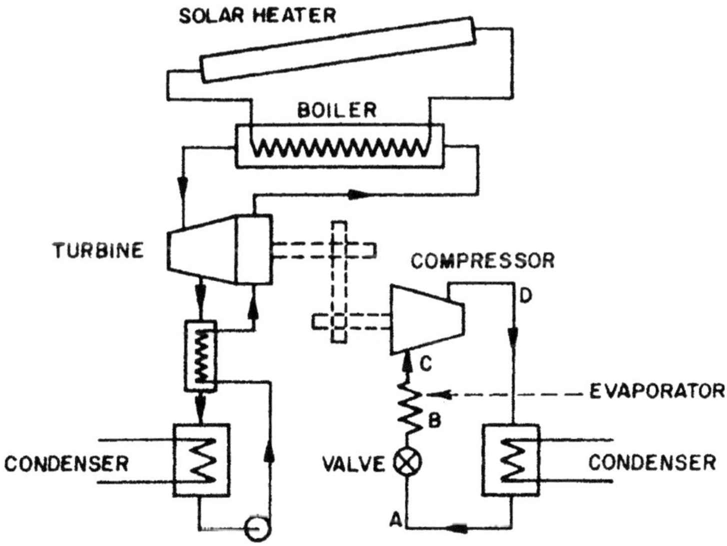

Fig. 5 - Solar Heated Rankine Cycle Compression Air Conditioning System

The coefficient of performance (the refrigeration rate through the evaporator divided by the power supplied to the compressor) may typically be of the order 5.

Solar energy used to operate such a system must produce mechanical power. This has been done by using solar heat to operate a Rankine cycle engine, and also by using solar photovoltaic cells to operate an electric motor.

Prigmore and Barber(9) have given the design considerations and test data for a Rankine cycle prototype in a "Mobile Solar Research Lab", which included a trailer with 58 m^2 of flat-plate solar collectors and a 3 ton (10.6 kW) Rankine cycle cooling system. Hot water at 101°C from the solar collectors was used to heat the working fluid R113 ($C\ Cl_2\ F\ C\ Cl\ F_2$) for a high speed turbine, and cooling water at 29°C was used in the condenser. The efficiency of the Rankine cycle was 11%. The shaft power output drove a conventional 12-piston compressor in the refrigeration unit through reduction gearing. R12 ($C\ Cl_2\ F_2$) was used as refrigerant with evaporator temperature 7°C. The coefficient of performance of the air conditioning unit was 7.4, and the cooling ratio (the room cooling divided by the heat absorbed from the solar collector water) was 0.5. If a collector efficiency of 0.4 is assumed, then the cooling effect obtained is 20% of the solar radiation incident on the collectors in fine weather.

This system illustrates a question that arises whenever solar energy is used to operate thermodynamic cycles. A high thermodynamic efficiency requires a high temperature. In flat-plate collectors, which collect both diffuse and direct solar radiation, 100°C is generally the highest temperature compatible with an acceptable collector efficiency, and the above system represents the best efficiency that can be obtained. The cooling ratio of an absorption refrigeration cycle is limited to about 0.6, and so there is little to choose between the compression and absorption cycles with flat-plate collectors.

However, the overall performance (refrigeration rate divided by the solar radiation flux incident on the collectors) of the Rankine cycle compression refrigeration system can reach 1.5 with collector temperatures as high as 300°C. To obtain these temperatures concentrating solar collectors with mirrors and tracking mechanisms are needed, which are more expensive than flat-plate collectors and (since they do not absorb diffuse solar radiation) are not good in hazy or cloudy weather. But for large installations over 100 tons (350 kW) the Rankine cycle system could become economically attractive(10) and outperform absorption systems in a suitable climate.

Another important consideration is the condensing temperature. If the water-cooled condensers at 30°C in a system with cooling ratio 0.5 are replaced by air-cooled condensers at 54°C, then the cooling ratio of the system will fall below 0.3. Therefore, to obtain the best results from a solar cooling system one should keep the condensing temperature as low as possible. This remark holds equally well for liquid absorption refrigration systems as for Rankine cycle systems.

Photovoltaic solar cells have been used to power domestic refrigerators in remote villages. In Arizona(10) an array of photovoltaic panels of area 88 m^2 had a peak power of 3.5 kW and was used to power a variety of appliances, including refrigerators. Used entirely for powering an air conditionining system this could provide a peak refrigeration rate of several tons, since at the temperatures employed in air conditioners the coefficient of performance can be relatively high. In practice the need to store the electricity and use it when solar radiation is weak would greatly reduce the air conditioning rate obtainable by this method.

The cost of electricity from a photovoltaic system in the United States has been estimated at about $1.00 per kWh for the early 1980's, and below $0.50 per kWh in the late 1980's(10). This compares favourably with the cost of electricity in remote areas , but not with the price of electricity in cities.

4. OPEN CYCLE DEHUMIDIFCATION

Another method of cooling employs the evaporation of water into the surrounding air. The process is governed by the temperature and humidity of the air. Possible ways of using it have been discussed by Szokolay(11) and Dunkle(12). The method is good in hot dry climates, and the question arises as to whether it can be used in tropical humid climates.

The properties of moist air are displayed on a psychrometric chart (Fig. 6) in which the horizontal axis represents the air temperature θ, and the vertical axis represents the humidity ratio W. The curves pass through points having the same relative humidity ϕ, while the sloping lines pass through points having the same adiabatic temperature and represent a process in which water is evaporated or condensed adiabatically exchanging latent heat for sensible heat of the air.

The purpose of an air conditioning system is to produce comfortable temperatures and relative humidities. One index of comfort is the temperature-humidity index defined by the formula

$$\theta - 0.55(1 - \phi)(\theta - 14.4),$$

where θ is the temperature in degrees Celsius, and is the relative humidity expressed as a decimal. The comfort zone normally embraces temperature humidity indices from 20°C to 30°C depending on individual opinion, clothing, and air movement.

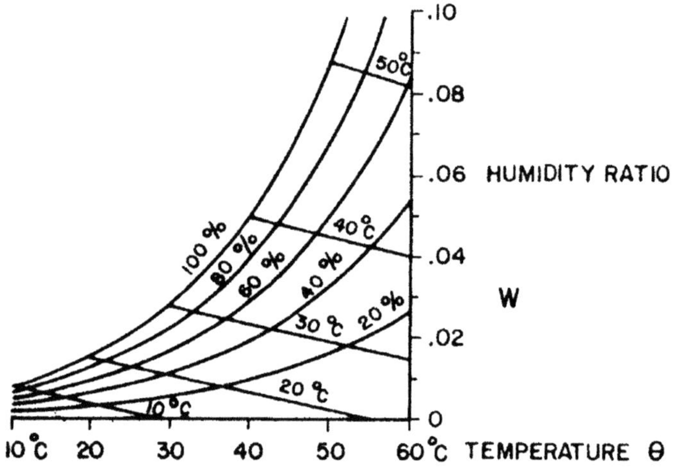

Fig. 6 - Psychrometric chart

Suppose the outside air has temperature 33°C and relative humidity 0.7. The humidity ratio is then 0.0222 and the temperature-humidity index is 30°C, which is barely comfortable. Our task is to reduce the temperature-humidity index to below 25°C by evaporative cooling.

The simplest process is direct evaporative cooling with humidification of the room air as shown in Fig. 7. If the humidifier produces 80% of the possible air temperature drop, then the temperature and relative humidity of the room air will be 29°C and 0.93 respectively, and the temperature-humidity index will be 28°C, which is too high.

To avoid the humidification of the room air one can apply indirect evaporative cooling as shown in Fig. 8. The outside air is cooled by humidification and this air, or the cooled water, is used to cool another air stream from the outside to the room via a heat exchanger.

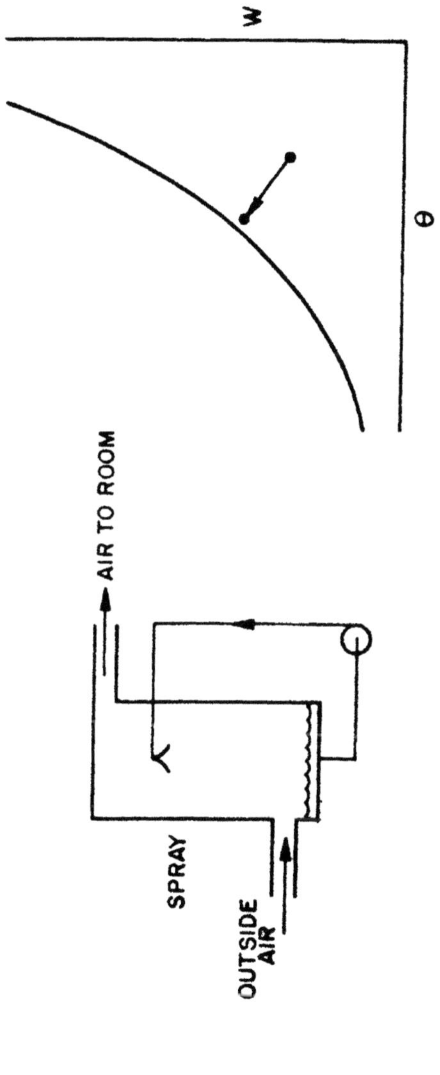

Fig. 7 - Direct Evaporative Cooling

The lowest possible temperature of the air entering the room is 29°C, but the humidity ratio is still 0.0222 so the relative humidity is only 0.87 and the temperature-humidity index is 28°C. This is still too high, but the lightly lower relative humidity is an improvement. Further reductions of the room air temperature can be made by humidification of both the outside air stream and the room air. The room air may also be returned to precool the incoming air, but none of these methods are effective enough for the humid tropics. The only way to use the method is to dehumidify the outside air before it enters the system.

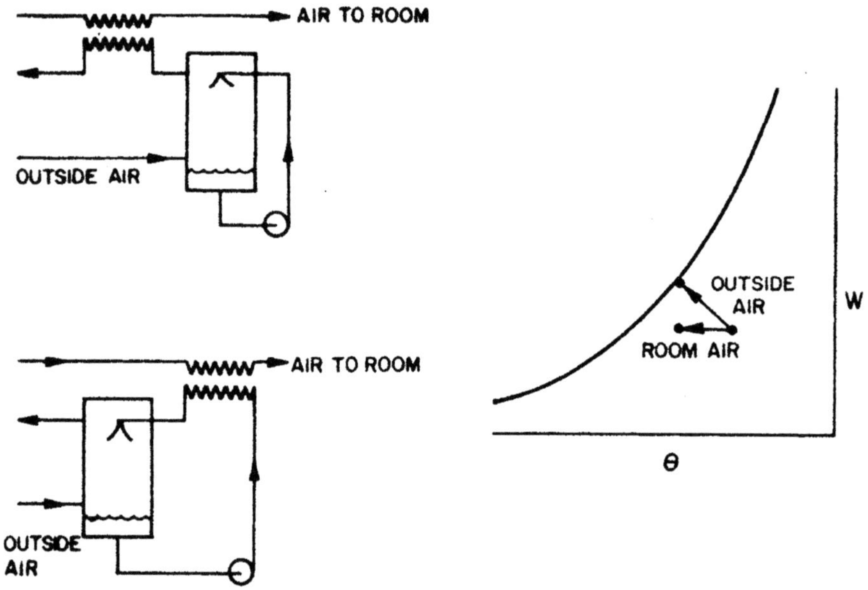

Fig. 8 - Indirect Evaporative Cooling

Fig. 9 shows the principle of a simple dehumidification method. Suppose that the outside air has temperature 33°C and humidity ratio 0.0222 (point 1 on the diagram). Let the desired condition of the room air have a temperature humidity index of 25°C and relative humidity 60%. Then the room temperature must be 28°C and the humidity ratio 0.0142 (point 4). Point 4 can be reached from point 1 by (1) dehumidifying the air with the help of a desiccant so as to reach point 2, (2) removing the latent heat of absorption of the water vapour so as to reach point 3 and (3) rehumidifying the air so as to reach point 4. If the air temperature at point 3 is the same as at point 1 (33°C), one finds that the humidity ratio at points 2 and 3 must be 0.0122, and the temperature at point 2 is 58°C.

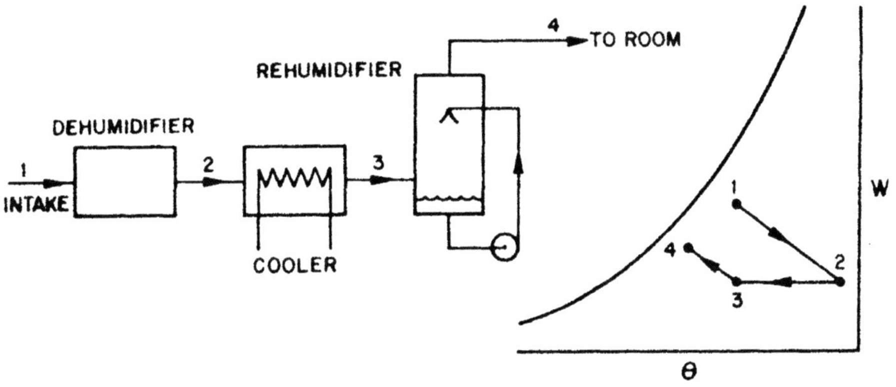

Fig. 9 - Simple Dehumidification

A variety of different systems all based on this principle have been proposed, some of them being rather complicated. An early system proposed by Lof(13) used triethylene glycol as the desiccant. Fig. 10 shows the system simplified by the omission of several heat exchangers. On a clear day, solar heated air from 10.2 m^2 of collector at temperatures from 60°C to 79°C could remove 2.7 to 6.4 kg of water per hour from glycol solutions of approximately 90% concentration, which is equivalent to a cooling capacity of 0.4 to 1.1 tons of refrigeration per 10 m^2 of solar collector for 8 hours. The total water evaporated per

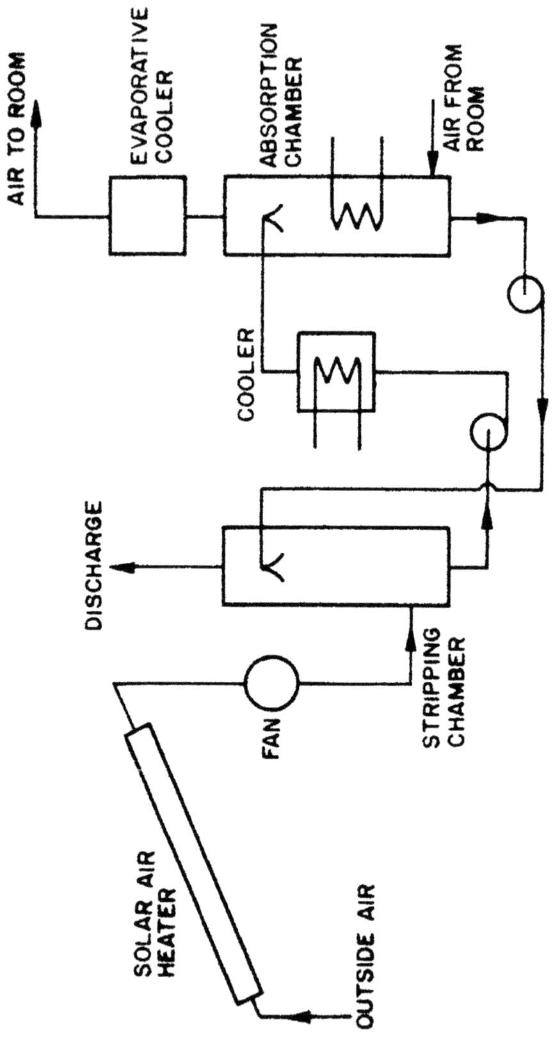

Fig. 10 - Solar Cooling by Dehumidification

day was about 32 kg, corresponding to about 70 MJ per day per $10m^2$ of solar collector. The potential average cooling capacity of a 56 m^2 collector in clear weather was, therefore, estimated at 14.7 KW, or 4 tons of refrigeration, for 8 hours.

It is possible to simplify the system by combining the solar collector and the stripper. Successful application of this idea in Turkmenia, U.S.S.R., has been reported by Baum and Kakabayev(14). An aqueous solution of lithium chloride is used as the desiccant. After picking up water in the absorption chamber the solution is pumped to the top of the building and allowed to flow down over the roof in a thin layer. There it releases water vapour to the outside air, the heat of evaporation being supplied by solar radiation, and the strong solution collected at the bottom of the roof is returned to the absorber. One of the installations built has a refrigeration capacity of 11.6 tons (40.7 kW) for cooling the rooms of a three-storey nine-apartment building with a floor area of 500 m^2. Chilled water from the evaporator is used for air conditioning.

This system, though satisfactory in the dry climate of Turkmenia, could not be used as it stands in the tropics because of the rain. However, investigations by Mullick and Gupta(15) in Madras using a calcium chloride solution as desiccant have shown that stripping can be achieved by allowing the solution to flow down over a blackened corrugated iron sheet inclined at 18° to the horizontal under a protective glass cover. Outside air heated by the sun rises by free convection through the space between the blackened sheet and the cover and carries away water vapour from the solution. It seems that dehumidification systems designed in this way should be applicable in equatorial regions. Fig. 11 shows how such a system might be arranged.

Another important line of research has been followed for some years using solid desiccants. From among the vaious systems studied two are selected for consideration here. The first, due to Close and Dunkle [16] and illustrated in Fig. 12, has two beds of desiccant (silica gel) that serve as energy stores. At any particular time one is being regenerated by solar heated air, as shown in the top diagram, and the other one is dehumidifying the room air, as shown in the bottom diagram. The following conditions are assumed: outside air temperature 35°C and humidity ratio 0.0223, a temperature of 25°C inside the building, an air conditioning load of 5.25 kW sensible heat and 0.5 kW latent heat (total 1.6 refrigeration tons), a building air flow rate of 0.87 kg/s. The temperature rise through the building is then 6°C. Assuming also an effectiveness of 0.8 for the heat exchanger, the evaporative cooler and the dehumidifier, it was found that the inside air humidity ratio would be 0.0094, and the temperature of the air leaving the solar collector would have to be 92°C.

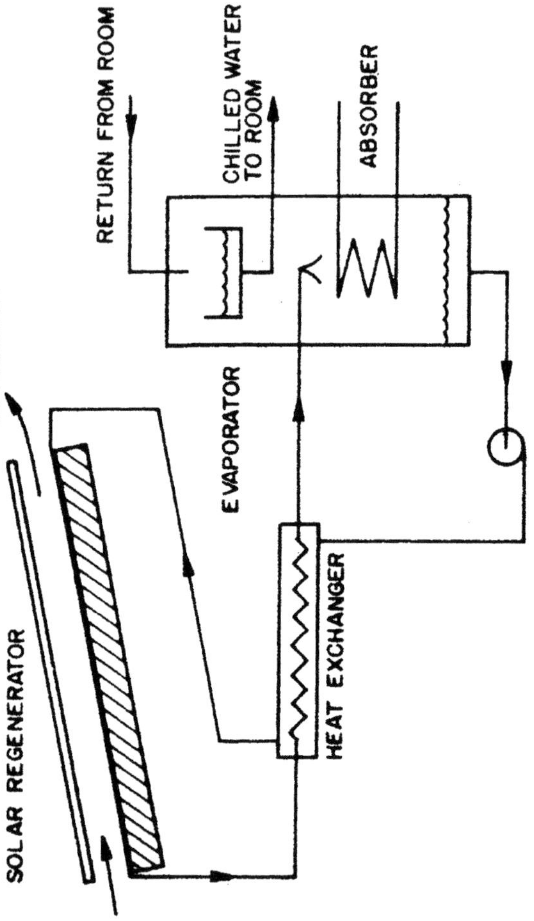

Fig. 11 - Stripping in a Solar Collector

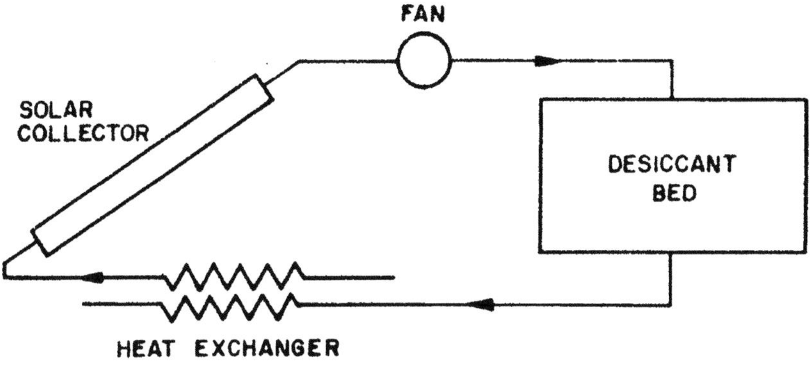

STORE REGENERATION

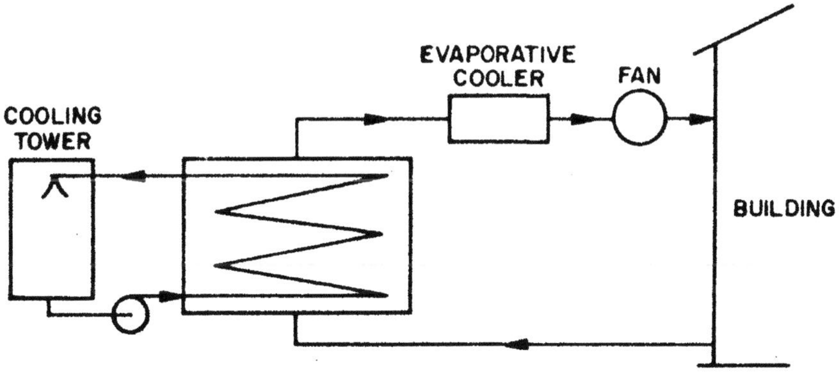

STORE COOLING AND BUILDING COOLING

Fig. 12 - Dehumidification Cooling with Dessicant Bed

Each of the dessicant beds would have to be 0.68 m^3 and contain 1000 kg of silica gel, and a solar collector with area 140 m^2 would have to have an efficiency of 0.26 if the insolation were 18.5 MJ/m^2 per day. Alternatively, if a collector with efficiency 0.48 were used, its area need be only 76m^2.

Finally, we describe one of the systems studied theoretically by Nelson et al.(17) using rotary generators. A rotary generator is a porous wheel so arranged that the porous matrix of the wheel passes in turn through two streams of air and transfers heat and moisture between them(18). The matrix of the wheel consists of ribbon wound spirally around the rotation axis on spacers. The material of the matrix may be a porous material carrying a desiccant such as silica gel, or a highly absorbent material such as wool.

Fig. 13 shows the system and Fig. 14 shows the state of the air at different points in the system on the psychrometric chart. The outside air at state 1 is dried by the dehumidifier wheel (state 2), cooled by the regenerator wheel (state 3), and humidified to state 4 for supply to the room. In the room the air picks up heat and moisture reaching state 5. It is then drawn out of the room, humidified and cooled to state 6, dried by the regenerator wheel (state 7), heated to state 8, and exhausted at state 9 after having given up its heat to dry the dehumidifier wheel. The heat input between states 7 and 8 may be from solar energy or auxiliary heating.

The simulations performed assumed a floor area 140 m^2 with a sensible heat load of 3.8 kW, a latent heat load of 1.0 kW, and a comfort zone from 22.2°C to 24.4°C with maximum humidity ratio 0.0112. Using past weather data for Miami, it was found that a 45 m^2 flat-plate solar collector having two glass covers and a non-selective surface with a 3.375 m^3 water heat storage tank would provide 95% of the energy required. The temperature required in the regeneration air stream would be less than 65°C for 60% of the time and above 80°C for only 5% of the time. The overall performance would be about the same as that of absorption refrigeration systems but, since the dehumidification cycle would require lower temperatures, the dehumidification system was better matched to a thermal solar energy supply.

5. AMMONIA-WATER ABSORPTION REFRIGERATION

The commonest ammonia absorption refrigeration system uses water as the absorbent, though a solution of sodium thiocyanate in ammonia has been successfully used(19). The principle of the method will be explained with the help of Fig. 15.

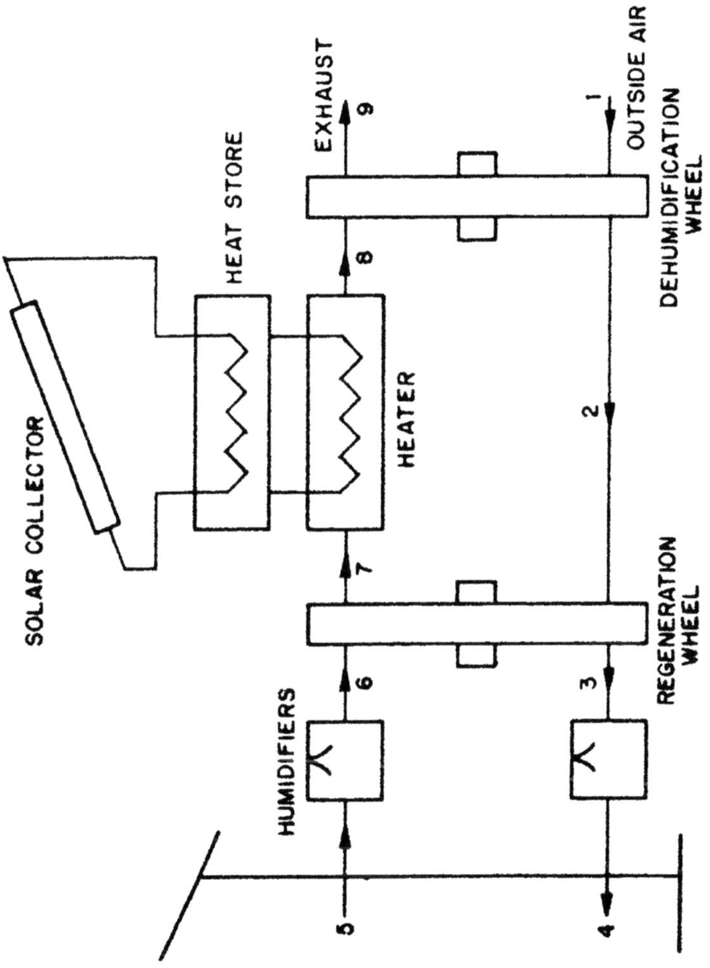

Fig. 13 - System with Rotary Generators

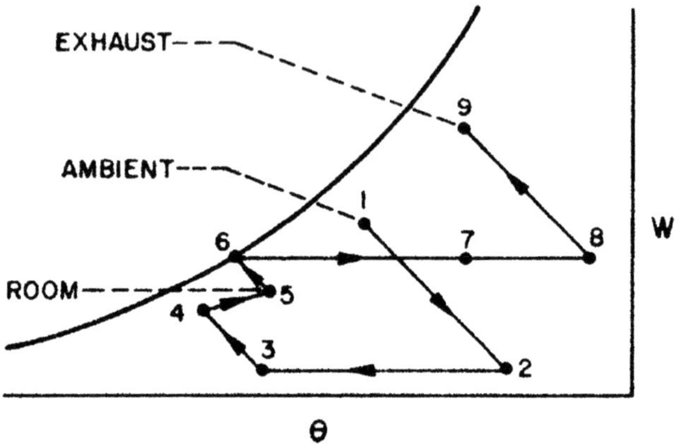

Fig. 14 - Psychrometric Chart for above System

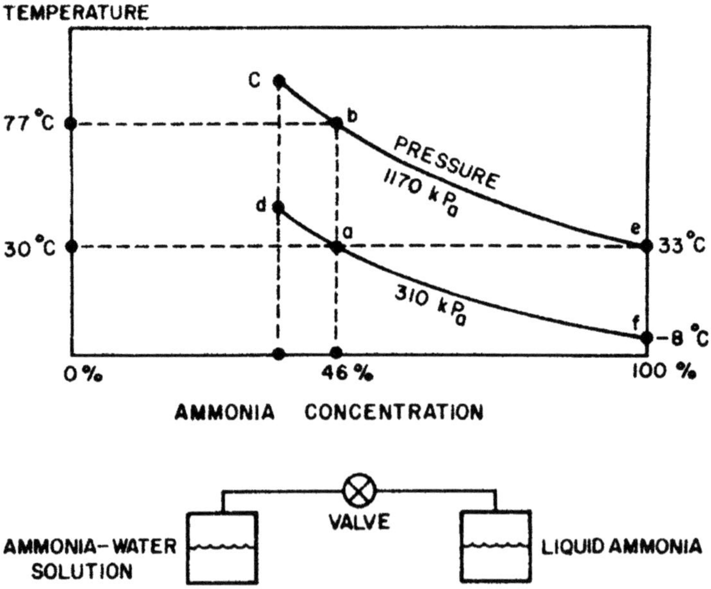

Fig. 15 - Ammonia-Water Absorption Cycle

An ideal cycle consists of the following processes.

(1) a $\rightarrow$ b : The heating of the ammonia-water solution with the valve closed, from ambient temperature T_a(30°C) and solution pressure P_a(317 kPa) to the generation temperature T_b(77°C) at which the pressure P_b(1170 kPa) is the saturation vapour pressure of pure ammonia at ambient temperature, the concentration remaining constant (0.46).

(2) b $\rightarrow$ c : The vaporization of ammonia from the solution, with the valve open, during which the pressure remains constant, ammonia condenses in the second vessel at ambient temperature (point e on the chart), the solution concentration is reduced (to 0.40), and the solution temperature rises from T_b to T_c(87°C).

(3) c $\rightarrow$ d : The cooling of the solution, with the valve closed, from temperature T_c to the temperature T_d (39°C) at which the pressure is P_a, the concentration remaining constant.

(4) d $\rightarrow$ a : The evaporation of the liquid ammonia, with the valve open, at constant pressure and temperature T_f(-8°C) (point f on the chart), during which ammonia is reabsorbed into the solution increasing its concentration while the solution temperature falls from T_d to T_a.

The points on this cycle have been chosen to produce freezing temperatures in a tropical climate. The processes a $\rightarrow$ b and b $\rightarrow$ c require a supply of heat, which can be obtained from solar energy. The generating temperatures T_b to T_c are within the range of flat-plate collectors in fine weather, though they are rather high for good efficiency if the black receiving surface is non-selective. One should note that with any absorption system the elevation (T_b - T_a) of the generation temperature above ambient temperature is always slightly higher than the depression (T_a - T_f) of the refrigeration temperature below ambient temperature.

The ideal cooling ratio of the cycle is defined to be the quotient Q_f/Q_g of the heat of refrigeration Q_f absorbed by the ammonia evaporating at temperature T_f and the heat of generation Q_g absorbed by the solution at temperature T_g during the distillation of the ammonia. At the beginning of the cycle shown in Fig. 15 we have T_g = T_b. Let Q_a be the total heat rejected to the environment at ambient temperature T_a during the distillation and reabsorption of ammonia.

Since in a complete cycle the system returns to its initial state, the total change in its internal energy is zero, thus

$$Q_g + Q_f - Q_a = 0.$$

Likewise, the total change in its entropy is zero, thus

$$(Q_g/T_g) + (Q_f/Q_f) - (Q_a/T_a) = 0,$$

the temperatures being on the absolute scale. Eliminating Q_a we obtain an upper limit for the cooling ratio

$$Q_f/Q_g = T_f(T_g - T_a)/T_g(T_a - T_f),$$

which for the above data comes to 0.93. In practice, because of heat losses and the need to heat the solution, the actual cooling ratio varies from 0.7 in large continuous systems to 0.3 in small intermittent systems.

If the generator is heated by solar energy, the overall performance of the system, defined to be the product of the cooling ratio and the efficiency of the solar collector, depends on the generator temperator T_g. This dependence has been studied by a number of authors. Satcunanathan and Kochhar(20) found, for example, that if a non-selective flat-plate collector with two glass covers is used to operate a continuous ammonia-water absorption system with incident solar radiation 1 kW/m^2, evaporator temperature 2.2°C, absorber temperature 30°C, and condenser temperature 32.2°C, the best generator temperature is 77°C giving an overall performance of 0.38.

Solar powered intermittent ammonia-water absorption refrigeration using flat-plate collectors was pioneered by Chinnappa(21). Numerous experimenters have since undertaken similar work. Fig. 16 shows the arrangement at present being studied in the Asian Institute of Technology (AIT)(22). The flat-plate solar collectors with total area 5 m^2 contain 67 kg of ammonia-water solution with concentration 0.46 from which ammonia is vaporized by solar heating during the day with valve A open and valve B closed. Flat auxiliary mirrors are used to enhance the solar heating. The ammonia is condensed and collected in a receiver in thermal equilibrium with a 3.5 m^3 tank of water kept at 3°C cooler than the mean ambient temperature (and about 8°C cooler than the ambient temperature on a hot afternoon) by thermally insulating the tank during the day and exposing it at night. Over 14 kg of liquid ammonia can be distilled on a bright day.

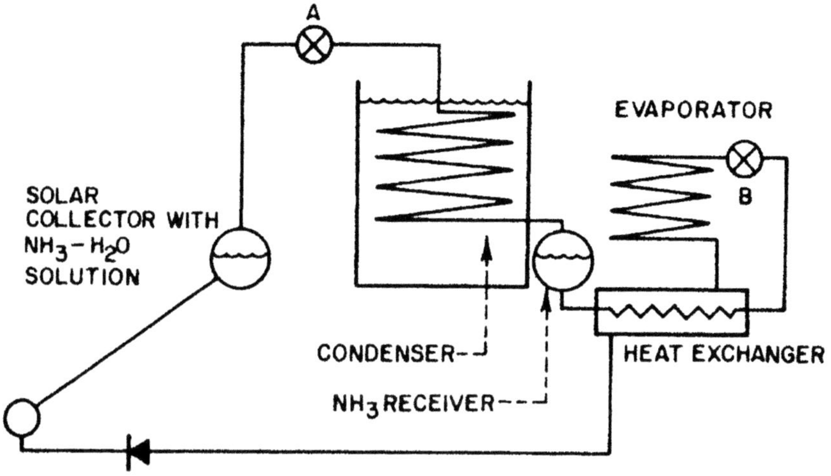

Fig. 16 - Experimental Unit in AIT

At sunset, valve A is closed and air is allowed inside the collector to cool the ammonia-water solution. Later in the evening, the expansion valve B is opened to produce refrigeration in the evaporator. The ammonia vapour is reabsorbed into the solution through inlets to the collector feeder, while the heat of absorption escapes from the collector surface. In one test, which did not use up all the liquid ammonia available, 25 kg of ice were produced from 32 kg of water initially at 28°C. If all the ammonia had been used all the water would have frozen.

This type of solar refrigerator, because it does not need any auxiliary power, is likely to be of use in ice making or cool storage for food or vaccines in remote mountain villages or small islands. However, the use of the intermittent cycle restricts the overall performance, which in the AIT experiments has been about 8%. If solar absorption refrigeration is to be used for food storage or air conditioning in or near cities, then continuous absorption cycles employing small amounts of electrical power for circulating the liquids should be used. Even for remote villages such systems could be designed with the help of small electrical circulation pumps powered by photovoltaic cells. It would be interesting to determine whether the increased overall performance would offset the extra cost of the electrical units.

Solar powered continuous ammonia-water absorption refrigeration using flat-plate collectors was pioneered by Farber(23, 24). Fig. 17 shows the arrangement used in an experimental 3 ton air conditioning unit(23). Solar heated water from an array of collectors having total area 37 m^2 enters the generator at a temperature between 60°C and 93°C, with the usual operation at about 79°C. Ammonia vapour from the generator is condensed and allowed to flow through the expansion valve into the evaporator to produce the cooling. The vapour then passes into the absorber where it is reabsorbed into the solution. A small pump returns the strong solution (concentration 0.58) from the low pressure generator through a heat exchanger where it takes up heat from the weak solution (concentration 0.39) passing from the generator to the absorber. The cooling ratio of the system is 0.57.

A smaller unit working in the same way, but designed for ice making, is shown in Fig. 18(24). The flat-plate collector area is 1.5 m^2, and the yield on a good day is 18 kg of ice from water initially at 24°C. This represents a cooling effect of 5.2 MJ per square metre of collector surface per day. In the case of the AIT experiments the cooling effect was 2.9 MJ per square metre of collector surface per day. These figures show the extent to which the continuous refrigeration system surpasses the intermittent system. An important requirement in solar cooling is the functioning of the system at night. If the excessive use of auxiliary power is to be avoided, then energy storage must be available. The most convenient form of energy storage in ammonia-water systems is in a supply of excess pure liquid ammonia. The intermittent system being developed by AIT (Fig. 16) uses this method. It has also been used in a continuous absorption refrigerator developed by Dornier-System for a food cooling project in Egypt. Fig. 19 shows a simplified diagram of the system.

The cooling house consists of an inner cooling chamber of 10 m^3 designed to hold 300 kg of food at temperatures in the range 6°C to 8°C, and a buffer room of 7.5 m^3 not actively cooled. The evaporator of the refrigerator is fixed under the ceiling of the cooling chamber. The double glazed flat-plate solar collector acting as a roof over the system has an area 25 m^2 and supplies heat directly to the generator at 90°C through aluminium heat pipes filled with R113. The system pressures are 1500 kPa in the generator and 450 kPa in the evaporator. The output of the cooling machine is 4 kW, but it has the special feature of being able to store a cooling capacity of 10 kWh. During the day the generator produces a surplus of liquid ammonia that is stored in a vessel prior to entering the evaporator. At the same time, a surplus

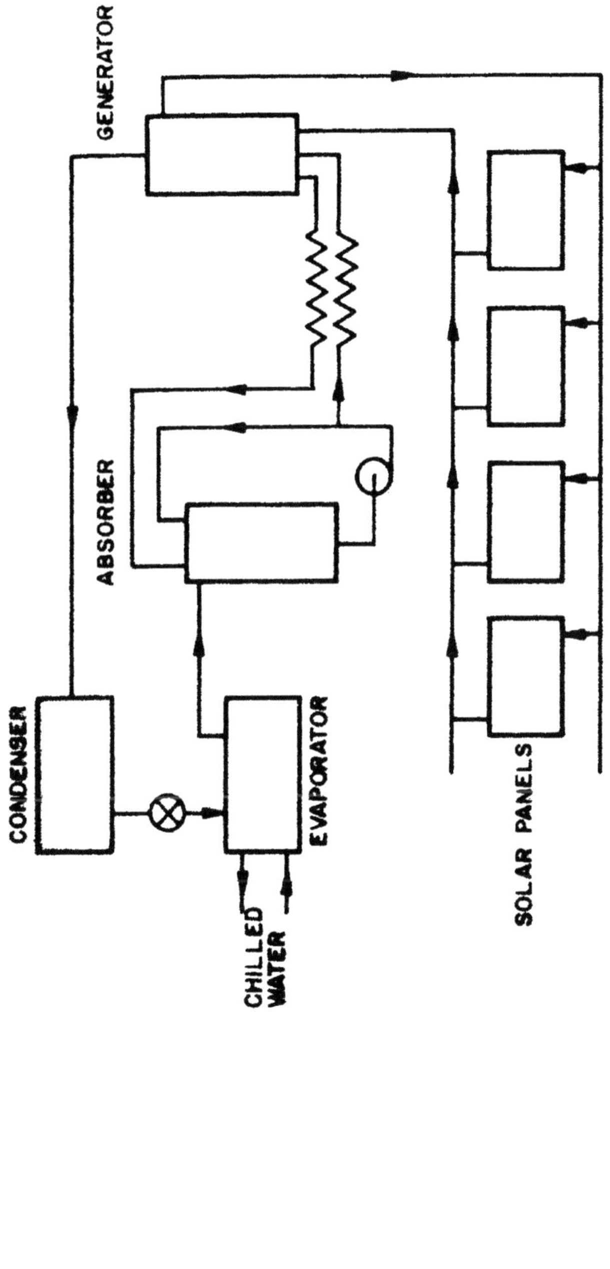

Fig. 17 - Three Ton Air Conditioning System

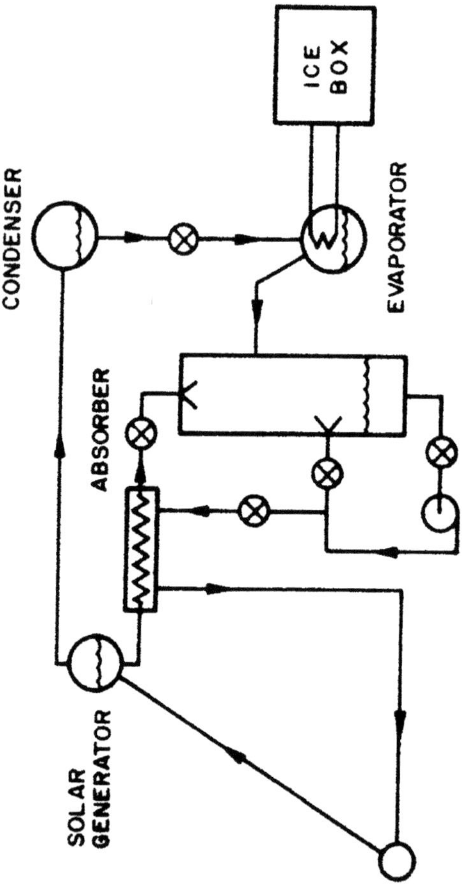

Fig. 18 - Small Refrigerator

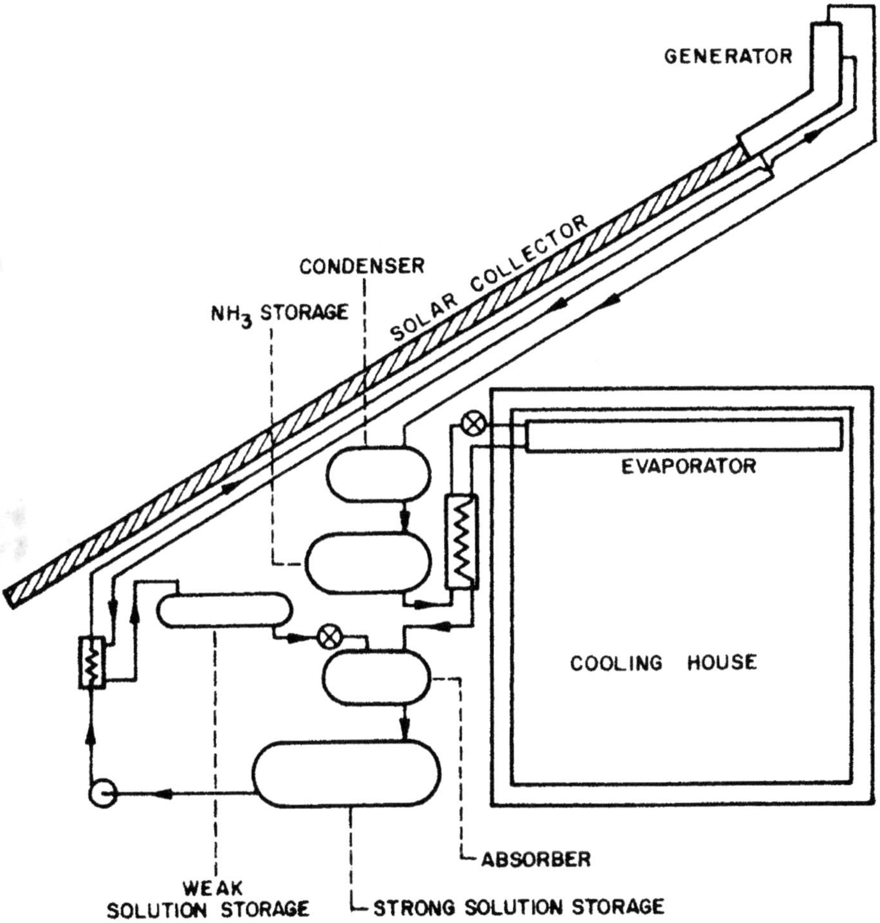

Fig. 19 - Cool Storage System

of weak ammonia-water solution is stored in another vessel to supply the absorber during operation in the absence of solar heating.

6. WATER-LITHIUM BROMIDE ABSORPTION

The most popular type of solar refrigerator for air conditioning is the water-lithium bromide chiller, and indeed it is the only type on the commercial market. The principle of operation of the system is exactly the same as that of the ammonia-water system. The refrigerant is water and the absorbent is a solution of lithium bromide in water. The cycle involves the distillation of water from the hot weak solution and the reabsorption of this water into the cold strong solution. Refrigeration is produced by the evaporating water during the reabsorption process.

Continuous systems cannot be operated below the freezing point of water. Consequently, the process is suitable only for air conditioning with evaporation temperature around 4°C. Operating pressures are very low (4 kPa t 0.8 kPa). One problem is the risk of crystallization of the lithium bromide if too much water is removed from the solution, either through high generation temperature or low condenser temperatures. However, the cooling ratio is typically 0.8 against 0.6 for the continuous ammonia-water system.

Fig. 20 shows the basic circuit of a water chiller that is available commercially(25) and is designed for operation by solar energy. Water at 75°C to 100°C obtained from single glazed flat-plate solar collectors with a selective surface is used to heat the generator. The cooling water is at 30°C, and chilled water for air conditioning emerges from the unit at 9°C. The cooling capacities available are from 2 tons to 50 tons depending on the size of the unit. A hot water storage tank is used for operation at night.

Many buildings have now had this system installed(26). A small test installation in Bangkok has a solar collector area 52 m^2 connected to a 2 ton unit for cooling a room of floor area 44 m^2. At present, the total cost of such a system is about twice the cost of a conventional electric compression system. The pay back period in terms of saved power cost is estimated at 8 to 10 years, while the life of the system is expected to be 20 years.

Another interesting possibility for a solar water-lithium bromide absorption system is the open cycle shown in Fig. 21(27). Instead of regenerating the absorbent solution by distilling water from it in a closed circuit, the water is driven off to the atmosphere in an open solar collector as in Fig. 21. The strong solution then passes through a liquid column where its pressure is reduced before entering the

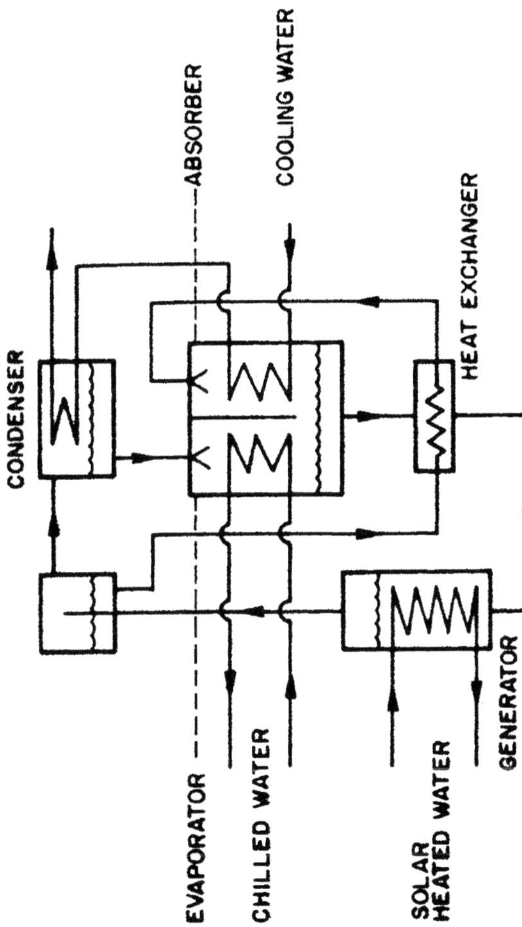

Fig. 20 - Water-Lithium Bromide Chiller

absorber/evaporator. The vacuum pump shown removes air dissolved in the solution during exposure in the solar collector.

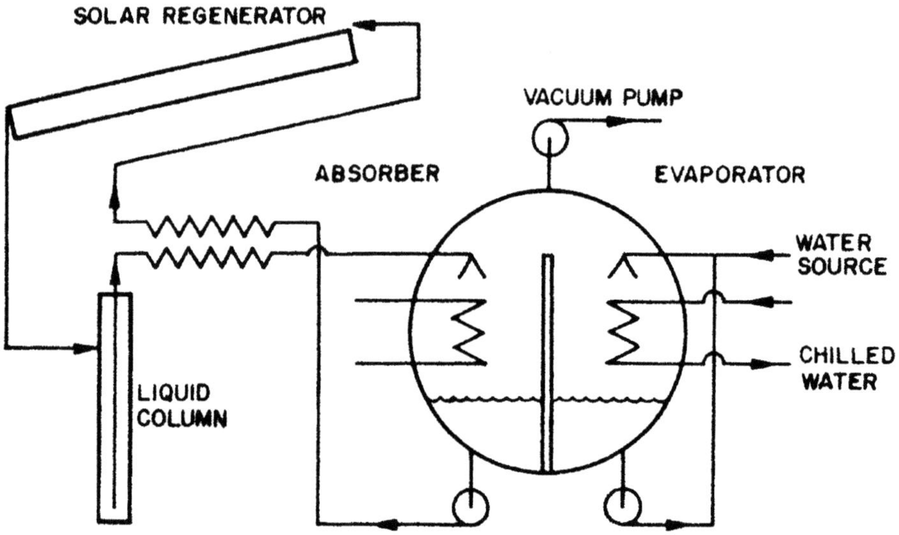

Fig. 21 - Open Cycle Absorption Chiller

The advantages of this system would include a simpler collector, which also acts as the generator, and a reduction in thermal losses, since the energy is transferred as chemical potential. On the other hand, auxiliary power to operate the pumps could be excessive and there is a risk of contaminants entering the system from the outside.

The overall performance of the system is governed entirely by the rate at which water is driven from the solution in the collector, since this determines the flow of water that can be introduced into the evaporator as refrigerant to maintain steady continuous operation. Studies have shown that it compares favourably with that of closed cycle solar powered absorption cooling, ranging from 0.18 to 0.45 depending on conditions. In humid climates rather long collectors (over 20 m) must be used to provide sufficient transfer of water from the solution to the outside air.

7. SOLID ABSORPTION REFRIGERATION

For intermittent refrigeration systems solid absorbents may be used instead of liquid absorbents. If ammonia is the refrigerant, a suitable solid absorbent is calcium chloride(28, 29). There are two reactions each with heats of absorption approximately twice the latent heat of vaporization of ammonia. Hence the theoretical cooling ratio is about 0.5, and in practice is near 0.4. For a condensation temperature of 28°C, the required generation temperatures for the two reactions are about 85°C and 96°C, while for an evaporator temperature of -10°C the absorption temperatures are 53°C and 65°C.

A porous solid made of calcium chloride and 15% Portland cement place in tubes in a flat-plate solar collector may be used as the generator. The generating temperatures are somewhat higher than those in ammonia-water systems. Consequently, the collector should have double glazing and a selective surface, and perhaps slight concentration by means of mirrors. On the other hand, the reabsorption of ammonia should be easier. During the refrigeration at night the back insulation of the collector may be removed to release the heat of absorption. Since the solid absorbent cannot evaporate, no rectification column is needed as in the case of the ammonia-water combination. But there seems to be some uncertainty as to whether the absorbent will continue to function efficiently after a very large number of cycles.

A very promising recent development in solar refrigeration is the use of zeolites as solid absorbents. In particular, the water-zeolite 13 x combination is being carefully studied. Besides providing a method of storing solar energy as chemical potential through the heat of absorption of water(30), it also provides a good intermittent refrigeration system with water as refrigerant(31).

Zeolite 13 x is a sodium alumino-silicate with a crystal structure having cavities capable of accepting water molecules by hydrogen bonding. The important property of zeolite for solar refrigeration is that the quantity of water absorbed in equilibrium is strongly dependent on temperature and only weakly dependent on the water vapour pressure (Fig. 22). Therefore, during the refrigeration process at night, when the zeolite is near ambient temperature and unsaturated, water vapour is readily absorbed keeping the pressure below 0.55 kPa and causing the refrigerant water to freeze, while during the day the efficiency of the regeneration process is only slightly affected by the condensation pressure.

The maximum theoretical cooling ratio is equal to the latent heat of vaporization of water (2.3 MJ/kg) divided by the heat of absorption of water on zeolite. This is about 4.2 MJ/kg for synthetic zeolite giving

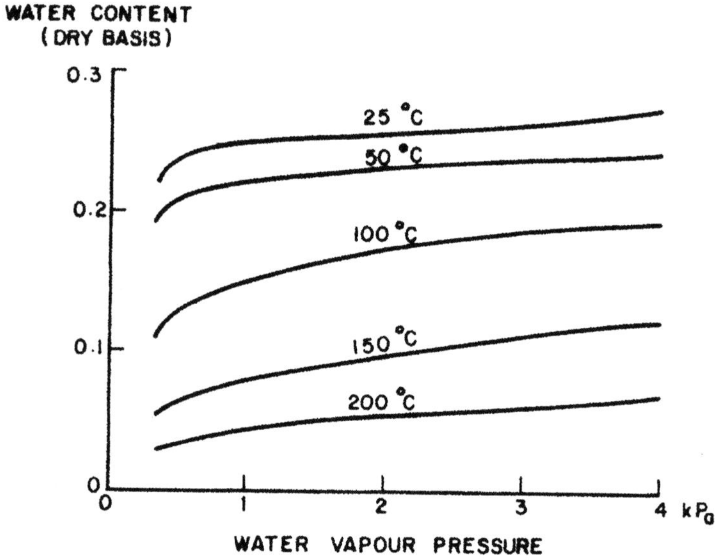

ADSORPTION ISOTHERMS : WATER VAPOUR ON ZEOLITE 13 X

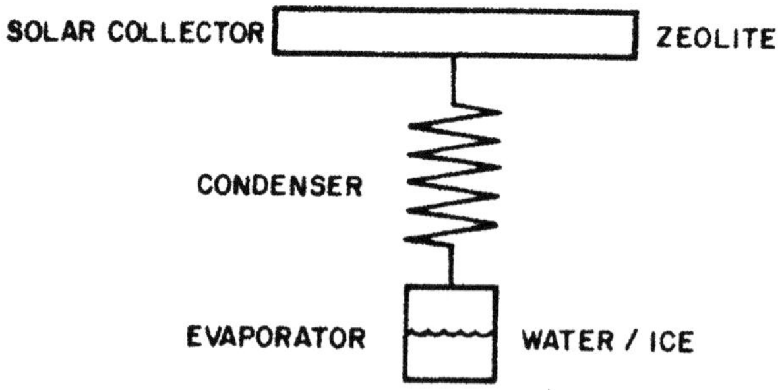

Fig. 22 - Solar Refrigeration with Zeolite

a cooling ratio of 0.55, and about 2.8 MJ/kg for natural zeolite giving the improved cooling ratio of 0.8.

A small commercially produced refrigerator by Tchernev(31) working on this principle with a flat-plate collector of area 0.74 m^2 containing 36 kg of natural zeolite makes 6.8 kg of ice after a sunny day (20 MJ/m^2). This ice, being contained in the hermetically sealed system, is not accessible to the user. It melts slowly during the day and cools an insulated box of volume 113 dm^3. In dull weather the system continues to produce cooling due to the desorption and absorption of water resulting from the diurnal temperature variations of the solar panel, though ice may not be formed.

REFERENCES

1. Hay, H.R. and Yellott, J.I. (1970) A naturally air conditioned building. Mechanical Engineering 92, pp 19-25

2. Niles, P.W.B. (1976) Thermal evaluation of a house using movable-insulation heating and cooling system. Solar Energy 18, pp 413-419

3. Lawrance, R. (1979) Private communication

4. Excell, R.H.B. (1978) The atmospheric radiation climate of Thailand. Solar Energy 21, pp 73-79

5. Johnson, T.E. (1975) Radiation cooling of structures with infra-red transparent wind screens. Solar Energy 17, pp 173-178

6. Catalanotti, S. (1975) The radiative cooling of selective surface. Solar Energy 17, pp 83-89

7. Addeo, A. et al. (1980) Light selective structures for large scale natural air conditioning . Solar energy 24, pp 93-98

8. Rao, K.R. (1979) Solar induced ventilation of buildings in warm humid climates. Seminar on the Applications of Solar Energy in South and Southeast Asia, 19-23 Feb. 1979, AIT, Bangkok

9. Prigmore, D. and Barber, R. (1975) Cooling with the sun's heat. Solar Energy 17, pp 185-192

10. Rosenblum, L. et al. (1979) Photovoltaic power systems for rural areas of developing countries. NASA Technical Memorandum 79097

11. Szokolary, S.V. (1975) Solar Cooling. Review 1, Architectural Science Unit, U. of Queensland

12. Dunkle, R.V. (1966) Regenerative cooling systems. Australian Refrigeration, Air Conditioning and Heating 20, pp 13-23

13. Lof, G.O.G. (1963) The heating and cooling of buildings with solar energy. Cht. 12 in Zarem, A.M. and Erway, D.D. Introduction to the Utilization of Solar Energy, McGraw Hill, N.Y.

14. Baum, V. and Kakabayev, A. (1978) Utilizing solar energy for creating comfortable room conditions in Turkmenia. Paper No. 1185, International Solar Energy Congress, 16-21 Jan. 1978, New Delhi

15. Mullick, S.C. and Gupta, M.C. (1974) Solar desorption of absorbent solutions. Solar Energy 16, pp 19-24

16. Close, D.J. and Dunkle, R.V. (1976) Open cycle solar energy cooling systems. Annual Engineering Conference, Townsville May 1976, The Institute of Engineers, Australia

17. Nelson, J.S. et al. (1978) Simulations of the performance of open cycle desiccant systems using solar energy. Solar Energy 21, pp 273-278

18. Dunkle, R.V., Banks, P.H. and Ellul, W.M.J. (1976) Regenerator research, development and applications in Australia. Paper 56, International Institute of Refrigeration meeting, 6-10 Sept. 1976, Melbourne, Australia.

19. Swartman, R.K., Ha, V. and Swaminathan, C. (1975) Comparison of ammonia-water and ammonia-sodium thiocyanate as the refrigerant-absorbent in a solar refrigeration system. Solar Energy 17, pp 123-127

20. Satcunanathan, S. and Kochhar, G.S. (1975) Optimum operating conditions of ammonia-water absorption refrigeration systems for flat plate solar collector temperatures. Paper 44/4, International Solar Energy Congress, U. of California, 1975

21. Chinnappa, J.C.V. (1962) Performance of an intermitter refrigerator operated by a flat-plate collector. Solar Energy 6, pp 143-150

22. Excell, R.H.B. et al. (1978) Experiments on a prototype solar powered refrigerator for remote areas. Paper No. 0371, International Solar Energy Congress, 16-21 Jan. 1978, New Delhi

23. Farber, E.A. et al. (1966) University of Florida solar air conditioning system. Solar Energy 10, pp 91-95

24. Farber, E.A. (1970) Design and performance of a compact solar refrigeration system. Paper No. 6/58, International Solar Energy Society Conference, 2-6 March, 1970, Melbourne, Australia.

25. Yazaki Solar Air Conditioning System (1979), Yazaki Comp., Tokyo

26. Yazaki News (Special Issue), 1 Jan. 1979, Yazaki Comp., Tokyo

27. Collier, R.K. (1979) The analysis and simulation of an open cycle absorption refrigeration system. Solar Energy 23, pp 357-366

28. Muradov, D. and Shadiev, O. (1971) Test of a solar absorption refrigerator. Heliotechnology 7 (3), pp 33-35

29. Wors e-Schmidt, P. (1978) Solar Refrigeration by Means of a Solid-Absorption Process, Report F 30-78.01, Technical University of Denmark

30. Shigeishi, R.A., Langford, C.H. and Hollebone, B.R. (1979) Solar energy storage using chemical potential changes associated with drying of zeolites. Solar Energy 23, pp 489-495

31. Tchernev, D.I. (1979) Solar refrigeration utilizing zeolites. Proc. 14th Intersociety Energy Conversion Engineering Conf., 2070-2073, Amer. Chem. Soc.

PERFORMANCE COMPARISON OF TWO TYPES OF SOLAR COLLECTORS

Bong Tet Yin,
P. Eng., Associate Professor in Mech. Eng.,
National University of Singapore.

1. INTRODUCTION

There are many types of solar collectors available on the market today, and they can be used in a variety of applications which require the conversion of solar energy into thermal energy. When selecting the solar collectors for a particular usage, it is necessary to compare the thermal performance of different solar collectors. Such comparison can be conveniently done if all solar collectors are tested and their performance presented in the same way. However, this is not always the case, since solar collectors are produced in different countries, and each manufacturer can also have his preference in presenting the thermal performance of his solar collector. As a case in point, the thermal performance of a flat plate collector and an evacuated tube collector are to be compared, where there is no uniformity in the performance presentation format.

2. DESCRIPTION OF SOLAR COLLECTORS

A flat plate collector and an evacuated tube collector are being considered for an application where it is necessary to produce hot water in the range of 80° - 90°. This temperature range is required, for instance, for some industrial processes and absorption chillers. The absorber plate of the flat plate collector is of the tube-in-chest construction, a longitudinal section of which is shown in Fig. 1. Each evacuated tube collector consists of 10 collector elements. A cross section of one collector element is shown in Fig. 2. A vacuum is maintained between the glass tube and the collector tube. The technical data of the two solar collectors is as follows:

Bill B. P. Lim (ed.), Solar Energy Applications in the Tropics, 129–137.

	Flat Plate	Evacuated Tube
Absorber area, m^2	1.88	1.75
Gross area, m^2	2.03	2.77
Absorber material	Stainless steel with selective coating	Aluminium plate with selective coating (Nickel - base)
Absorptance	0.91 - 0.94	0.91
Emittance	0.09 - 0.12	0.12
Insulation	50mm fibre glass at the back	Vacuum, $^{-4}$ torr
Cover material	3mm tempered glass	2mm Soda glass
Transmittance	0.876	0.92
Flow rate, l/h unit	25 - 85	50 - 300

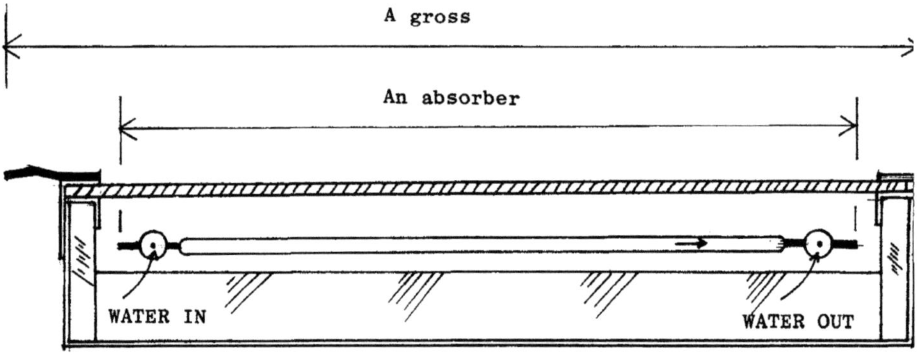

Fig. 1 - Longitudinal Section of Flat-plate Collector

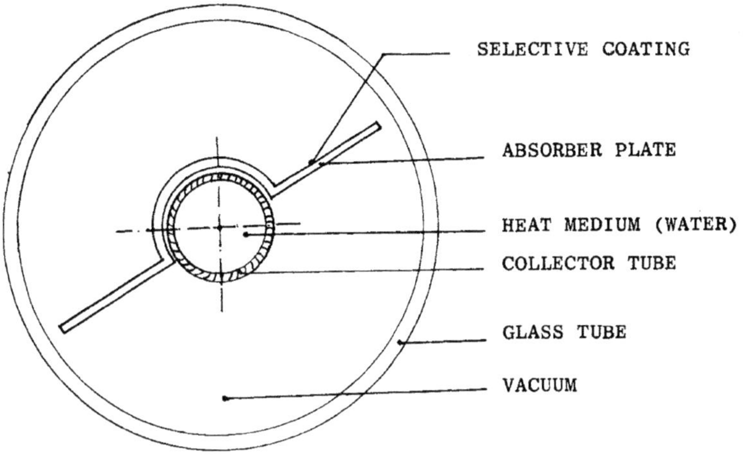

Fig. 2 - Cross Section of Collector Element of Evacuated Tube Collector

3. THERMAL PERFORMANCE

The thermal performance of a solar collector is well presented by its efficiency curve. Efficiency is the ratio of the useful energy collected to the solar energy incident on the collector. According to ASHRAE Standard 93-77, at a particular flow rate, the efficiency of a flat plate collector can be written in the form of

$$\eta = (A_a/A_g)(F_R) \quad [(\tau\alpha)_e - U_L(T_{f,i} - T_a)/I_t] \qquad - \quad (1)$$

This efficiency is based on the gross area of the collector. F_R is the heat removal factor, defined as the actual useful energy collected to the useful energy that could be collected if the entire collector surface were at the inlet fluid temperature instead of its actual temperature.

The manufacturer of the flat plate collector under consideration provides the following information regarding the efficiency of the collector:

$$\eta = 0.684 - 3.929P - 10.507P2 \qquad - \quad (2)$$

where $P = (T_{f,i} - T_a)/I_t$

This collector has been tested according to ASHRAE Standard 93-77. The flow rate used for the testing was 0.010 kg/s m^2. The information provided by the manufacturer of the evacuated tube collector is as follows

$$\eta_a = 0.913 - 2.782X - 2.578x^2 \qquad - (3)$$

where $X = [(T_{f,i} + T_{f,o})/2 - T_a] / I_t$

This efficiency is based on the absorber area and not the gross area as in the previous case. Also the parameter X is different from the parameter P in that the average water temperature in the collector is used instead of the inlet water temperature. As such, the two efficiencies cannot be directly compared.

The efficiency of the evacuated tube collector based on the gross collector can be obtained by multiplying the right hand side of equation (3) by the ratio of its absorber area to gross area. This yields

$$\eta = 0.577 - 1.757X - 1.629X^2 \qquad - (4)$$

Since $\eta = \dfrac{M\,C_p(T_{f,o} - T_{f,i})}{A_g I_t}$

It can be shown that

$$P = x - \frac{A_g}{2M\,C_p}$$

Using the same flow rate of 0.010 kg/s as used for the flat plate when it was tested, the efficiency of the evacuated tube collector is plotted in Fig. 3 and compared to the efficiency of the flat plate collector. It can be seen that the efficiencies of the two collectors are equal to 48.5% when $(T_{f,i} - T_a)/I_t$ is equal to 0.045. For a value of $(T_{f,i} - T_a)/I_t$ higher than this, the efficiency of the evacuated tube collector is higher than that of the flat plate collector. This is due to the fact that, when operating at high water temperature, the evacuated tube collector has less heat loss compared to the flat place collector. The heat loss from a solar collector is determined by its heat loss coefficient, U_L, and it can be determined from the efficiency curve. From equation (1), which is the equation for the efficiency curve:

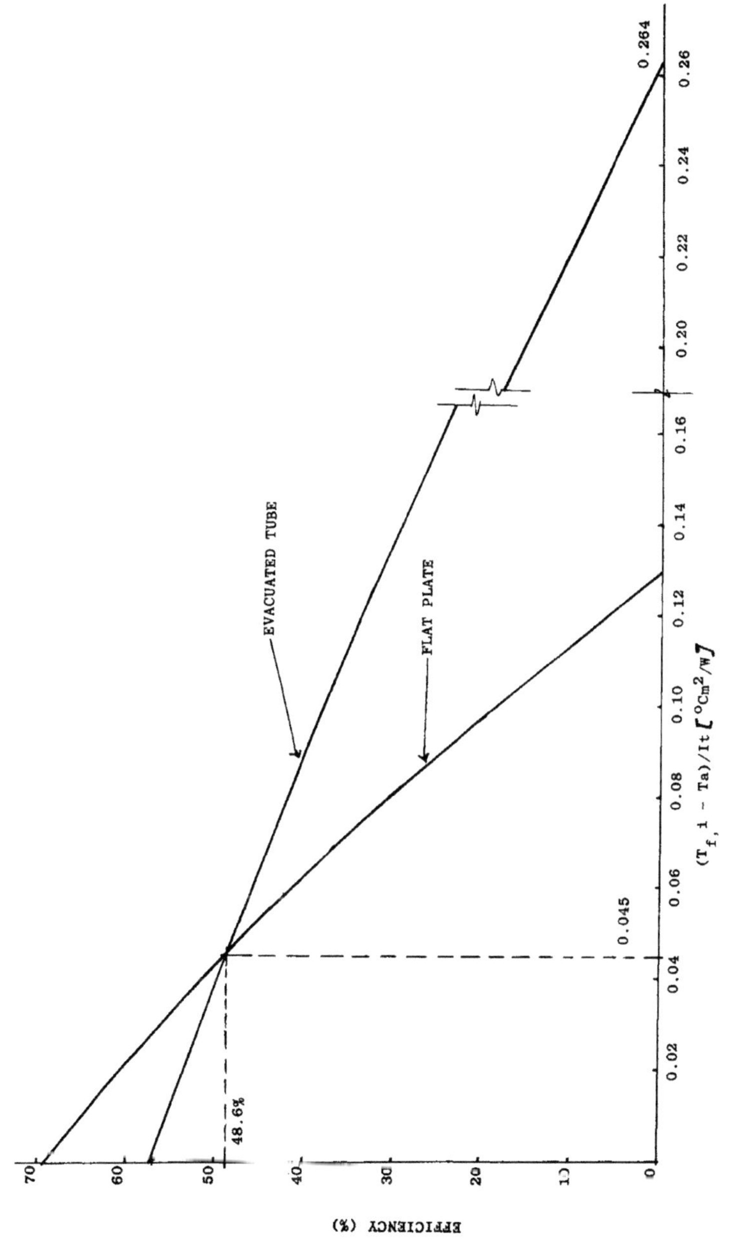

Fig. 3 - Performance Comparison between flat plate and evacuated tube solar collectors

$$\text{slope} = -\frac{Aa}{Ag} F_R U_L$$

$$\eta - \text{intercept} = \frac{Aa}{Ag} F_R (\tau\alpha)_e$$

The η - intercept is the efficiency value when $(T_{f,i} - T_a)/I_t$ is equal to zero. It can be shown that the collector loss coefficient

$$U_L = -(\text{slope})\ (\tau\alpha)_e\ /(\ \eta - \text{intercept})$$

For the flat plate collectors:

$$\text{slope} = -3.929 - 10.507P$$
$$(\tau\alpha)_e = 0.81$$
$$\eta - \text{intercept} = 0.684$$

If the variation of the value of $(\tau\alpha)_e$ with solar radiation incident angle is neglected, the collector loss coefficient becomes

$$U_L = 4.65 + 12.44P$$

Therefore for the whole range of operation, U_L varies from 4.65 $w/m^2\ °C$ to 6.26 $w/m^2\ °C$. Assuming that equation (1) is also valid for an evacuated tube collector loss coefficient varies from 2.59 $w/m^2\ °C$ to 3.66 $w/m^2\ °C$.

Under the operating condition when the collector loss coefficient of the flat plate collector is 6.26 $w/m^2\ °C$ (P = 0.1295), the corresponding value for the evacuated tube is 3.17 $w/m^2\ °C$. This lower collector loss coefficient is due to the elimination of convective heat loss from the collector tube to the glass tube cover by having a vacuum between the two.

The value of $(T_{f,i} - T_a)/I_t$ at which the efficiency of a solar collector is zero determines the theoretical maximum inlet water temperature at which the solar collector can operate without losing heat.

$$[(T_{f,i})_{max} - T_a]\ /I_t = 0.129 \text{ for flat plate collector}$$
$$[(T_{f,i})_{max} - T_a]\ /I_t = 0.264 \text{ for evacuated tube collector}$$

At the average local daytime ambient temperature of 30°C and average daily solar radiation of 350 w/m2, the maximum inlet water temperature are:

$$(T_{f,i})_{max} = 75°C \text{ for flat plate collector}$$
$$(T_{f,i})_{max} = 122°C \text{ for evacuated tube collector}$$

4. EXPERIMENT

Temperature readings were taken from two hot water systems, one using flat plate collectors and another one using evacuated tube collectors. Each system consisted of 10 solar collectors connected in parallel, and operated in a closed loop circulation with a hot water storage tank. The flat plate collectors were inclined at 5° to the horizontal plane and facing the direction of 25°E of S. The evacuated tube collectors were installed horizontally, with their axes parallel to the same direction which was faced by the flat plate collectors. The amount of water in the storage tank and the water flow rate were respectively 510 l and 510 l/h (51 l/h unit) for the system using flat plate collectors, and 690 l and 690 l/h (69 l/h unit) for the system using evacuated tube collectors. The storage volumes and flow rates were proportional to the gross solar collector area in each system. The supply and return hot water pipes were of ¾-inch diameter insulation with 1" Rockwool, and the storage tank was insulated with polyurethane foam. A reverse supply system was used so that the water could be distributed as uniformly as possible to all the solar collectors in the two systems. Both systems had practically the same length of piping.

Fig. 4 shows the variation of the water temperatures with the existing radiation and ambient temperature. The values presented were the average values over 50-minute periods. Initially, the temperature in both storage tanks was at 26.0°C. During the experiment, no hot water was used. It can be seen that the hot water temperature entering the storage tank from the evacuated tube collectors was always higher than that from the flat plate collectors. However, the thermal performance of both systems fell short of the predicted values when the efficiency curves were used and the discrepancy could not be explained by the heat loss from the piping systems. Such discrepancy in performance could only be explained by the existence of vapor locks in both systems, that some solar collectors received very little water, if any, causing them to be practically out of operation.

5. CONCLUSION

For the convenience of comparing the thermal performance of different solar collectors, it is desirable that manufacturers make use of a standard format in representing the efficiency of their collectors. Of the two types of solar collectors being compared, it was found that the evacuated tube collector performed better under high temperature and low radiation conditions, but for any solar water heating system to perform well, good plumbing is a prerequisite.

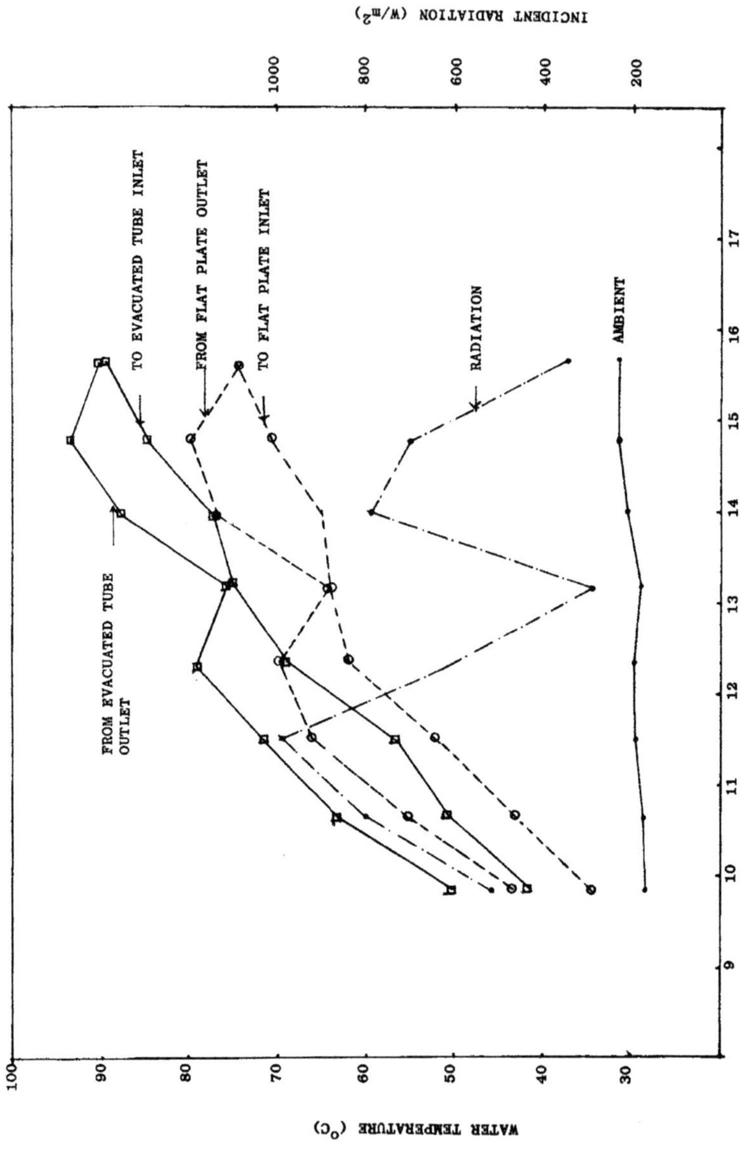

Fig. 4 - Storage Tank inlet and outlet water temperatures

NOMENCLATURE

A_a	=	Absorber area, m^2
A_g	=	Gross area m^2
C_p	=	Specific heat of water, J/kg°C
F_R	=	Heat removal factor
I_t	=	Total solar radiation incident upon the aperture plane of collector, W/m^2
U_L	=	Loss coefficient based on absorber area, W/m^2°C
M	=	Water flow rate, kg/s
$T_{f,i}$	=	Inlet fluid temperature, °C
$T_{f,o}$	=	Outlet fluid temperature, °C
$(\tau\alpha)_e$	=	Effective transmittance-absorptance product
η	=	Collector efficiency based on gross collector area
η_a	=	Collector efficiency based on absorber area

DISCUSSION

Question:

Dr. Bong, what is the capacity of the storage tank and what is the flow rate? How long and over what period of time did you carry out this work, and does it still need any modification?

Assoc. Prof. T. Y. Bong:

Each storage tank is about $1.2m^3$. However, I did not fill them up because I wanted to make the storage volume proportional to the area of the collector. For the flat plate, it was 510 litres, and for the evacuated tube, 670 litres. This is about 25 litres per square meter of collector area. These rather small storage volumes were used in order to get a high temperature.

The flow rate was 51 litres per hour per unit for the flat plate and 67 litres per hour per unit for the evacuated tube. They were within the range recommended by the manufacturers. The system is being modified at the moment.

Question:

Prof. Bong, I just want to ask you a question about your calculation. You have done a very complex calculation based on the solar radiation on a daily factor and hourly factor. Now, how accurate are these calculations? If you take into consideration the atmospheric conditions that might fluctuate, for example, in your university area, which is so near to the Jurong Industrial area, how will atmospheric pollution affect your calculations.

Assoc. Prof. T. Y. Bong:

I took the average. Now averaging will be right only if the line is straight. If it is linear, it does not matter - you can average it out; but this curve is not really linear. It is second order. However, it is quite close to being linear. Many institutions which test collectors, for instance the DISET, always present it in a second order equation, and this shows the variation in the loss coefficient. So this loss coefficient is not constant, it varies.

Long term performance is more difficult to predict. According to the manufacturer, the transmissivity can be assumed to be 5% less in order to take care of the loss. But the surface properties may also change and this will be more serious than the pollution, I believe. In the case of the evacuated tube, what happens if the vacuum is gone? This collector costs US$600 a piece, and there is no guarantee.

Question:

Based on the two collector systems that you described just now, if you recommend a system of evacuated tubes for a hot water system in a hotel, then a problem of maintenance arises - the changing of tubes if they are broken, etc.

Assoc. Prof. T. Y. Bong:

I agree that this should be one of the considerations, if we are to go for the more sophisticated systems. For instance, with our absorption chiller, when something is wrong with it, we do not really know how to go about fixing it. Nobody is really familiar with absorption chillers. So it is with the evacuated tube. No one is sure how to maintain it, if something goes wrong.

Question:

Does this mean that if we are talking about marketing and putting it into practical application, we would have to sacrifice the greater efficiency of the evacuated tube for the flat plate?

Assoc. Prof. T. Y. Bong:

That is really up to the professional engineer - to make his best judgement.

Prof K. R. Rao:

I would request any of our foreign delegates who have more experience on the evacuative tube over a longer period to tell us more about the breakage, leakage and other maintenance problems.

Comment:

These problems are being tested now, because we developed these collectors only 5 or 6 years ago. At this moment, the evacuated tube can last at least 15 years without leakage.

Question:

Prof. Bong mentioned in his paper that in order to calculate the optimum solar system, one would require detailed data on hourly radiation rather than the average radiation. Then, I think in that case it is impossible to design an optimum solar system because the hourly data change from day to day and from year to year.

Assoc. Prof. T. Y. Bong:

The reference year method is used. The radiation data of several years is analysed for the average and the variance of each month. The most representative monthly data is used to build up the reference year. For instance, the value for January is January 1972, and then for February, it is the data of 1977 and so on. This reference year consists of data of various years. There are also slight variations to this approach. In the University, we use the auto-regression moving average model to predict the radiation from one hour to the next. It makes use of the transmittance matrix, which is the ratio of radiation received to that outside the atmosphere.

With this prediction technique, one may work out, for instance, how big a boiler is needed to take care of all eventualities within the following 20 years.

CENTRAL SOLAR HOT WATER APPLICATION

Chiam Joon Tong
Ph.D., M.I.E.S., P. Eng. (Singapore),
Engineering Consultant - Energy Management.

ABSTRACT

Central Solar Hot Water installations whose main components are of multi-solar collectors and storage capacity were introduced into Singapore only recently in 1978. Such systems supply hot water to sports/social clubs, hotels, restaurants, hostels, hospitals, other commercial/institutional buildings and industry. Against the background of rising fuel and electricity prices, central solar installations are currently proving to be economically acceptable. Recovery of capital ranges from 5 to 7 years whilst Discounted Cash Flow computations give an earning power ranging from 16 to 25%.

Experience during the past two and a half years has shown that such installations have, in general, functioned satisfactorily. The demand has grown steadily and with the active encouragement of the Government, particularly in specifying solar hot water installations in some development projects, the future for such systems appears promising.

1. INTRODUCTION

1.1. Central Solar Hot Water Installations in Singapore

In April 1978, two central solar hot water systems were first installed in Singapore - at the Changi Golf Club and the Tanglin Club. The total solar collector areas installed were relatively small, namely, 16 and 40 sq. metres respectively. Within a short space of two and a half years, the demand for such installations has grown steadily, culminating at the time of writing, in the proposed giant solar plant at the Passenger Terminal Building of Changi International Airport. The total specified solar collector area of this solar plant is 1.152 sq. metres. Assuming a standard collector area of 2 sq. metres, the design calls for 576 solar collectors. Table I gives a list of some major solar installations currently at the completion/development stage.

Bill B. P. Lim (ed.), Solar Energy Applications in the Tropics, 143–162.

Table I - Major Central Solar Installations at the Completion/ Development Stage

	Location	Solar Collector Area m^2	Target Completion Date
1.	Changi International Airport Passenger Terminal Building	1152	31 Mar 81
2.	S'pore International Airlines (SIA) Clubhouse	120	31 Dec 81
3.	Orchid Inn (Hotel) Phase II	240	31 Dec 81
4.	S'pore Island Country Club (Island location)	120	31 Oct 80
5.	Tanglin Club (Phase III Development Project)	144	31 Oct 80
6.	Jurong Country Club	80	31 Oct 80

Table II indicates major solar plants which have been installed and are in operation.

Table II - Major Singapore Central Solar Installations in operation

	Location	Solar Collector Area m^2	Target Completion Date
1.	Changi Golf Club	16	April 1978
2.	Tanglin Club (Phase I Development Project)	40	April 1978
3.	S'pore Island Country Club (Bukit)	64	Dec. 1978
4.	Shangri-la Hotel Phase I	32	May 1979
5.	Orchid Inn (Hotel) Phase I	96	June 1980

These lists are by no means exhaustive since solar installations with collector areas below 16 sq. metres have been omitted, including large systems which are in the planning stage.

However, from these tables, it is evident that central solar hot water systems are considered an economical and practical means of conserving energy by the managements concerned which have taken the decision to invest.

Typical solar installations in Singapore are given in Fig. 1a and 1b respectively.

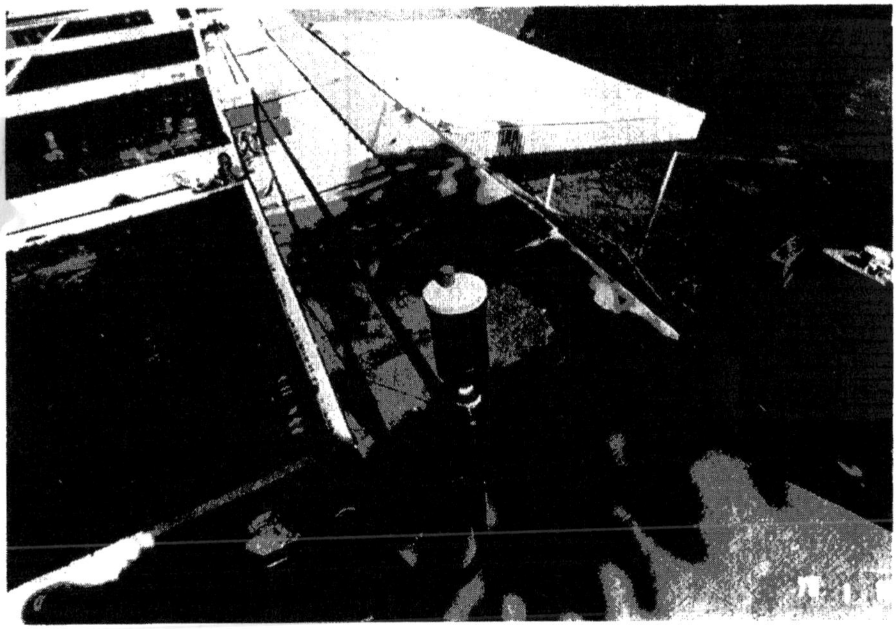

Fig 1a - Solar Hot Water Heating System, Changi Golf Club - showing end connection of solar panel series with ventilation valve allowing ventilation through the panels

Fig. 1b - Solar Hot Water Heating System, Changi Golf Club - showing a series of solar panels directly placed on roof of Club House

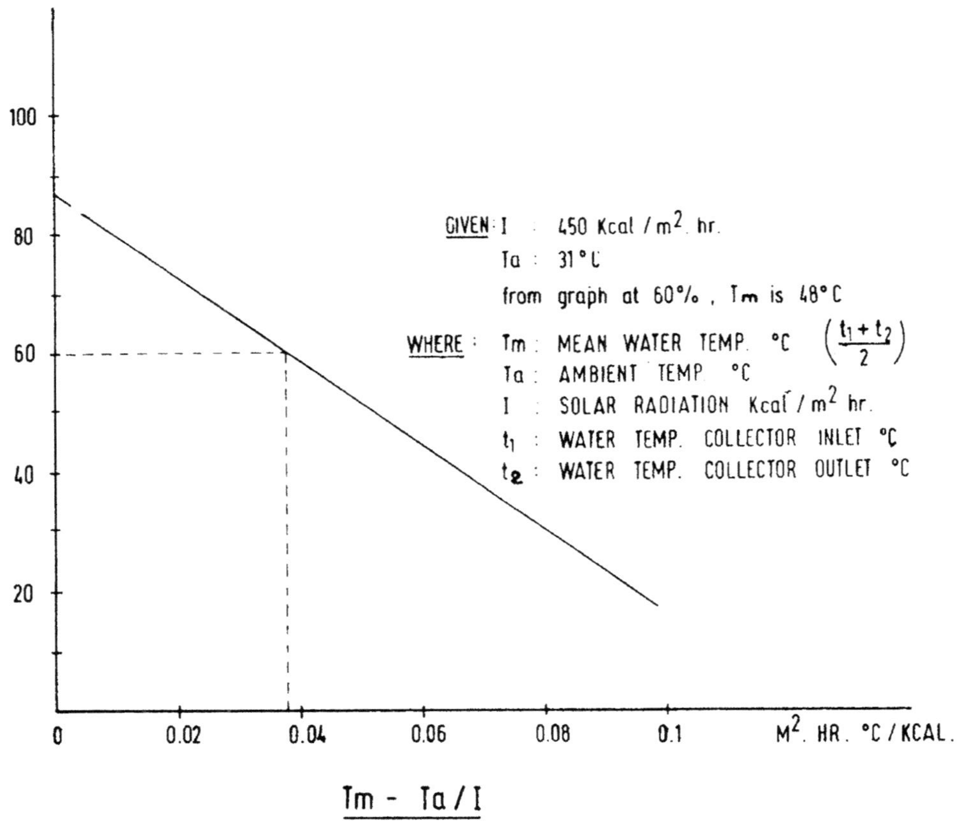

Fig 2 - Efficiency Performance of a Typical Finned Tube Solar Collector

1.2. Types of Solar Collectors

In general, solar collectors applied in Central Installations are of the simple "flat-plate type" with an exposure area of about 2 sq. metres. Except for one local firm which manufactures this type of collector, most of these collectors are imported.

Flat plate collectors installed in Singapore are equipped with either selective or non-selective coatings.

Several installations in Singapore have utilised the finned tube solar absorber. The design of this type of collector is radically different from the conventional flat-plate type as it is based on contoured aluminium fins, working on the "heat trap" principle as shown in Fig. 3.

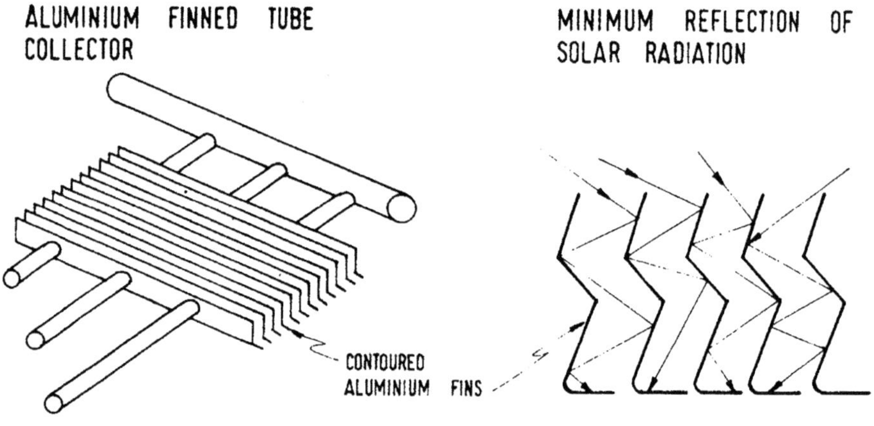

Fig. 3 - Typical Finned Tube Collector

The aluminium fins inhibit convectional losses by trapping heated air between adjoining fins thereby permitting maximum heat transfer to the operating liquid (water). Operating on this principle the collector achieves relatively high efficiencies. As shown on the performance curve in Fig. 2, finned tube collectors operating at a mean temperature of 48°C at an insolation of 450 kcal/m^2hr. are projected to achieve an efficiency of 60%.

1.3. Incentive for Central Solar Installation

Since the energy crisis of 1973, the consistent and sizeable price increases in petroleum fuel, gas and electricity (generated from petroleum fuel oil) have been strong incentives in Singapore to prompt the thermal application of solar energy for residential, commercial and institutional buildings.

Since 1977, the demand for solar heaters, both of the domestic and central type, has increased steadily. With unabated increases in petroleum fuel prices, the consensus is that this demand will continue to grow. The graph in Fig. 4 shows that within a period of 15 months from January 1979 to March 1980 the price of industrial diesel oil escalated from 28 cents/litre to 68 cents/litre, an increase of 143%.

The demand for central Solar Hot Water installations is, therefore promising in the context of Singapore because of the following factors:

(a) Against the background of spiralling petroleum fuel prices and Singapore's total reliance on petroleum fuel, the economic viability of central solar projects will potentially become more attractive.

(b) Apart from solar, other forms of renewable energy such as tidal, wind, waves and geothermal are either non-existent/negligible or impracticable to harness at this point in time.

(c) The predominance of high rise buildings both in existence and under development tend to favour the application of central solar installations rather than simple domestic solar units.

(d) Nuclear energy with its high capital investment, problems in waste disposal, operational hazards and environmental pollution is an unlikely alternative in Singapore.

1.4. Objective

Singapore with its high and constant solar insolation throughout the year is well poised to exploit this renewable energy.

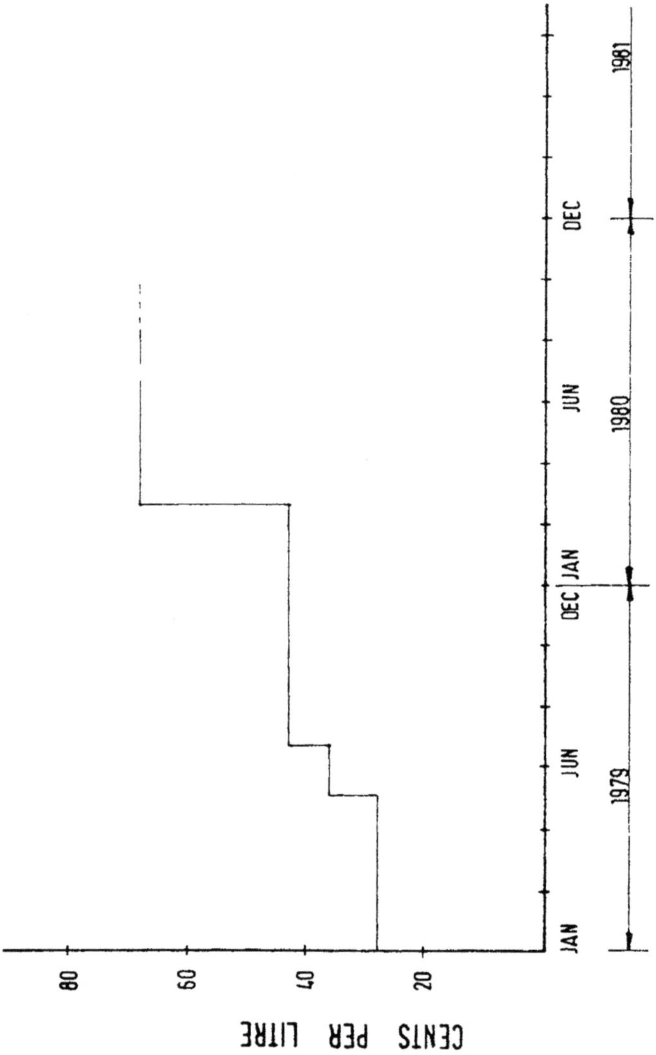

Fig. 4 - Increase in Singapore Diesel Prices

It is important at the outset that efficient and cost-effecive solar thermal technology is applied to ensure that such investments achieve the optimum returns. It is equally important that such projects which demand fundamental Engineering skills and involve considerable capital outlay are professionally designed and supervised by experienced engineers.

This paper discusses some of the basic design parameters of a central solar installation and makes an assessment of its economic viability.

2. OPTIMISATION OF COLLECTORS

One of the first problems confronting the designer of a central solar installation is the number of collectors required.

A simple solution is to assess the peak demand and then calculate the number of collectors to meet this requirement. This method is acceptable only if hot water demand during the normal periods does not deviate significantly from the peak.

In sports/social clubs for example, when the peak demand during the week-ends is considerably more than week-days, designing to peak demand will lead to over-sizing of capacity and unattractive economies. A more prudent method of assessing the optimum collector area is to take into account two major parameters viz:-

(a) The increasing cost relative to the number of collectors installed.

(b) The effect of the collectors on fuel savings over the life span of the plant.

Fig. 5 illustrates the method of achieving an optimum solution.

The life span of the collectors is conservatively assumed to be 10 years (in practice it is more). The cost of fuel over a period of 10 years is computed, taking into account savings given by the collectors. For example, in the extreme case when solar collectors are nil, the fuel cost is maximum, and as collectors are increased, the fuel cost decreases correspondingly reflecting savings by solar energy.

Further fuel costs over the 10-year period are discounted at 12% to Present Values to obtain the same basis of comparison with collector costs.

From the plot of the total cost (fuel and collectors), as shown in Fig. 5, the optimum number of collectors is determined. In this case the optimum is 60.

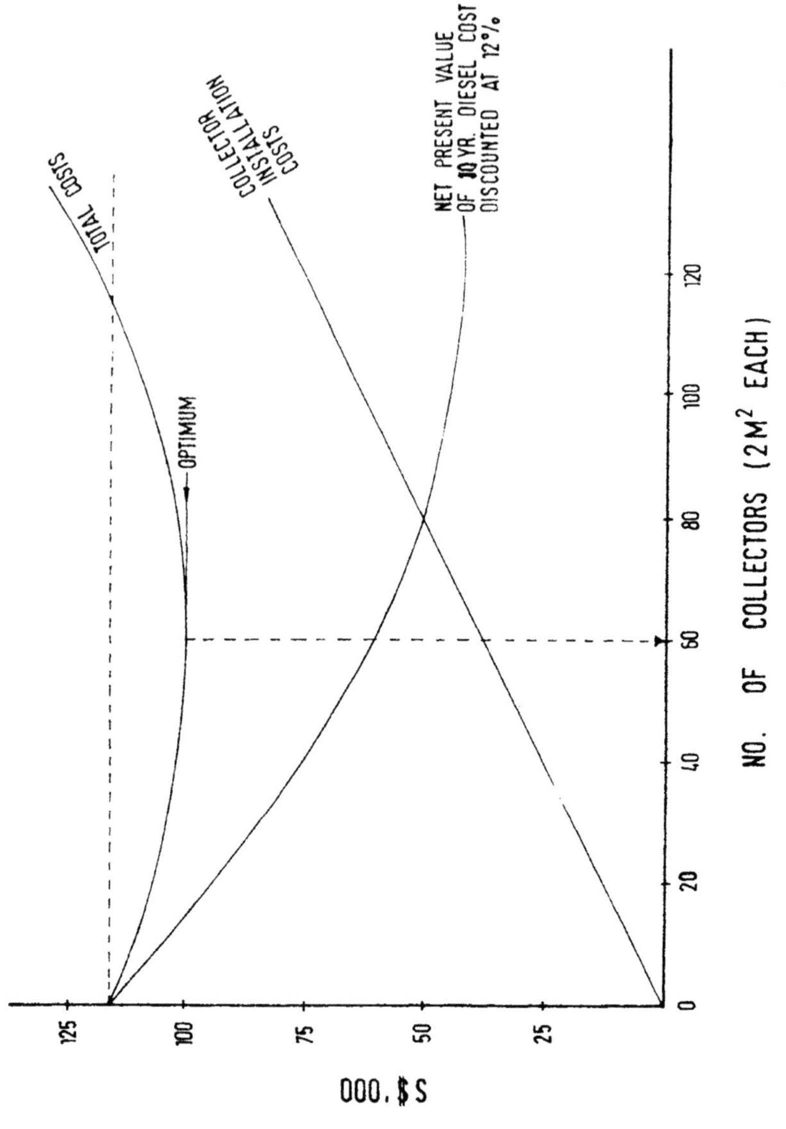

Fig. 5 - Optimisation of Collectors (Based on Diesel Price, Aug '79)

3. FLOW SCHEME

A simplified layout of an ideal central solar hot water installation is given in Fig. 6.

Cold water, either from the city mains or an overhead storage tank, is fed to the main pressurised tank. The cold water is circulated to an array of solar collectors where it is heated and returned to the tank.

The solar heated water from the main tank then flows to a smaller secondary tank. Hot water from the secondary tank is then circulated by a pump to users.

Attached to the secondary tank is the ancilliary heating source which, in this case, is a diesel fired hot water heater. Alternative conventional back-up heating sources are electricity and gas.

4. CONTROLS - MAIN SOLAR ENERGY CIRCUIT

The overall efficiency of a central solar heater depends largely on the control of the circulation pump in the main solar energy circuit.

Generally, there are three types of controllers, viz:-

4.1. Time Switch Control

This simple controller activates/stops the pump within a pre-set period of the day (e.g. from 8 a.m. to 5 p.m.) but discounts weather variations. Since solar collectors are basically heat exchangers, pumping warm water from the storage tank to the collectors during a rainy or cloudy day is counter-productive. Weather variations in Singapore, therefore, do not favour this type of control.

4.2. Electronic device using solar cells

This control system relies on silicon cells, sensing solar radiation. The electric potential generated by the cells activates an electronic switch, which in turn energizes the pump motor from the electrical mains. The drawback of this controller is that it does not take into account the system temperature.

4.3. Differential temperature controller

The most efficient method of controlling the circulation pump is by the differential temperature controller. Two temperature inputs are fed into the controller, one from the collector and the other from the tank.

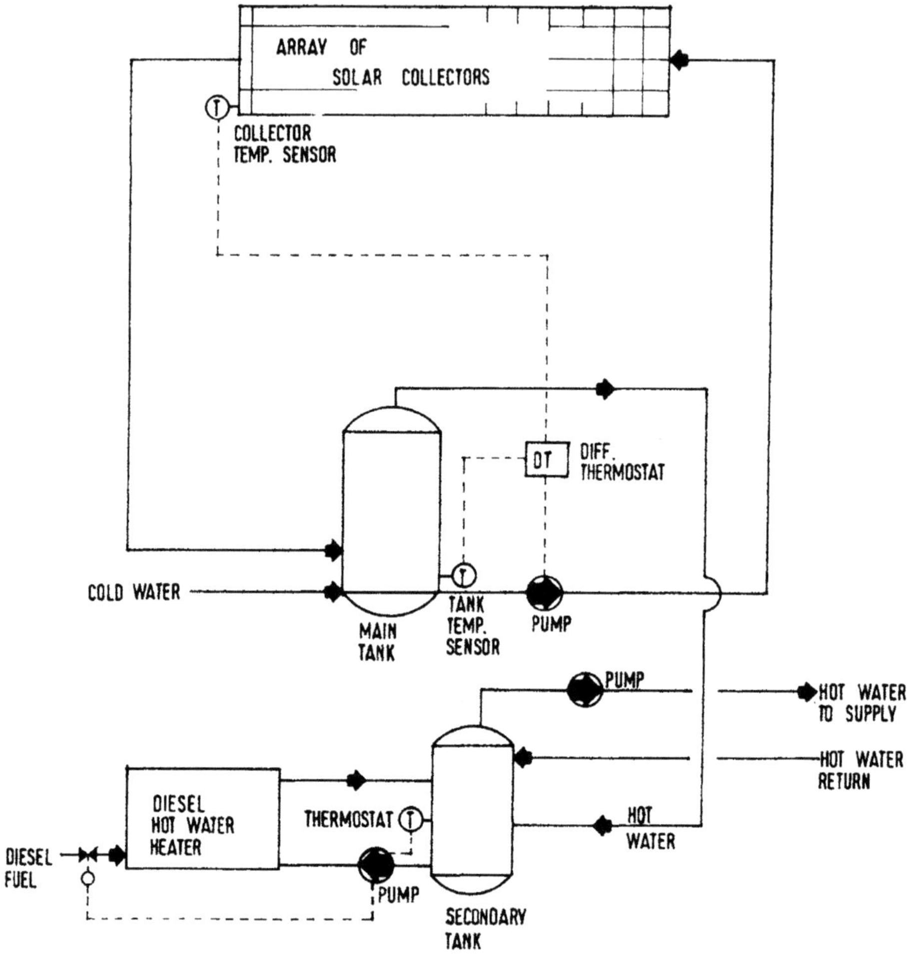

Fig. 6 - Simplified Schematic Flow Diagram - Central Solar Plant Integrated to Diesel Heater

At a pre-set temperature differential, (for example, when the collector outlet temperature exceeds tank temperature by, say, 3°C) the differential controller will automatically energize the circulating pump. On the other hand, when the differential temperature falls to the present cut-off differential of, say, 1°C, the pump stops operating.

To achieve maximum efficiency in the solar energy circuit, it is essential that temperature differentials are set as low as possible.

5. CONTROLS IN THE SECONDARY TANK

To maximise overall efficiency in the system, auxiliary heating input is sited in the secondary tank.

This basically ensures that the main tank is freed from the interference of auxiliary heat input, thereby allowing the solar collectors to operate at maximum efficiency.

The controls in the secondary tank involve a cut-in/cut-out thermostat which, in turn, governs the pump in the diesel energy circuit.

On a sunny day, the secondary tank thermostat, sensing the temperature of the solar heated water from the main tank, automatically cuts off the pump in the diesel energy circuit, consequently deactivating the diesel heater.

During inclement weather, the thermostat sensing low water temperature from the main tank automatically switches on the pump which, in turn, activates the diesel heater. The pump/diesel heater will automatically be switched off again when the water is heated to the high temperature set point.

6. CONTROLS IN THE COLLECTORS

As heated water releases air, it is important to ensure adequate automatic venting facilities at the high points of the collectors installed.

7. OPTIMUM FLOW RATES IN COLLECTORS

The flow rate of water in the conduits of the solar collectors is an important parameter in determining optimum heat transfer. From theoretical heat transfer considerations, it is essential to ensure that the Reynold's number related to flowrate rate in the collector conduits exceeds 2100.

In the state-of-art practice, operating flow rates range from 1.0 to 2.0 litres/min per sq. metre of collector area.

8. ECONOMICS OF CENTRAL SOLAR INSTALLATIONS

8.1. Basis for economic assessment

The economic viability of a central solar hot water installation depends largely on the basic assumptions on which the assessment is made. Broadly there are two assumptions, viz:

(a) The first assumption is that the capital invested covers the entire installation which takes into account both the conventional heating facilities together with the solar energy circuit which involves essentially the solar collectors and the associated controls, pumps and piping systems.

(b) The second assumption is that the economic evaluation is confined essentially to the solar energy circuit and any incremental storage capacity directly related to the solar installation.

The argument in favour of the second assumption is that, given the need to supply hot water in the first instance, it is mandatory, in any case, to install the conventional hot water heating facilities. Following this, it is logical then to consider the feasibility of incremental investment in a solar hot water system compared with the potential fuel/energy savings it will achieve.

The second premise is considered to be the more realistic of the two, and is adopted in the economic appraisal which follows.

8.2. Economic Appraisal

An economic appraisal is made of the accumulated fuel savings over the life span of the solar hot water installation based on the following assumptions:

(a)	Capital Investment	:	incremental cost of installing the solar energy circuit
(b)	Annual insolation	:	$5822 MJ/m^2$
(c)	Diesel fuel price increase	:	15% per annum
(d)	Interest rate	:	12%

(e) Annual Inflation rate : 6%

(f) Price of diesel fuel : 68 cents/litre

(g) Nett heating value diesel fuel : 35MJ/litre

(h) Diesel heater efficiency : 70%

(i) Overall installation efficiency factor : 0.495

(j) Life-span of solar installation : 15 years

Results of the computation are given in Fig. 7, reflecting accumulated fuel savings over a period of 15 years compared with capital investment.

The capital investment ranges from $\$400/m^2$ to $\$600/m^2$ of installed solar collectors. The lower investment range represents the case where ancilliary equipment is minimal whilst the higher range takes into account marginal incremental storage capacity and/or sophisticated equipment specified.

Results in Fig. 7 indicate, on the basis of present diesel fuel prices (68 cents/litre), that central hot water solar installations would achieve a payback period ranging from 4.7 to 6.7 years.

A computation on the discounted cash flow method based on data represented in Fig. 7 indicated that the central solar hot water system achieves an earning power varying from 16% to 25% depending on the initial capital involved. As a general rule, it is considered that capital projects with an earning power in excess of 15% are acceptable.

9. PERFORMANCE OF A CENTRAL SOLAR HOT WATER SYSTEM

9.1. Design Specification and Test Parameters

Test runs were conducted on a central solar hot water installation of $32m^2$ collector area with a storage capacity of 2500 litres.

The main design specification calls for heating 2500 litres of water by the sun to 50°C per day.

The collectors installed were the finned tube type as discussed

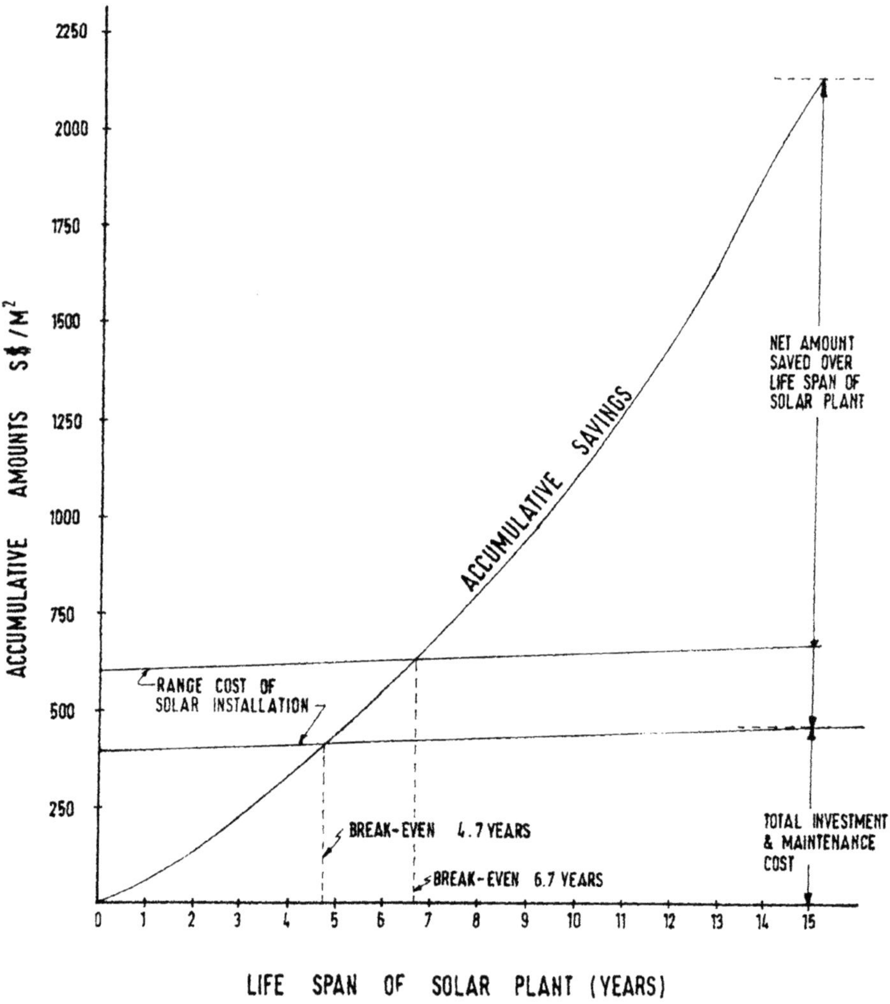

Fig. 7 - Economics of Typical Central Solar Installation Based on Diesel Fuel Savings earlier. Sixteen collectors each of $2m^2$ were utilized, giving a ratio of storage capacity to collector area of 78 litres/m^2.

The flow rate during the test runs was set at 45 litres/min.

9.2. Discussion of results

Two series of test-runs were conducted (Fig. 8). The period of both trials was from 10.00 a.m. to 3.00 p.m.

In the first series, 2500 litres of water were heated from 29°C to 51°C. The aggregated insolation recorded during the test period was 3.31 kwh/m^2. Solar energy absorbed was computed at 64 kwh whilst solar insolation incident to the total collector surface was 106 kwh. This gives an overall efficiency for the system of 60%.

In the second series, as a result of relatively lower insolation, 2500 litres of water were heated from 31° to 49°C. Aggregated insolation of 2.84 kwh/m^2 falling on $32m^2$ of collector surface gives a total available solar energy of 90.9 kwh. On the other hand, 52.3 kwh was absorbed, thereby giving a system efficiency of 58%.

It was concluded from the test results that the performance of the solar installation was satisfactory. The average temperature of 50°C for the two series covering a period of 5 hours indicated that the performance of this system met the main specification.

10. FUTURE PROSPECTS

10.1. Initiative by Government

From both the economic and technical viewpoints, there are no serious factors which would inhibit the demand of solar hot water installations. On the contrary, increasing petroleum fuel prices and a high year-round availability of solar radiation in Singapore are positive factors which will promote the shift towards increasing demand for central solar installations purely out of economic considerations.

On a national scale, central solar installations will assist not only in energy conservation and pollution abatement but also in the conservation of foreign exchange.

The initiative taken by the Government in encouraging large solar hot water projects augurs well for future prospects of such systems. As an example, the tender called by the Public Works Department for the solar hot water project at Changi International Airport mentioned earlier has injected renewed interest in the fledgling solar industry.

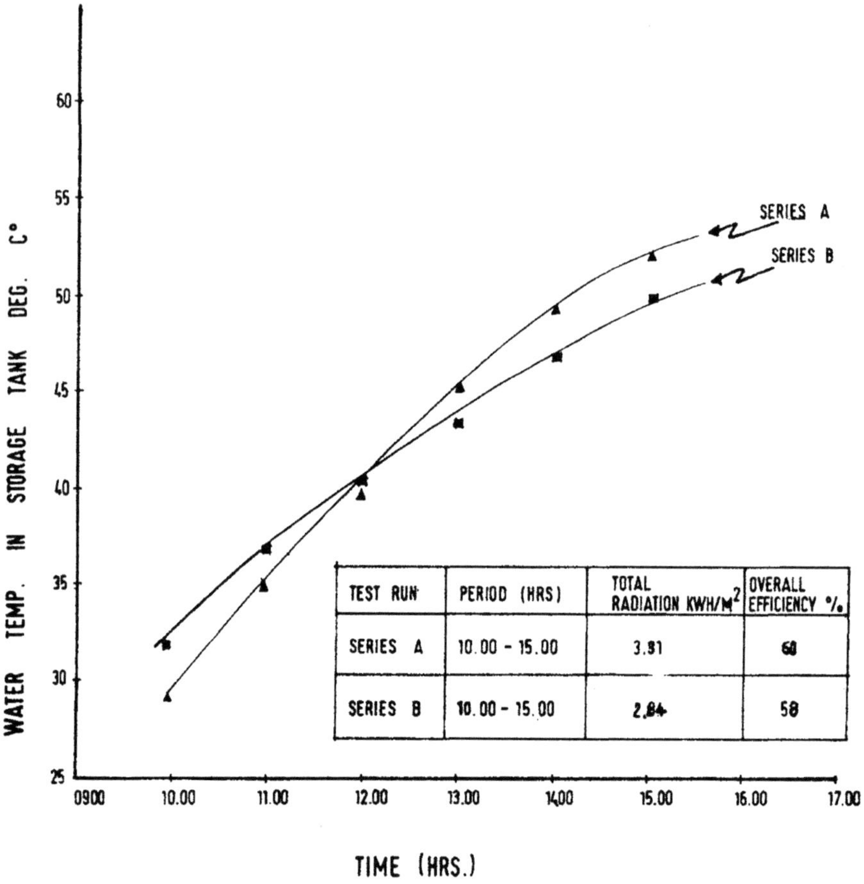

TEST RUN	PERIOD (HRS)	TOTAL RADIATION KWH/M^2	OVERALL EFFICIENCY %
SERIES A	10.00 - 15.00	3.31	60
SERIES B	10.00 - 15.00	2.84	58

Fig. 8 - A Typical Central Solar Plant Heating of 2500 litres water using collectors (Total area - 32 sq metres)

No less than 14 local companies registered to participate in this tender, although eventually only 8 submitted quotations. Of these companies, except for one local firm which manufactures its own solar collectors, most represent manufacturers from countries with well established solar technology.

Thus, the transfer of solar technology through expatriate organisations/representatives in Singapore will be an important factor in getting the industry off the ground.

10.2. Application in Industry

Most of the central solar hot water installations in Singapore are confined to applications in hospitals, schools, canteens, clubs, commercial and institutional buildings. The application in industry has been generally ignored.

It must be noted that extensive studies in countries with established solar technology have shown that the prospects of applying thermal solar technology in industry are promising.

The U.S. Department of Energy (DOE) concluded that about 35 per cent of industrial process heat could be supplied by solar energy by the year 2000 if there were sufficient tax incentives, progressive improvements in the cost-effectiveness of solar energy collecting systems, and petroleum fuels were made available at competitive prices(6).

An Australian study indicated that, by applying established technology, a minimum of 50% of the heat requirements in the food processing industry could be met by solar collectors(7).

A U.K. study(5) showed that low temperature heat requirements (below 80 °C thermal heat) constitute a sizeable proportion in the following industrial sectors:-

Industrial Sector	% Total energy in the Sector
(a) Food, Drink and Tobacco	32
(b) Textile, leather and clothing	13
(c) Engineering and other metal trades	13

The low temperature energy which is supplied largely by conventional fuels or electricity could be replaced by solar energy.

From this brief survey, it is apparent that the prospects of applying central solar heating systems in industry, particularly within the food, drink and tobacco industries in Singapore, are promising.

11. CONCLUSION

Central solar installations have generally operated to expectations in Singapore for the past two and a half years. Some cost-effectiveness procedures are recommended with the aim of improving the performance of such system in the design.

An assessment shows that the economies of solar hot water systems are reasonably attractive and, at least, are not deterring potential investors at this point in time.

Actual trials on a central solar installation showed that its performance in a tropical environment is satisfactory.

Future prospects for the development of the solar industry are promising, particulary with the interest shown by the Government and the exploitation of the untapped market in the industrial sector.

REFERENCES

1. J. T. Czarnecki and W. R. W. Read. Advances in solar water heating for domestic use in Australia. Solar Energy Vol. 20, p 75 (1978)

2. R. N. Morse. Solar industrial process heating for can warming at Queanbeyan, N.S.W. Australia. CSIRO SES Report No. 10 (April 1978)

3. H. Salt. Progress report on the performance of three Australian Solar Hot Water Systems. CSIRO SES Report No. 8 (September 1976)

4. K. S. Ong. Utilisation of Solar Energy in Malaysia - Some prospective applications. IES/IEM Eng. Convention (April 1977)

5. C. W. Lewis. The prospects for Solar Energy use in Industry within the United Kingdom. Solar Energy Vol. 24, p 48 (1980)

6. R. C. Spongberg. A U.S. Govt. appraisal of potential of solar energy for industry. Solar Energy for Industry UK-ISES London (1978)

7. D. Proctor and R. N. Morse. Solar Energy for the Australian food processing industry. Solar energy Vol. 19, p 63 (1977)

DISCUSSION

Question:

May I refer to your Fig. 2 which shows that the maximum efficiency at the level for this flat plate collector is about 88%? Could you tell us what sort of absorption and transmission co-efficients have been used in this particular application.

Dr J. T. Chiam:

This is a black coating type, the absorptance factor is about 0.9. Because it is a non-selective coating, the emittance factor is relatively high. It is not as low as 0.1 as in selective coatings but, off hand, I would say it is about 0.8. However, the heat transfer mechanism of the fin-tube collector is different from the flat plate with its selective coating in that heat is trapped between the fins and convectional currents are minimized. On the other hand, there is a relatively larger amount of reflection in the conventional flat plate collector. As a result of this phenomenon, in the case of the fin-tube collector, because there is relatively less reflection, more solar energy is trapped within the finned configuration. Therefore, higher levels of efficiency have been achieved.

Question:

As you have mentioned, the absorptivity is about 0.9, (is that right?) and the transmittance of the glass is, say, around 80%. That amounts to saying the maximum efficiency from a flat plate is around 70 to 72%. It appeared to me that Figure 2 is exagggerated.

Dr J. T. Chiam:

This efficiency curve shown in Fig. 2 is actually derived from the Weizmann Institute of Science in Tel Aviv and tests were conducted, I think, on 3 fin-tube collectors. This Institute is basically an independent body. The results produced can be verified.

Prof K. R. Rao:

Another way of looking at this is to assume that the absorptivity of the black fin may be 0.9 if it is a flat one. If it is a fin tube, it will, (more or less), be close to that of a black body. Probably, in this case, the total absorption will be more than 0.9.

Assoc. Prof. T. Y. Bong:

Yes, if I may add. I am going to look at it from a different point of view. The curve given by Dr. Chiam cannot be compared to the first curve that I showed. He is making use of T average minus TA over IT. If this is converted into TI minus TA over IT which is according to ASHRAE Standards 93-77, then the curve will be lower. I showed you the second example of the evacuated tube. For instance, instead of cutting the vertical axis at 0.577, it cuts the vertical axis at 0.565, which is actually lower. We hope that all manufacturers will give us the data in the same format so that we can compare. Another thing is the unit. One cannot compare Dr Chiam's figures with my figures. I used watts per square meter, Dr. Chiam used kilo calories.

Question:

Being someone from industry, I am very glad to hear Dr. Chiam mention that there is a lot of potential for the application of solar systems in industry, but I also understand that the Building Control Department in Singapore recently required new buildings, such as factories, to install non-reflective roofs. I wonder whether anybody in Singapore who is connected with the solar industry has checked with the Building Control Department, whether the panels that are available at the moment conform to that requirement of non-reflective materials.

Prof K. R. Rao:

Clear glass is not normally branded as reflective. Its reflectance will be below 0.1. The glass which reflects 70% to 80% of the total solar radiation is not acceptable. The use of clear glass should not be disputed.

Dr J. T. Chiam:

I am sure that the PWD must have done their homework when they specified solar panels on the roof of the terminal building at Changi International Airport. Presumably, they would have checked with the Building Control Department that the panels do not reflect unacceptable light levels. If one flies into Tel Aviv Airport, a large proportion of the roofs of the surrounding buildings are covered with solar collectors, and as far as I know there have been no complaints by the pilots or by the aviation authorities in Tel Aviv. As Professor Rao has just said, the amount of reflective light from the clear glass panel is relatively minimal. Furthermore, solar collectors are generally coated with a non-reflective black paint which minimises reflected light.

DEVELOPMENT OF PHOTOVOLTAIC SOLAR SYSTEMS FOR URBAN APPLICATIONS

Donald G. S. Chuah,
B.Sc., M.Sc., Ph.D., M.Inst.P., F.Inst.P., M.I.E.E.E.,
Chevalier de 1 Ordere des Palmes Academiques,
Associate Professor, School of Physics,
University Sains Malaysia, Penang,
Malaysia.

ABSTRACT

A photovoltaic system converts sunlight into electricity. The elements of such a system are briefly described. Performance characteristics of these systems under local atmospheric conditions are presented. Urban applications of photovoltaic solar cells and cost goals are indicated.

1. INTRODUCTION

The escalating price of oil since 1973 and its possible shortage have fuelled interest in the development of alternative energy sources such as solar energy. Todate, the development of solar energy applications is aimed primarily in the rural areas owing to its special features such as cost effectiveness compared to conventional fuel. However, 85% of the world power consumption goes to the rich while only 15% goes to the 2.4 billion poor(1). It appears that the urban area is where the wealth is concentrated. In this context, data is presented to discuss the various proposals that are being put forward to introduce the applications of photovoltaic solar electricity to urban areas. Such a deliberation is consistent with a workshop paper. The development work going on so far in this region, hopefully, for the eventual application of photovoltaic solar electricity, will be indicated.

2. WHAT IS PHOTOVOLTAIC SOLAR ELECTRICITY?

Photovoltaic solar electricity is the conversion of sunlight into electricity by efficient photovoltaic silicon solar cells. High conversion efficiency cells became commercially practical in the mid-1950s for space applications. Vanguard I, launched in 1958, was the first solar-powered satellite. Since that time, photovoltaic solar cells have been used to supply electrical power to almost every manned space vehicle, space laboratory and communications satellite(2). Numerous terrestrial applications, currently available, are summarised in Table 1(3).

Bill B. P. Lim (ed.), Solar Energy Applications in the Tropics, 165–185.

Table I - Examples of Present Terrestrial Application of Photovoltaic Power Units

POWER SUPPLY APPLICATION	PEAK RATING, W
Warning lights:	
airport light beacon	39
marine light beacon	90
railroad signals	
highway barrier flashers	1.2
tall structure beacon	
lighthouse	
Communications systems	
remote repeater stations for	
- microwaves	50
- radio	109
- TV	78
remote communications station	3500
mobile telephone communication station	2400
portable radio	50
emergency locator transmitter	
Water systems:	
pumps in desert regions	400
water purification	10800
Scientific instrumentation:	
telemetry - collection and transmission platforms for environmental, geological, hydrological and seismic data	
anemometer	100
remote pollution detectors - H_2S	
- noise	3
Industrial:	
remote machinery and processes, e.g., copper electrolysis installation	1500
cathodic protection of underground pipeline	30
electric fence charger	
domestic water meter	20
off-shore drilling platforms	
forest fire lookout posts	
Battery charging:	
boats, mobile homes and campers, golf carts	6-12
construction site equipment	
Ni-Cd-powered military equipment	74

POWER SUPPLY APPLICATION	PEAK RATING, W
Recreational and educational:	
educational TV	35
vacation home	
- lighting, TV	
- refrigerators	
sailboats	200
- lighting, ship-shore communication	
- automatic pilot	
portable TV camera	66
camping lighting	
electronic watches, calculators	
recreational center sanitary facility	168
Security systems:	
closed circuit TV surveillance	150
intrusion alarms	6

Pulfrey (3).

3. A PHOTOVOLTAIC POWER SYSTEM

A photovoltaic (PV) system typically consists of a solar cell array, energy storage and regulation and control devices.

A solar cell array is made up of several photovoltaic modules, each of which encapsulates an assembly of solar cells which convert sunlight into electricity. In its simplest form, a cross-section of a silicon solar cell is shown in Fig. 1a. Basically, it is a p-n junction semiconductor. When light falls on the diffused layer (emitter) the right quantum of photon creates an electron-hole pair in the p-n junction. The electron-hole pair drift towards a region of built-in electric field in the p-n junction and there the electron is separated from the hole by the high electric field. The electrons then flow to the metallic fingers in a flow pattern as depicted in Fig. 1b. The electric current is then collected from the bus bar(4, 5). Thus, the PV module is the basic building block of the solar array, from which electrical power is collected. The electrical energy obtained from sunlight is typically stored in a number of lead acid cells which are connected in series and/or parallel to provide the required power. Sufficient storage capacity must be provided to meet specific load requirements and to account for diurnal and seasonal variations in solar insolation. Alternatively, it can be stored in the form of water in water-pumping application. Voltage regulation is provided to protect the batteries from overcharge and excessive discharge, and to protect the loads from voltage extremes. The load can be an electric motor/pump assembly, a bank of fluorescent lamps or some mechanisms/devices which are electrically powered. Fig. 2 illustrates the basic elements making up a photovoltaic system. A large scale PV system is illustrated in Fig. 3. In this system a power conditioner is incorporated to convert the output DC power from the array to AC power, and to track the input power so that the solar cell module can operate continually at its maximum power point (Fig. 4).

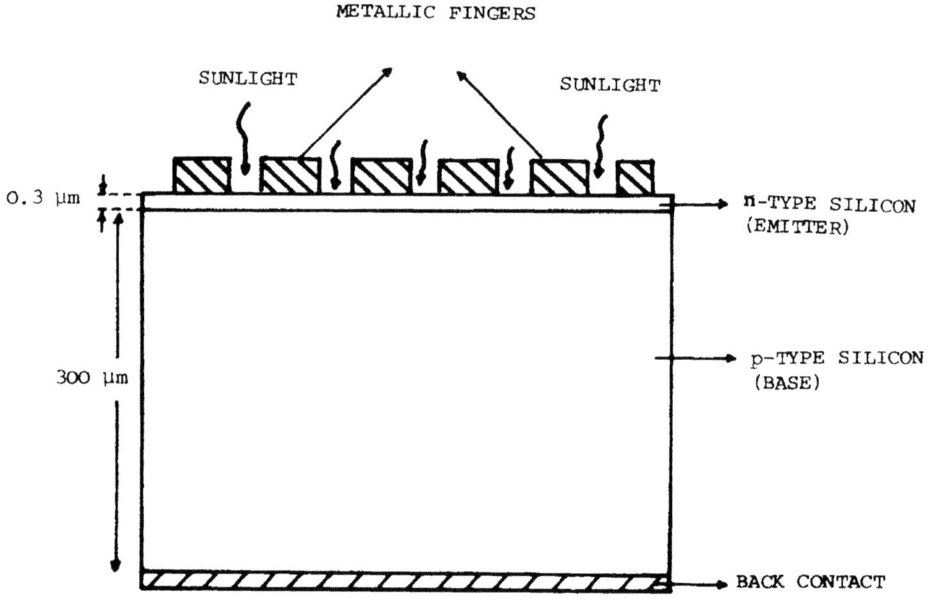

Fig. 1a - Cross sectional view of silicon solar cell

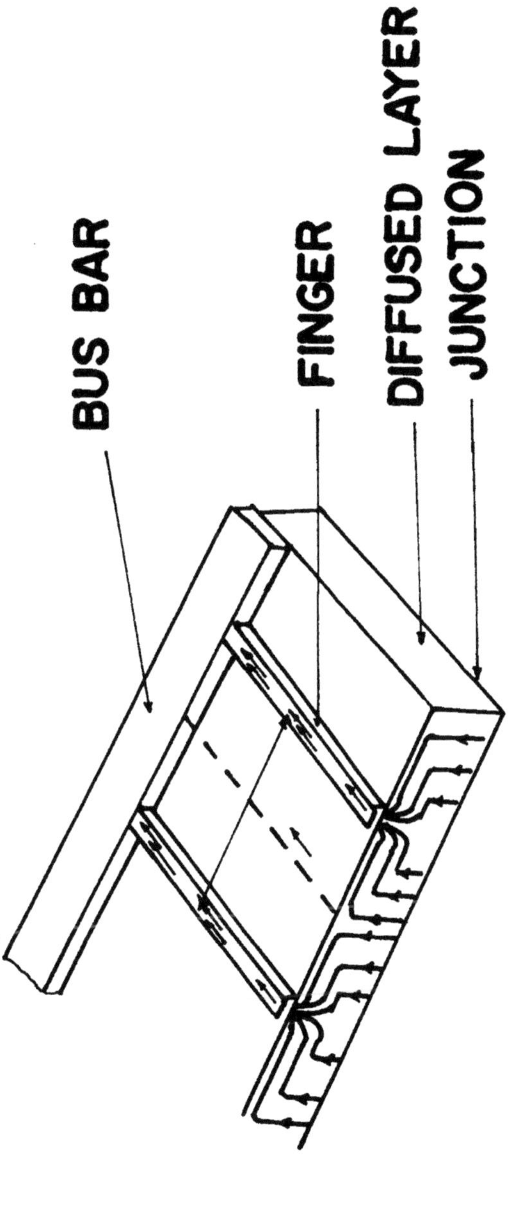

Fig. 1b - Current Flow Pattern in the surface sheet and grid lines for the linear grid configuration

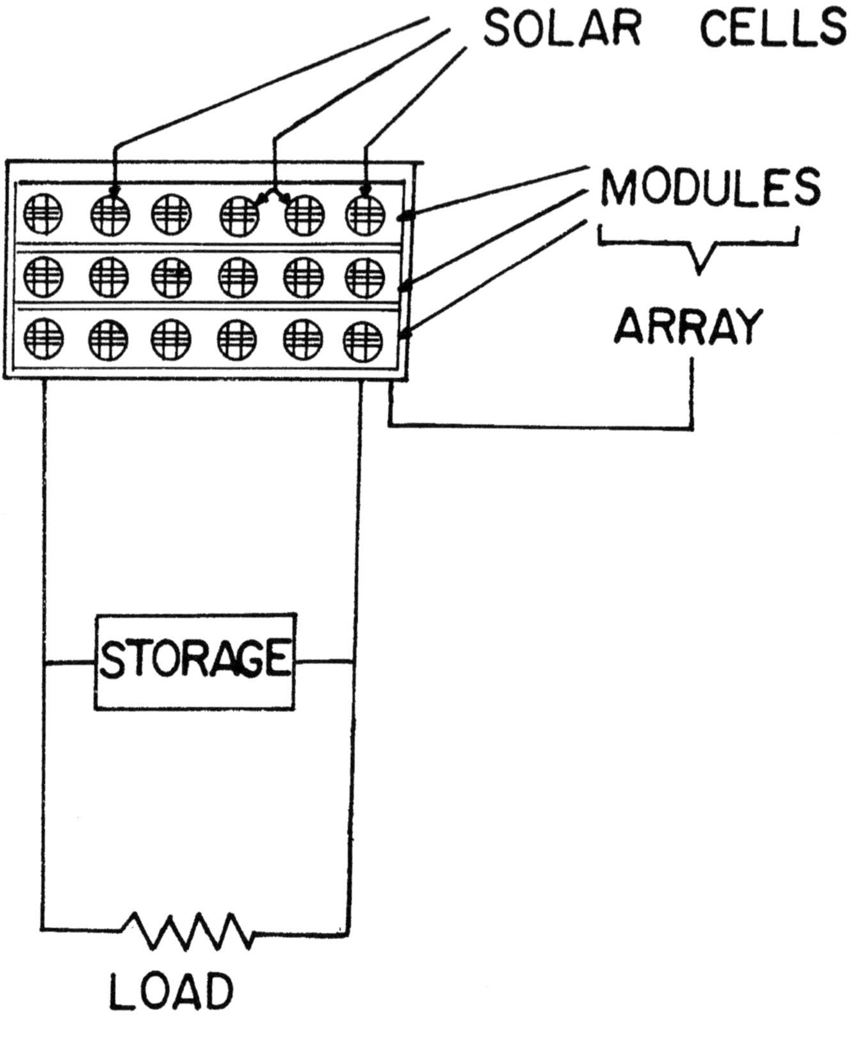

Fig. 2 - Elements of a simple Photovoltaic system

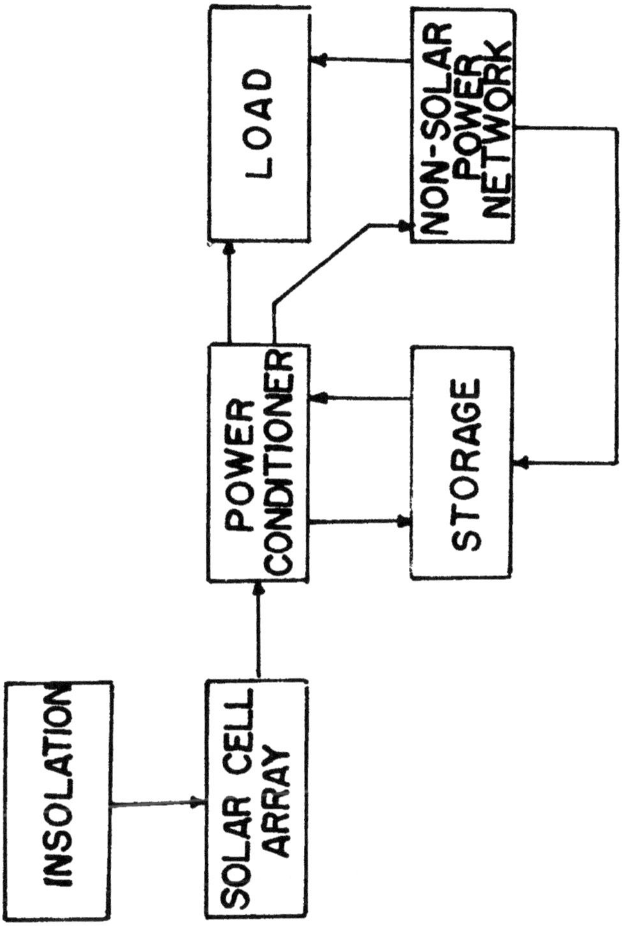

Fig. 3 - Elements of a large-scale Photovoltaic system

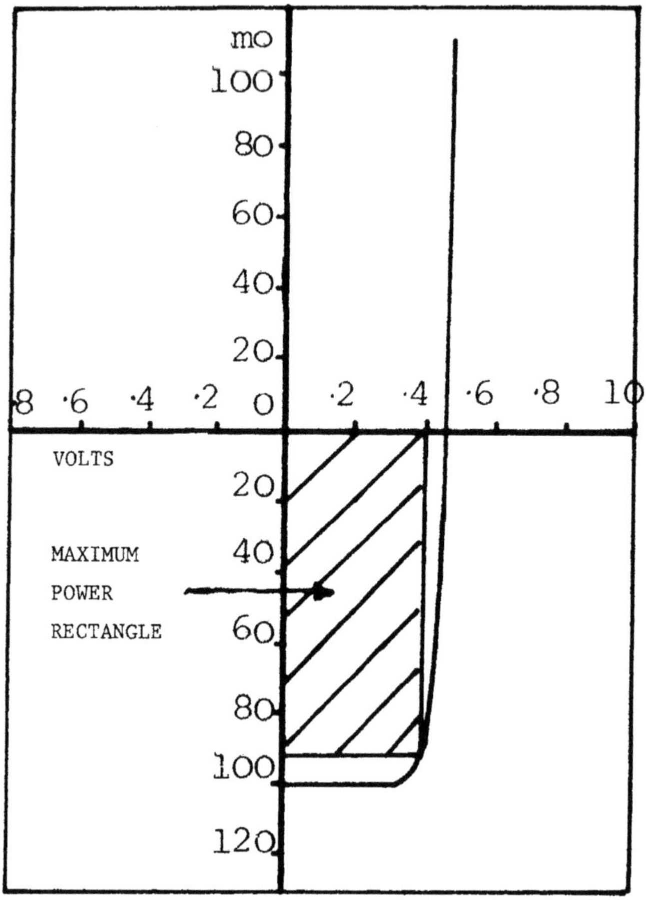

Fig. 4 - Ideal I - V characteristic of a photocell or solar energy converter

4. WHY DO WE CHOOSE PHOTOVOLTAIC SOLAR ENERGY?

Solar electricity, apart from photovoltaic conversion, can be obtained by thermal conversion as demonstrated in the solar tower plant. However, photovoltaic conversion is a direct conversion compared to the thermal conversion, which is indirect. Other advantages(6) of photovoltaic conversion of solar energy into electricity are:

(a) the life span of a solar cell is approximately 20 years provided encapsulation of the solar cells is done properly;

(b) once a photovoltaic solar generator is installed, it can ensure regular power supply;

(c) little, if any, maintenance is needed;

(d) it is absolutely pollution free;

(e) except for the initial capital cost, the production cost of solar electricity is relatively low since the source energy is free;

(f) moreover, the system, which does not include sophisticated instruments, is very simple and thus very easy and safe to operate.

5. THE SOURCE ENERGY

The incident power of the Sun is about 1.2×10^{17} watts falling on the earth surface, that is, about 10^{21} watt hour per year, which is fantastically high. This energy is equivalent to 3.3×10^{8} MTCE (Million Tons of Coal Equivalent) or 8000Q (1Q = 10^{18} BTU or about 40,000 MTCE).

Comparing the 8,000Q with the world consumption in 1970 of 0.18Q and that forcast for 1980 world consumption of 0.3Q, the solar energy hitting the earth amounts to about 30,000 times human needs(7, 8). But the incident energy is subject to climatic hindrances, and day and night cycle.

In order to utilise the sun energy in a particular location it is essential to know its average power, peak power, the ratio of diffuse to direct radiation and the statistical distribution of each class of power. Work in this area has been done by Exell(9) (Thailand), Chia(10) and Goh(11) (Singapore), Malik and Kurisman(12) (Indonesia) and Chuah and Lee(13, 14) (Malaysia). (This is the list known to me so far and my apologies for not knowing similar work of others).

At most, the incident power is 1kW m^{-2}, which necessitates large collection areas to capture the energy. This is a factor which characterises a solar power system - large and capital intensive. However, this characteristic could be reduced by concentration of sunlight. Various types of concentration systems are well described in the literature. High concentration could be achieved by using linear Fresnel lens and low concentration with booster reflectors(15).

6. LOCAL PERFORMANCE OF SOLAR CELLS

The silicon solar cells currently available commercially are rated 10% efficient at standard conditions of 25°C and incident peak solar power of 100 mW cm^{-2}. The average local ambient temperature is 30°C, which is 5°C above the standard operating temperature; thus, the efficiency is expected to drop under local conditions. Furthermore, the peak solar power of 100 mW cm^{-2} is seldom attainable even during excellent clear sky conditions around noon time.

During the period of investigation, the average maximum cell efficiency was found to be 7.0% and the average photovoltaic conversion efficiency was 5%(16). Cell temperature up to 60°C was also observed and this caused a deterioration of the output current(17). However, the ampere-hour efficiency for two generators of 25W and 1.4W respectively appear to be comparable to that of nickel-iron cells though below that of lead-acid cells as tabulated in Table II.

Table II - Performance Characteristics(18)

Characteristics	25W, 12 Generator	1.4W, 12V Generator	Lead Acid Cells	Nickel-Iron Cells
Sunshine hours	214.0(1977)	189(1976)	-	-
Charging power (AH)	72.0	8.65	-	-
Discharging power (AH)	51.0	7.11	-	-
Ampere-hour efficiency	71.0%	82.0%	90%	70-75%

Further improvement could be expected with encapsulation suitable to a high ambient temperature. Nevertheless, these data show that photovoltaic solar generators under local atmospheric conditions are capable of providing dc output power electricity.

On average, the solar radiation intensity in Malaysia is about 50%

direct and 50% diffuse(13). A suitable concentrator for this type of weather is a trapezoidal groove. Preliminary work on a concentrator generator with a concentration factor of 2 gives the current-voltage characteristics as plotted in Fig. 5 and Fig. 6. In Fig. 5, the current output is higher with concentration compared to that without concentration at 1000 hours and 1200 hours when the solar radiation intensities are 79.89 mW cm^{-2} and 49.84 mW cm^{-2} respectively. At other times, when the solar radiation intensities are low, owing to diffuse conditions, the current output is nearly the same, indicating that the solar cells can function normally. This is an advantage of a low concentration device.

Sufficient amount of current is also obtainable during poor solar radiation intensity as shown in Fig. 6. Such current can be used to trickle-charge battery for electric lighting. The results are consistent with the manufacturer's specifications under various solar radiation and cell temperatures. The typical photovoltaic conversion efficiency obtained was about 5%(19).

7. PROPOSED URBAN APPLICATIONS

7.1. Utility Grid Feedback

Kleiner(20) and Sater-Goradin(21) have suggested a combined photovoltaic/ invertor system to provide peak load sharing using utility grid feedback. This can be achieved by incorporating an invertor synchronized to the utility grid so that the invertor can directly transform the solar photovoltaic array generated energy to local load (residence) demand with any excess energy feedback to the utility grid. Where the solar energy is insufficient to supply the local load, it is supplemented. The objective of such a proposal is to (a) minimize initial capital cost, (b) maximize energy conversion, and (c) defer utility capital commitments. Such a utility grid feedback photovoltaic system would omit storage batteries. The flow diagram in Fig. 7 illustrates how an ideal system of this type might function, namely, a photovoltaic array interfaces through a power conditioning unit to the local load and the utility grid. The introduction of large numbers of systems would result in a displacement of fuel consumption and also result in considerable energy conservation.

7.2. "Buy Back Power"

With the above proposed utility grid feedback system, the Massachusetts Institute of Technology (MIT) Energy Laboratory(22) proposes that the home owner using PV systems be allowed to provide power back to the utility when he has excess capacity relative to his own demand.

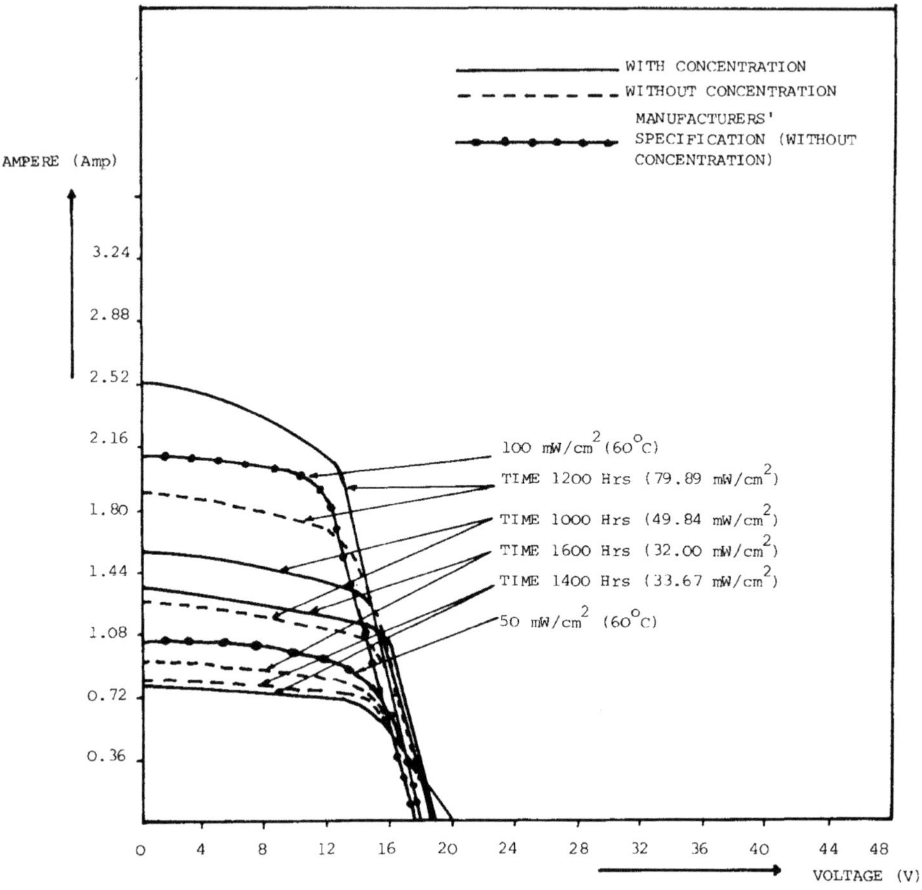

Fig. 5 - Result of current-voltage characteristics on a concentrator generator with a concentration factor of 2

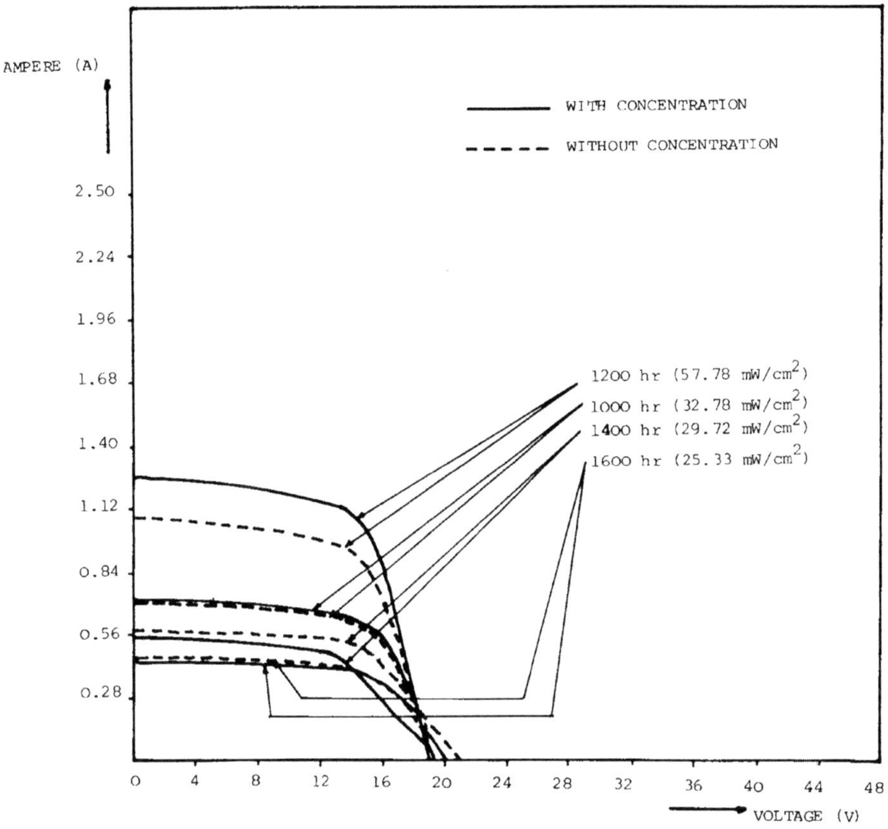

Fig. 6 - Result of current-voltage characteristics on a concentrator generator with a higher concentration factor

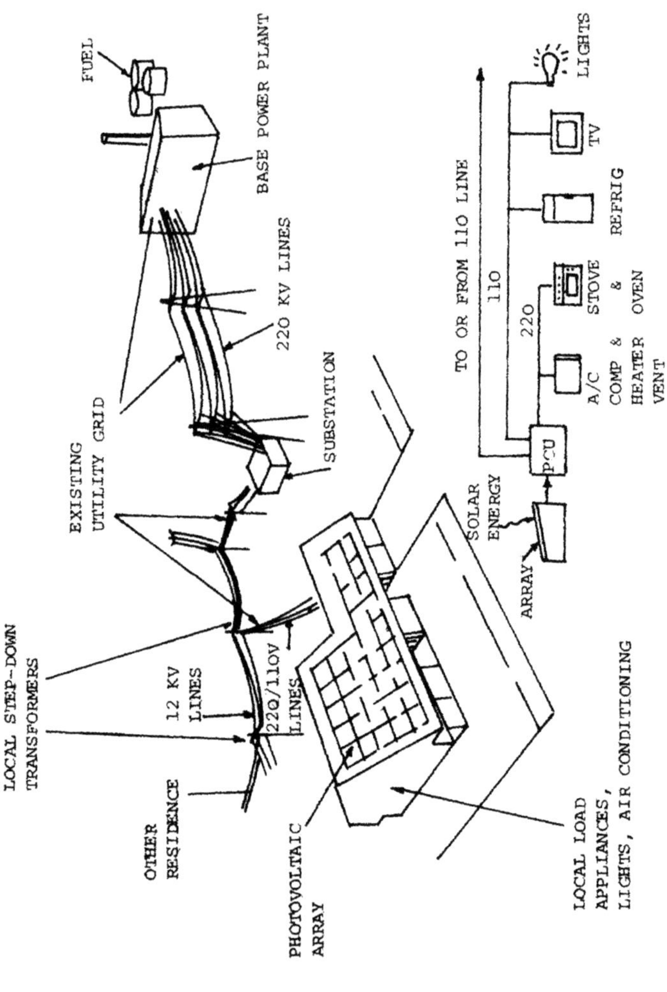

Fig. 7 - Illustration of Residential System with Existing Utility Grid Incorporating a Photovoltaic Array and Power Conditioning Sub-system

Two alternative rates at which the utility would buy back power from the home owner were considered. The first rate is 50% buyback (that is, the utility is willing to buy from the home-owner at half the time-specific price that the utility charges). The second possibility is a 100% buyback (that is, the utility is willing to pay the home owner exactly what it charges, effectively providing free storage).

The MIT studies are based on three different settings, namely, single family residences, institutional buildings and central facilities. The results of these studies indicate a high breakeven system price of US\$2.60 Wp^{-1} under a 100% buyback rate to about US\$1.00 Wp^{-1} at 50% buyback depending on the region studied. This result appears attractive compared to a module cost of US\$10.00 Wp^{-1} assumed for 1980(23).

7.3. Partial Energy to Electrical Vehicle

An interesting application to study is the charging of an electric vehicle, which has renewed consideration recently. This resurgent interest has come about due to three events, (1) concern over the environmental quality in cities, (2) the increase in oil consumption to power automobiles as indicated in Fig. 8(24) and (3) the recent breakthrough in battery technology(25). Thus, there is a likelihood that in the not-too-distant future electric vehicles may be competitive with internal combustion vehicle in an urban environment.

The potential of incorporating arrays of solar cells to supply partial energy to electric vehicles has been evaluated by Varde(26). The vehicle is primarily used for commuting to work or short distance travel and is powered by batteries. Solar cells supply the energy to maintain the batteries charged during travel and rest periods. The specifications of the vehicle studied by Varde are quite similar to that of a utility car as specified by Arthur D. Little, Inc. in Table III. For such a design the maximum area of solar cells required was 2.4 m^2. This could be reduced by minimizing the total weight of the vehicle. The chassis and body are to be of light weight structure. The battery plates could be housed in blown polypropylene(27). Solar cells matching the discharge rate of the battery and motor speed controls using electronic circuitry for swift and even acceleration and minimum energy loss are the criteria of a matched power system(28).

7.4. Price Goals of Photovoltaic Systems

Fig. 9 shows the development of the price goals for a photovoltaic module/array(29). The present array price is US\$7.00 to US\$14.00 Wp^{-1} (1978). Based on the current silicon technology the lowest array price attainable appears to be US\$2.50 Wp^{-1} by 1982.

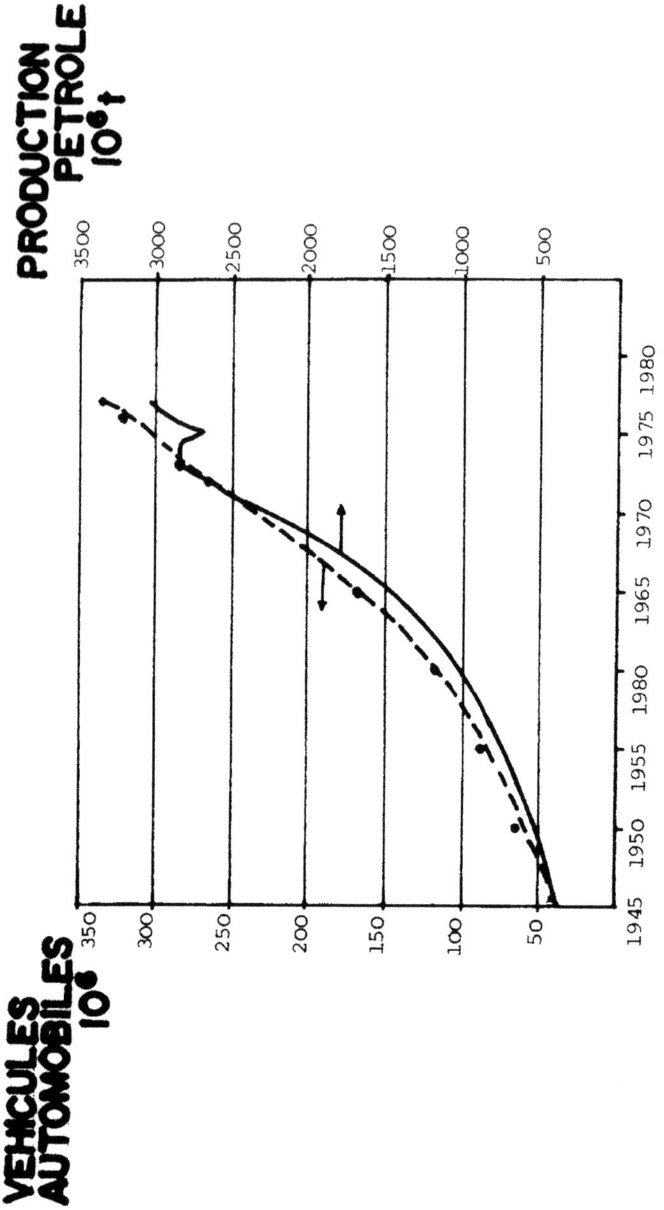

Fig. 8 - Increase in oil consumption to power automobiles

Table III - Electric Vehicle Specification

		Family car	Commuter car	Utility car	Delivery van	City taxi	City bus
Assumptions							
1. Acceleration to	(mph)	60	60	30	40	40	30
2. in	(sec)	15	30	10	20	15	15
3. Range	(mi)	200	100	50	60	150	120
4. Seats or payload	(lb)	6	4	2	2,500	6	10,000
5. Loaded weight	(lb)	4,000	2,500	1,700	7,000	4,000	30,000
6. Curb weight	(lb)	3,500	2,100	1,400	4,500	3,500	20,000
7. Weight assignable to propulsion, energy storage, controls	(lb)						
a. Conventional construction		1,250	750	500	1,400	1,250	5,000
b. Lightweight construction		1,750	1,050	700	2,000	1,750	7,000
8. Frontal area	(ft^2)	25	18	18	42	25	80
9. Drag coefficient		0.35	0.25	0.25	0.85	0.35	0.85
10. Elec transmission efficiency	(%)	82	77	72	79	76	85
Derived parameters							
11. Max. power delivered by motors	(k W)	70	22	12	49	36	135
	(hp)	94	30	16	66	48	180
12. Max. output of power source	(k W)	85	29	17	62	47	159
13. Max. velocity	(mph)	100	80	65	56	77	55
14. Delivered energy	(k Wh)	100	20	8	45	75	300
15. Stored energy	(k Wh)	122	26	11	57	99	353
16. Weight of motors, transmission, and controls	(lb)	348	118	82	259	210	615
17. Weight assignable to power source	(lb)						
a. Conventional construction		902	632	418	1,141	1,040	4,385
b. Lightweight construction		1,402	932	618	1,741	1,540	6,385
Power source requirements							
18. With conventional construction							
Energy density	(Wh/lb)	135	41	26	50	96	81
Power density	(W/lb)	94	46	40	55	45	36
19. With lightweight construction							
Energy density	(Wh/lb)	87	28	18	33	64	55
Power density	(W/lb)	60	31	28	36	30	25

* The loaded weights given for the cars and taxis are not the maximum which they are capable of carrying but rather reflect typical usage.

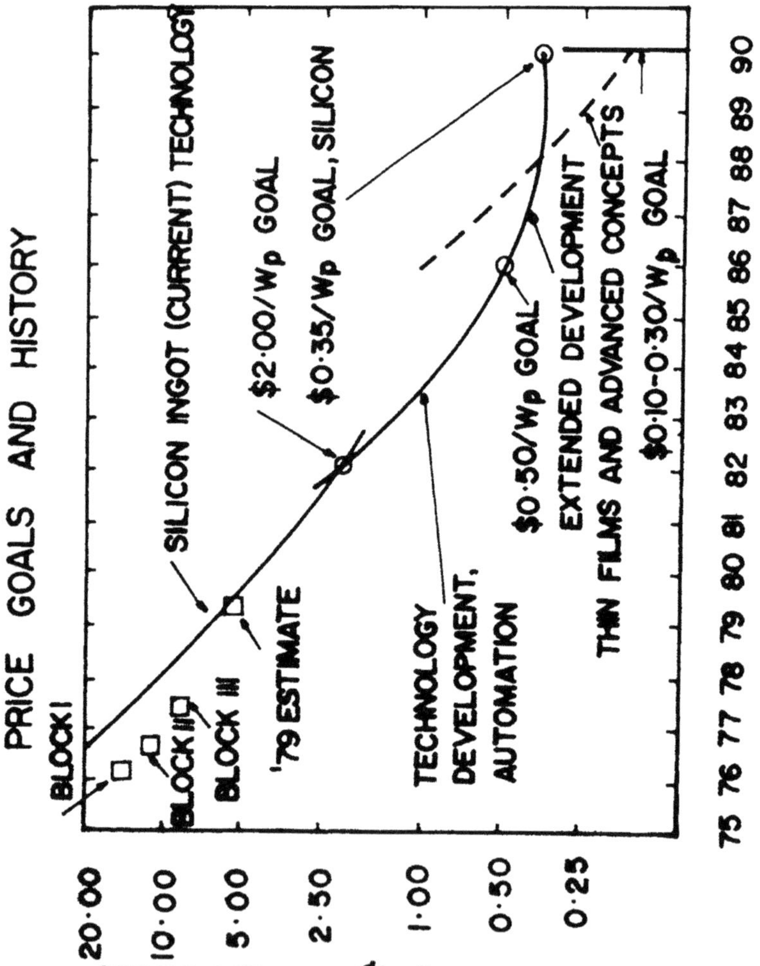

Fig. 9 - Photovoltaic Module/Array

However, if the "buyback" proposal were to be implemented, this array price could be brought forward from 1982. But it is still away from the price goal of US$0.50 Wp^{-1} set by ERDA (U.S.A.), at which price, the photovoltaic system would be competitive to conventional electricity. It is hoped that this goal could be reached by introducing automation and other technological development to the current silicon technology. The last stage of using thin film solar cells and advanced concepts could bring the price below the ERDA (U.S.A.) competitive price goal to US$0.10 - US$0.30 Wp^{-1} in 1990. Thus, we note that there is a steady effort to decrease price to meet the escalating oil price.

8. CONCLUSION

Experiences reported with installed photovoltaic systems indicate their viability in converting solar energy into electricity. The reliability of such systems also appears satisfactory. The price is being steadily decreased to make photovoltaic solar electricity competitive with conventional power supply. The development of the technology of utility grid feedback and the study of its economics give rise to the hope of a solarized society with homes completely powered by photovoltaic solar electric arrays and electric cars run off the sun's energy.

9. ACKNOWLEDGEMENT

I thank the Organisers of this Seminar and Workshop and the Universiti Sains Malaysia for giving me the opportunity to participate in it.

REFERENCES

1. B. T. Feld, "A Crude But Plausible Model For World-Wide Energy Needs Into The Next Century", Private Communication, 12.3.73

2. Robert R. Davis, "The Photovoltaic Market: An Overview", Arco Solar

3. David L. Pulfrey, "Photovoltaic Power Generation", Van Nostrand Reinhold Co. (1978)

4. Harold J. Hovel, "Semiconductor and Semi-metals", Vol. 11, Solar Cells, Academic press (1975)

5. A.R. Moore, "An Optimized Grid Design of a Sun-Concentrator Solar Cell", Vol. 40, No. 2, pp.140-165, RCA Review (1979)

6. Wolfgang Palz, "Solar Electricity - An Economic Approach To Solar Energy", Butterworths (1978)

7. H. Durand, "The Use of Photovoltaic Energy Conversion: A Technological and Economical Challenge", Laboratoires d'Electronique et de Physique Appliquee, Paris, April (1974)

8. OECD - Statistics, October (1973)

9. R. H. B. Exell, "The Fluctuation of Solar Radiation in Thailand", Solar Energy 18, p.549 (1976)

10. Chia Lin Sien, "Sunshine and Solar Radiation in Singapore", The Meteorological Magazine # 1166, 98, pp.265-274 (1969)

11. T. N. Goh, "Statistical Study of Solar Radiation Information in an Equatorial Region (Singapore)", Solar Energy, Vol. 22, p.105 (1979)

12. Fauzi Malik and Kurisman, "Studi Persiapan Pemataan Intensitas Radiasi Surya di Indonesia", National Physics Symposium, Universiti Kebangsaan Malaysia, 28th-30th June, Kuala Lumpur (1979)

13. Donald G. S. Chuah and S. L. Lee, "Solar Radiation in Peninsular Malaysia", presented in the same Conference as 12

14. Donald G. S. Chuah and S. L. Lee, "Solar Radiation in Peninsular Malaysia - Statistical Presentations", submitted to Energy Conversion, Pergamon Press, May (1980)

15. Meinel and Meinel, "Applied Solar Energy - An Introduction", Addison-Wesley (1977)

16. Donald G. S. Chuah, "Solar Cell Efficiencies in the Tropics", Int. Conf, Solar Energy Dev., Varese, Italy, 26-29 March (1979)

17. Donald G. S. Chuah et al., Proc. 1977 Photovoltaic Solar Energy Conf., pp.1253-1260, Luxembourg, 27-30 September (1977)

18. Donald G. S. Chuah, "The Ampere-Hour Efficiency of photovoltaic Solar Generators", Energy Conversion, Vol. 19, No. 3, pp.177-180, Pergamon Press (1979)

19. D. G. S. Chuah, S. L. Lee and C. K. Koh, "Water Pumping by Photovoltaic Solar Energy", Conference on Tenaga 80, The Malaysian National Committee of the World Energy Conference, Hilton Hotel, Kuala Lumpur, 21-23 August (1980)

20. C. T. Kleiner, "Advanced Semiconductor Technology for Alternative Energy Sources: D.C. to A.C. Invertors", Alternative Energy Sources, Vol. 9, pp.4097-4090 (1977)

21. B. L. Sater and C. Goradin, "An Integrated Photovoltaic/Thermal High Intensity Solar Energy System (HISES) Concept for Residential Applications", 11th Intersociety Energy Conversion Conference, Vol. II, pp.1316-1323, 12-17 September (1976)

22. P. Crupeuter and G. Taylor, "An Economic Analysis of Grid-Connected Residential Solar Photovoltaic Power Systems", Massachusetts Institute of Technology, Energy Laboratory, Cambridge, Massachusetts (1978)

23. Dennis Costello and David Posner, "An Overview of Photovoltaic Market Research", Solar Cells, Vol. 1, pp.37-53 (1979/80)

24. Michel Grenon, "A Propos des Carburauts du futur", Premiere partie, Revue de L'Energie, 311, Janvier (1979)

25. Peter D. Bliar, "Modelling Energy nd Power Requirements of Electric Vehicles", Energy Conversion, Vol. 18, pp.127-134, Pergamon Press (1978)

26. K. S. Varde, "Partial Energy Supply to Electric Vehicles Through Solar Cell Systems", Vol. 3, pp.1091-1096, Alternative Energy Sources, McGraw-Hill (1978)

27. Karl V. Kordesch, "Lead-Acid Batteries and Electric Vehicles", Vol. 2, Marcell Dekker, Inc. (1977)

28. Gould, Inc., "Technical Bulletin on Electric Vehicle Power Systems", Illinois, U.S.A.

29. P. Rappaport, "Prospects and Opportunities in Photovoltaics", Proceedings of the Photovoltaic Solar Energy Conference, Berlin (West), 23-26 April (1979)

UTILIZATION OF SOLAR ENERGY - THE JAPANESE EXPERIENCE

Ken-ichi Kimura,
Professor, Department of Architecture,
Waseda University,
Okubo 3-4-1, Shinjuku, Tokyo 160,
Japan.

ABSTRACT

Recent research and development on low temperature utilization of solar energy with an emphasis on solar cooling are described. Selected projects carried out by Project Sunshine set up by the Japanese government are introduced in three categories: collectors with selective surfaces, refrigeration machines driven by solar energy and solar systems applied to four different buildings. Studies on the Soka Solar House project of the Science of Technology Agency are summarized with energy analysis of solar cooling made at the design stage and some test results, followed by an economic analysis. Important accomplishments in developing solar systems made by some companies include collectors and absorption chillers, and their features of current commercialization are explained. Some of the results of the experimental studies on the open cycle cooling and dehumidification system conducted at the author's laboratory are shown to have been only partially successful. Significant aspects of passive cooling devices seen in traditional Japanese houses are summarized, suggesting deliberate consideration of the natural energy in designing, which is applicable to modern buildings.

1. INTRODUCTION

Statistics show that 3.48×10^{15}kcal(14.57×10^{5}kJ) of energy was consumed in Japan during the year of 1975 and that 10.4% of it was used for heating and cooling and 3.1% for hot water supply(1). As for the annual energy consumption in residential buildings, it is generally understood that 40% of the total is used for heating, 30% for hot water supply and only 1% for cooling, the rest being for cooking, lighting, etc.. By contrast, the cooling period is much longer than the heating period in commercial buildings because of the large amount of internal heat gain. As living standards rise, people tend to like to have space cooling, and this brings the high peak demand for electricity in the middle of summer afternoons, which, in turn, gives rise to a serious problem necessitating the construction of new power plants.

Bill B. P. Lim (ed.), Solar Energy Applications in the Tropics, 187–216.

It is natural to consider that low temperature utilization of solar energy should be more economically effective than higher temperature utilization. The effectiveness of solar energy utilization also depends on the length of energy requirements during the year. For example, as hot water supply is needed throughout the year, it is economically feasible, whereas a solar heating system is idle during the non-heating season and thus takes longer to pay off the initial cost than a solar hot water supply system. The energy requirements for cooling in households is so small that energy saved with solar cooling could be much less than solar heating. The economic feasibility of solar cooling could only be justified when cooling requirements range over the year.

It is regarded reasonable, therefore, in Japan for solar cooling, heating and hot water supply system to be installed in rather large scale buildings but it is seldom applied to small single family residences. In fact, the amount of energy saved by solar cooling in large buildings is rather small compared to that with solar heating; and the economic feasibility is evaluated on the overall basis of solar hot water supply, heating and cooling system and not solar cooling only.

It is important, therefore, to find other means of solar cooling for small houses without very sophisticated components and systems. In this respect traditional passive devices are good examples of solar cooling and certainly useful if properly combined with modern technology. Open-cycle cooling and dehumidification may be another approach if simple home-made systems are developed.

2. RESEARCH AND DEVELOPMENT OF SOLAR COOLING BY SUNSHINE PROJECT

In July 1973, just before the oil crisis in October 1973, a plan to do research and develop solar energy utilization was disclosed under the sponsorship of MITI (Ministry of International Trade and Industry) of the Japanese government, and the implementation started in April 1974. This was known as Project Sunshine and the aim was to tackle the most difficult developments in solar energy. Solar cooling was considered one of the most important tasks because solar heating and hot water supply were regarded as rather simple.

It was then considered that there were three major problems to be solved for effective use of solar energy for cooling:

(a) development of efficient solar collectors,
(b) development of efficient solar refrigeration machines,
(c) development of efficient systems incorporating collectors, refrigeration machines and storage with conventional energy distribution systems.

Research into and development of this hardware were assigned to large companies with excellent research capabilities. The following is a selection of some of the accomplishments.

The Showa Aluminium Company developed a good selective surface by electrolysis on aluminium plates. Fig. 1 shows the characteristics which imply a=0.933, =0.10 and a/ =9.33[2]) The durability was found to be sufficient during a test with a temperature of under 300°C for 1000 hours. A flat-plate collector made of copper tubing clamped with extruded aluminium plates with this selective surface and one sheet of glass cover is manufactured and commercialized as shown in Fig. 2 and the performance is shown in Fig.3(2).

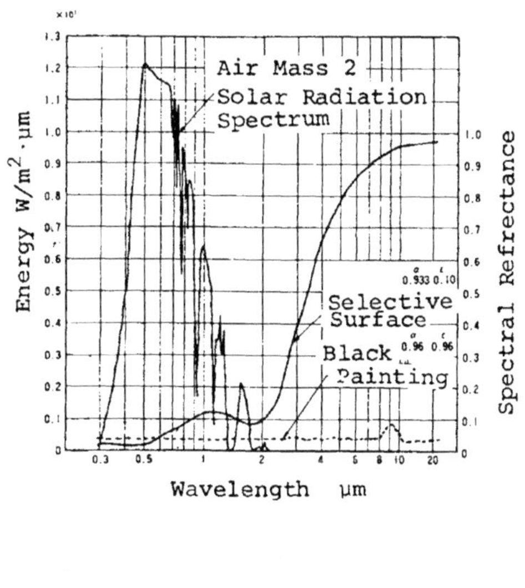

Fig. 1 - AM2 Solar Radiation Spectrum and Spectral Reflectances of Selective surface of Showa Collector Absorber and Black Painting

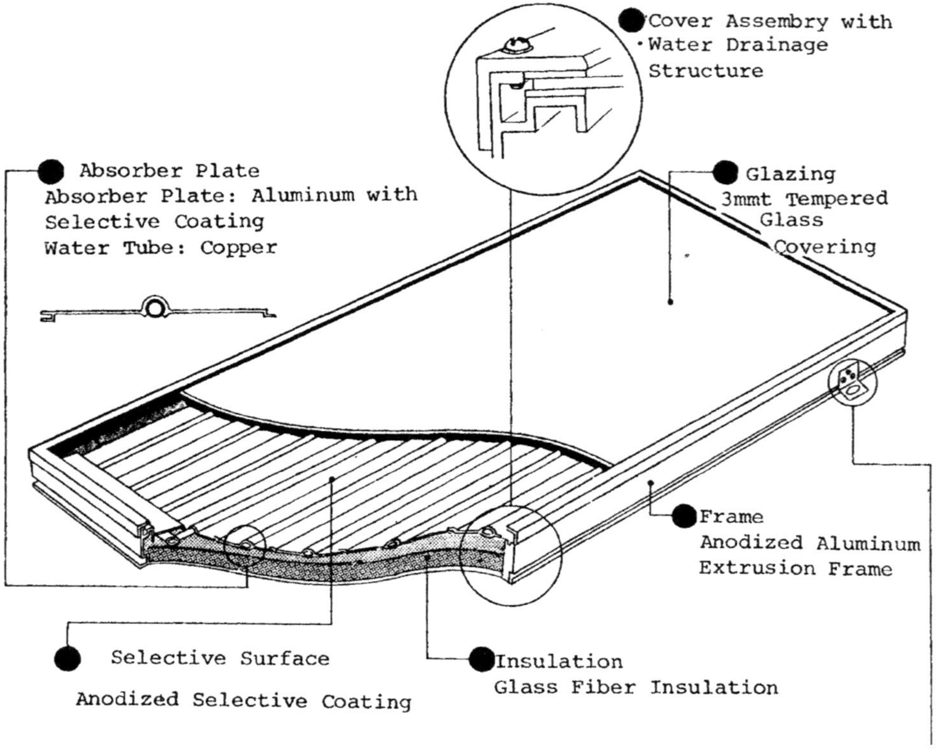

Fig. 2 - The Showa Collector

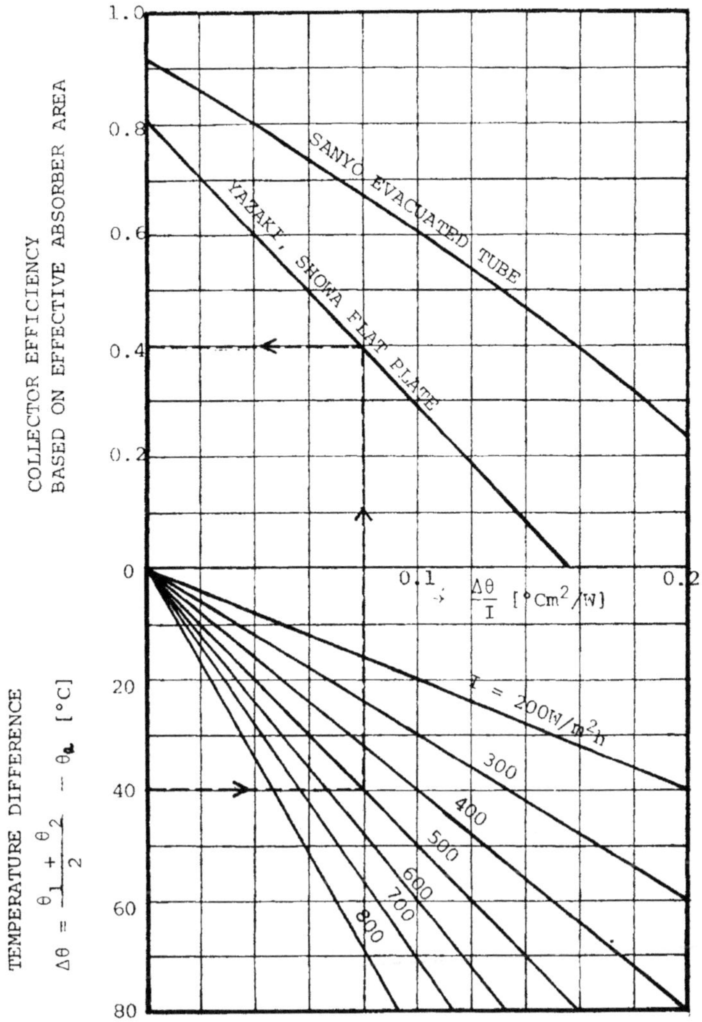

θ_1 : COLLECTOR INLET TEMPERATURE

θ_2 : COLLECTOR OUTLET TEMPERATURE

I : SOLAR RADIATION ON THE SURFACE

Fig. 3 - Collector Performance

The Sanyo Company developed an evacuated tubular collector. This holds Showa's collector plate within the glass tube and the end cap supports the glass tube with resilient sealing and bellows which allow for the difference in expansion between the glass tube and the collector tube. The collector is assembled into a rectangular shape composed of eight standard tubes connected in series as shown in Fig. 4, and the performance is shown in Fig.3(3). Note that the efficiency is based on the unit surface area of the collector plate. The efficiency looks much higher than that of the flat-plate collector, but the net collection per unit area of installation, including headers and casings, turns out almost close to that of flat plate collectors with good selective surfaces.

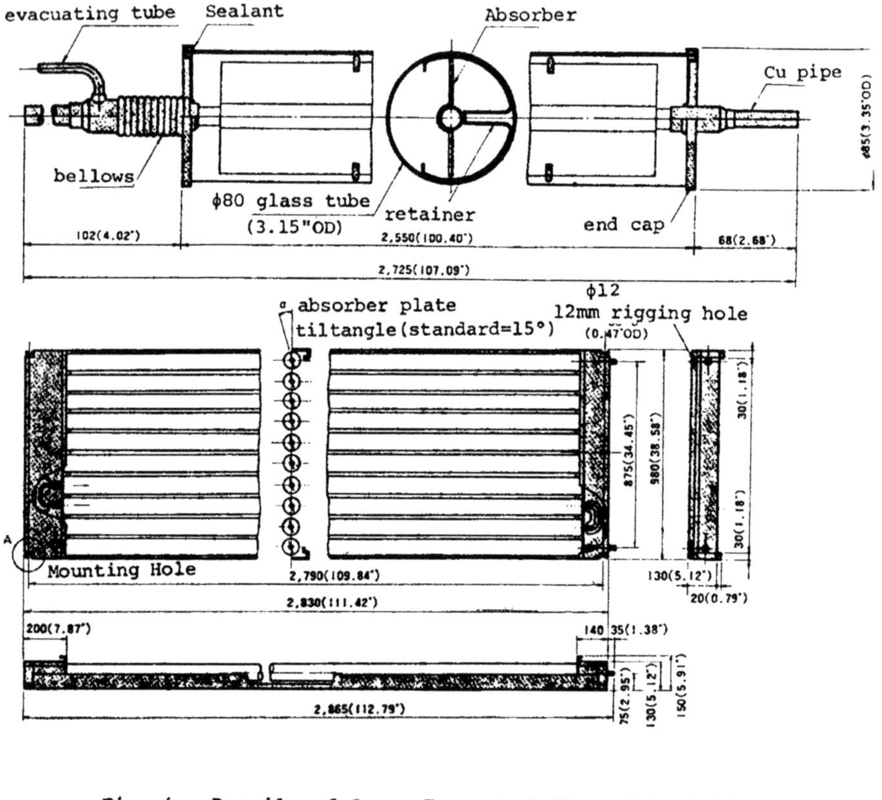

Fig. 4 - Details of Sanyo Evacuated Glass Tube Collector

A special type of absorption refrigeration machine for solar cooling was developed by the Kawasaki Heavy Industry Company. This machine was designed to operate with single effect by solar hot water at COP of 0.7 and with double effect by auxiliary conventional energy at a COP of 1.1. A unit of 30RT was installed in the solar school building of the Department of Energy Engineering of Oita University. The test results during the summer of 1978 show the average COP values of 0.538 with solar in single effect and 1.181 with auxiliary steam in double effect. This was designed and constructed by the Kawasaki Heavy Industry Company, the Kajima Construction Company and Toyo Netsu Kogyo Kaisha (Thermal Engineering Company). Fig. 5 shows the system diagram(4). The system performance during the summer of 1978 was analysed and evaluated using a seasonal performance factor (SPF) as defined in the following expression(5):

$$SPF = \frac{Q_c}{Q_E + Q^F}$$

where Q_c = seasonal total cooling output [Mcal]

Q_E = electrical power consumption by pumps to collector chiller and cooling tower and control system associated with solar system [kWh x 0.86]

Q_F = fossil fuel consumption as auxiliary input to absorption chiller [Mcal x 0.351]

The average value of SPF during the cooling season from June 1 to September 30, 1978 was reported to be as follows:

$$SPF = \frac{43{,}032}{12{,}089 \times 0.86 + 28{,}860 \times 0.351} = 2.1$$

This shows that the solar cooling system was operated rather poorly compared with the conventional cooling system with a centrifugal chiller with SPF = 2-3, and considerable improvements would be necessary in future.

Three other experimental systems of solar cooling, heating and hot water supply were installed by the Project Sunshine Fund, and their performances were tested for a few years.

Hirakata Solar House was built near Osaka by the Sanyo Company and the Ohbayashi-Gumi Construction Company as a new single family residence of two storeys. The system diagram, together with the seasonal energy balance, is shown in Fig. 6(5). Sanyo's glass tubular collector was mounted on the roof and the south facade. It was reported that the net

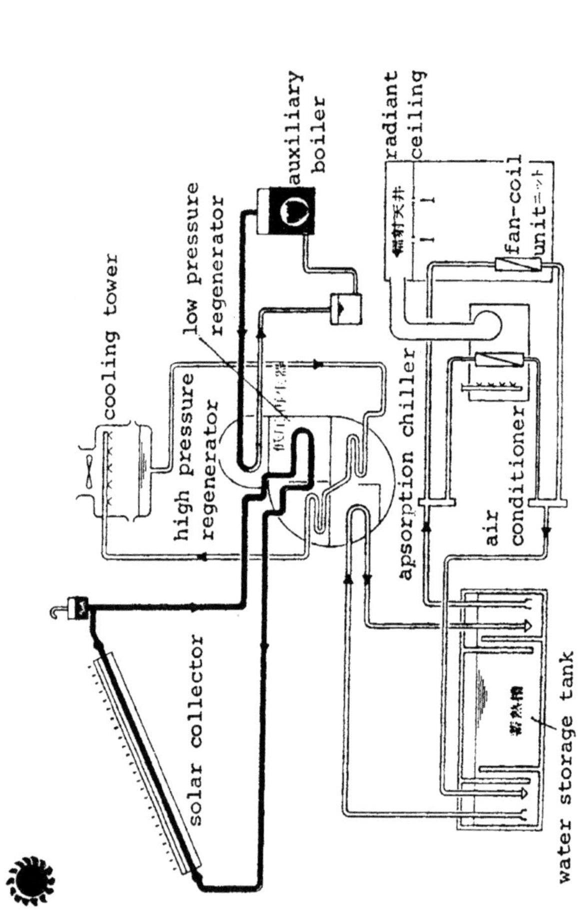

Fig . 5 - Solar Cooling, Heating and Hot Water Supply System of Oita University

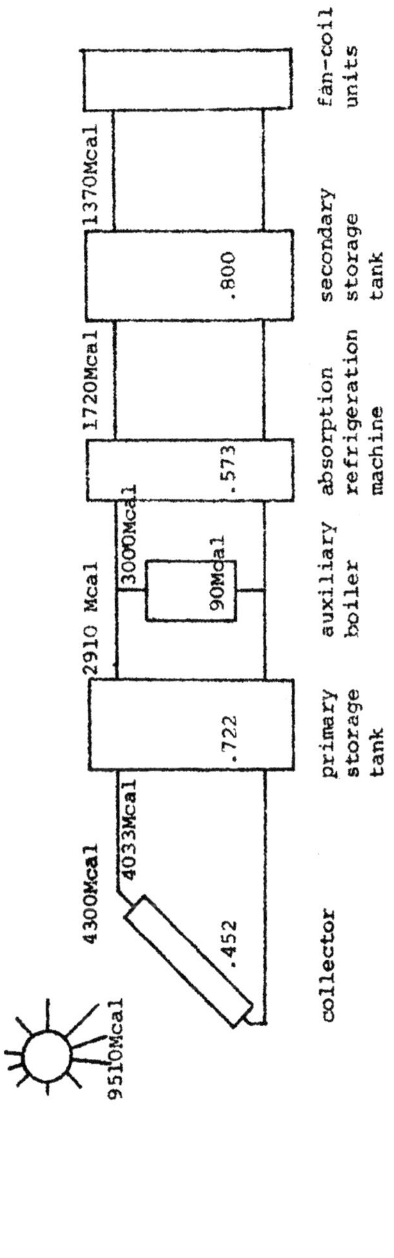

Fig. 6 - Energy Balance of Hirataka Solar House in Cooling season, July 17 - Sept 3, 1978 (except July 21)

loss through pipings connecting two storage tanks and the absorption chiller was fairly large, and this effected the performance of solar cooling considerably.

Ayase Solar House was built near Tokyo by the Mitsubishi Electric Company and the Taisei Construction Company as a retrofit application to an existing single family residence. Fig. 7 shows the system diagram. A 1RT Rankine cycle driven compressor developed by the Mitsubishi Company was installed. The operation of the solar cooling system was reported as unsatisfactory on the seasonal basis. This machine was designed to be driven by the heat source of 90°C, but, in fact, it was operated only on a very fine day when COP as cooling output vs. energy input was found to be 0.4. It seems rather difficult at this stage to expect the Rankine cycle-engine driven compressor to be successful from the viewpoint of energy saving, though it is possible to generate electricity from the Rankine cycle engine during the off-cooling season.

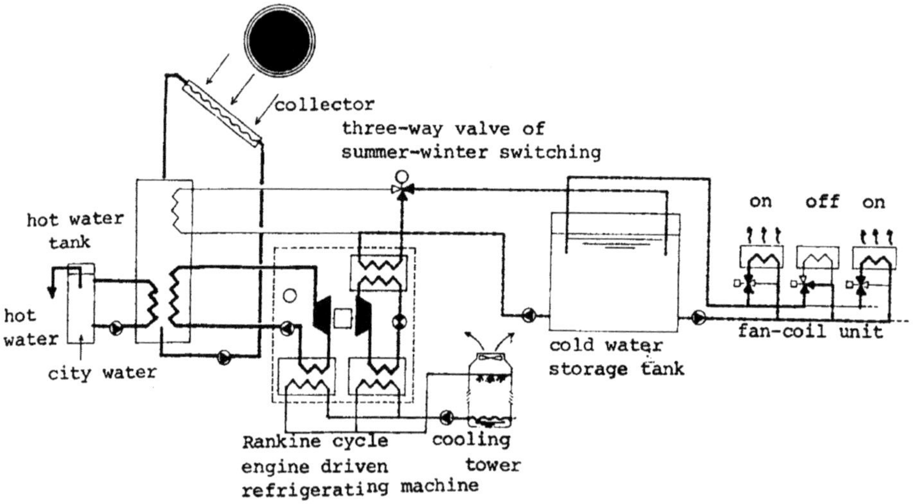

Fig. 7 - Solar Cooling, Heating and Hot Water Supply System of Ayase Solar House

Kikunodai Apartment Building of the Tokyo Electric Company was built in Chofu City in Tokyo. The Takenaka Komuten Company designed the building and the solar cooling, heating and the hot water supply system. The Ishikawajima Harima Heavy Industry Company developed a 20RT refrigeration machine driven by Rankine cycle engine and this was installed together with a heat pump. The test results showed that COP was 0.41.

Cost performance is not analyzed with these four solar buildings built by Project Sunshine, because these are all projects on an experimental basis with various different purposes. From an economic point of view, solar energy systems must be viable in comparison with conventional cooling systems, including possible savings due to future development and mass production of solar components. Special features of these four buildings are shown in Table I(6).

3. SOKA SOLAR HOUSE BY SCIENCE TECHNOLOGY AGENCY

The Science and Technology Agency of the Japanese Government initiated its own research project of a solar house with solar heating, cooling and hot water supply system in 1974. The purpose of this project was to verify the possibility of solar heating, cooling and hot water supply for a residential building through a proper combination of various existing technologies. Soka Solar House was completed in March 1975 with the solar system as shown in Fig. 8(6).

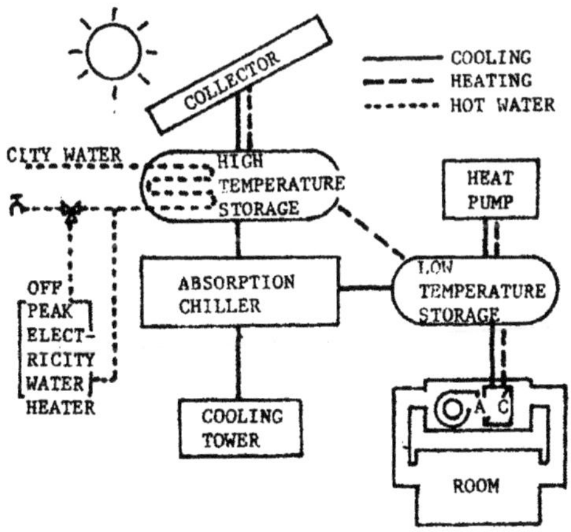

Fig. 8 - System Diagram of Soka Solar House

Table I - Special Features of Solar Houses under Project Sunshine of the Japanese Government

Category of Building	New Single Family House	Retrofitted Single Family House	Multi-Family Apartment	Large Scale Building
Project Executed by	Sanyo Electric Ohbayashi-gumi	Mitsubishi Elec. Taisei Construc.	IHI Takenaka Construc.	Kawasaki Heavy Ind. Kajima Corporation Toyo Thermal Eng.
Location	Hirakata City Osaka Pref.	Ayase Town Kanagawa Pref.	Chofu City Tokyo Pref.	Oita City Oita Pref.
Date Completed	Mar. 1977	Mar. 1977	Dec. 1978	Mar. 1978
Bldg Administrator	Sanyo Co.	Taisei Co.	Tokyo Elec. Power	Oita University
Total Floor Area	118.52 m^2	97.52 m^2	2240 m^2	1860 m^2
Construction	RC, 2 storey	wooden, 2 storey	RC, 3 storey	RC, 3 storey
Collector	Evacuated Tube	Flat-Plate	Flate Plate	Evacuated Flat-plate
Absorber	Al plate, selective	Al plate, selective	Al, Selec. paraboilc	Steel pl., black
Glazing	Glass tube	2 pane glass	2 pane cyclinder	2 pane glass
Unit Size	100 mm, l=1.2m	1500x2000 mm	0.95x3.6m 1.0x1.7m	1865x7520x225 mm
Total Area, effec	40.6 m^2 (roof)	48m^2	366, roof 19, roof	508 m^2 (roof)
Tilt Angle	0°	17°	20° 20°	10°
Storage (high T)				
Volume	1 m^3	1040x840x680	10 m^3	None
Media	water	$NH_4Al(SO_4)_2.12H_2O$	water	
Temperature S/W	60°C	94°C	90-95/10-40°C	
Storage (low T)				
Volume	8 m^3	5.5 m^3	80 m^3	45 m^3 x 2
Media	water	water+sodium acetate	water	water
Temperature S/W	10°C/45°C	5 - 10°C/50-60°C	8-13/40-60°C	9-15°C/40-46°C
Refrigeration Ma.	Absorption	Rankine/Comp. Ref.	Rankine/Heat Pump	Absorption, Double Eff.
Capacity	6000 kcal/h	3100 kcal/h	20 USRt	30 USRt
Input Temperat.	85°C	90°C	90°C	85°C
Cond/Evap Temp		38/5°C	35/7°C	
Expander		Sliding vane 750W, 2800rpm	Rotary multi-vane 12kW, 1500rpm	
Working Media	H_2O-LiBr	R 114/R 22	R 11/R 22	H_2O-LiBr
Chilled Water T	15-10°C	10-7°C	12-7°C	14-9°C
Cooling Water T	31-35°C	33-35°C	30-33°C	30-38°C
COP	0.67 at 85°C	0.4 (overall)	0.41 (overall)	0.65 (solar/1.0 (aux.)

The reason why this system was proposed was derived from the energy analysis of solar cooling with auxiliary energy on a primary energy basis, taking account of energy consumed for pumps and fans. Suppose that one tries to obtain one unit of cooling output from a solar absorption cooling system with back-up by oil to the generator and from a conventional system with a centrifugal chiller. Fig. 9 shows the comparison, which identifies the amounts of primary energy needed in both cases when percentage solar amounts to 72%(7). It was considered rational, therefore, to install both a solar absorption chiller and a heat pump that could work as an auxiliary heat source for both cooling and heating as shown in Fig. 8(6).

The operation results for three years with the occupants (a family of three) were not as satisfactory as expected for many reasons. Table II shows the annual average values of loads, loads shared by solar energy and percentage solar for heating, cooling hot water and their total(8).

Table II - Annual Load Pattern of Soka Solar House

	Heating	Cooling	Hot Water	Total
Load (Mcal)	2912	2168	3901	8981
Load Shared by Solar (Mcal)	2017	561	1851	4429
% Solar	69.3	25.9	47.4	49.3

Economic analysis was done based on the results and some assumptions. Taking ten units of town houses with central solar cooling, heating and hot water supply system, the total initial cost of the system was estimated and the difference of the cost between the solar system and the conventional system turned out to be ¥ 1,043,000.

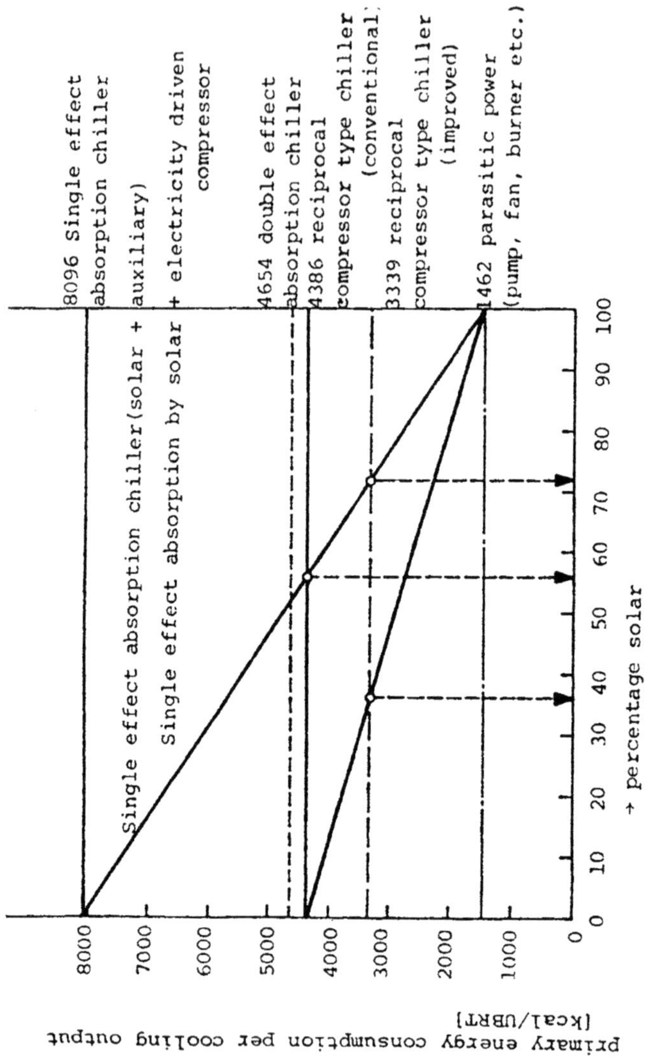

Fig. 9 - Relationship between Percentage Solar of Cooling and Primary Energy Consumption per Unit Cooling Output including Parasitic Power

Table III - Power Consumption at the Soka House and Conventional House for the year 1977-78

	Soka House	Conventional House
Space Heating		
heat pump	1652.3	1549.0
air conditioner	278.8	278.8
enthalpy exchanger	6.7	6.7
controls	591.6	349.9
collector pump	280.6	
heat source pump	64.1	
Sub-total	2874.4	2184.4
Space Cooling		
heat pump	1554.0	1598.0
air conditioner	369.0	369.0
enthalpy exchanger	3.6	3.6
controls	320.0	132.0
collector pump	107.0	
heat source pump	35.0	
cooling tower	65.0	
sub-total	2453.6	2102.6
Hot Water Supply	3755.0	4506.0
night-time heat loss		1854.0
sub-total	3755.0	6360.0
Total	9083.0	10647.0

Table IV - Annual Energy Consumption and Cost Savings?

		Solar House	Conventional House	Savings
1975-6	Energy	8165.8 kWh	11764.3 kWh	
	Cost	181,038	259,821	78,783
1976-7	Energy	9058.7 kWh	11105.9 kWh	
	Cost	201,103	246,884	45,781
1977-8	Energy	9083.0 kWh	10647.0 kWh	
	Cost	199,826	234,234	34,408
Average				52,990

The equation for estimating the pay back period of N years for the initial solar cost S(¥) against the annual savings M(¥/year) at the interest rate r and the annual rate of energy cost increase ℓ can be expressed by

$$S(1+r)^N \leq M \sum_{K=1}^{N} (1+r)^{N-K} (1+\ell)^{K-1}$$

Using this formula, the relationship between these values was obtained as shown in Fig. 10. This means that if r = 6% and ℓ = 9% the initial cost of ¥1.043.000 could be paid back in 17 years and thereafter profit could be obtained owing to the solar system.

4. EFFORTS OF PRIVATE COMPANIES FOR SOLAR COOLING

It is greatly appreciated that the Yazaki Company developed the components and systems of solar cooling through their own efforts within a very short period of time. After the International Congress "The Sun in the Service of Mankind" held in Paris in 1973, they started research and development of solar cooling and built a solar house called Yazaki Solar House One in Kosai City where the solar cooling operation began on July 21, 1974. Mr Ishibashi, leader of the solar group who formerly designed a gas-fired absorption chiller, tried to remodel this unit into one suitable for solar energy use. After several improvements, a solar absorption machine was commercialized to be capable of operating with hot water at a temperature of 75 - 95°C at a COP of 0.6. The cooling capacity of the smallest one is 1.3RT, i.e. 4000kcal/h and the largest 10RT i.e. 30000 kcal/h. For large scale application, multiple units may be installed and connected together to be operated in accordance with the amount of cooling load required by modular control. This machine is widely used in many solar buildings in Japan. The Yazaki Company also developed a good flat-plate collector. The absorber plate is made of two sheets of 0.4 mm thick stainless steel which is seam-welded and finished with a selective surface as shown in Fig. 11(9) and the performance is shown in Fig. 3.

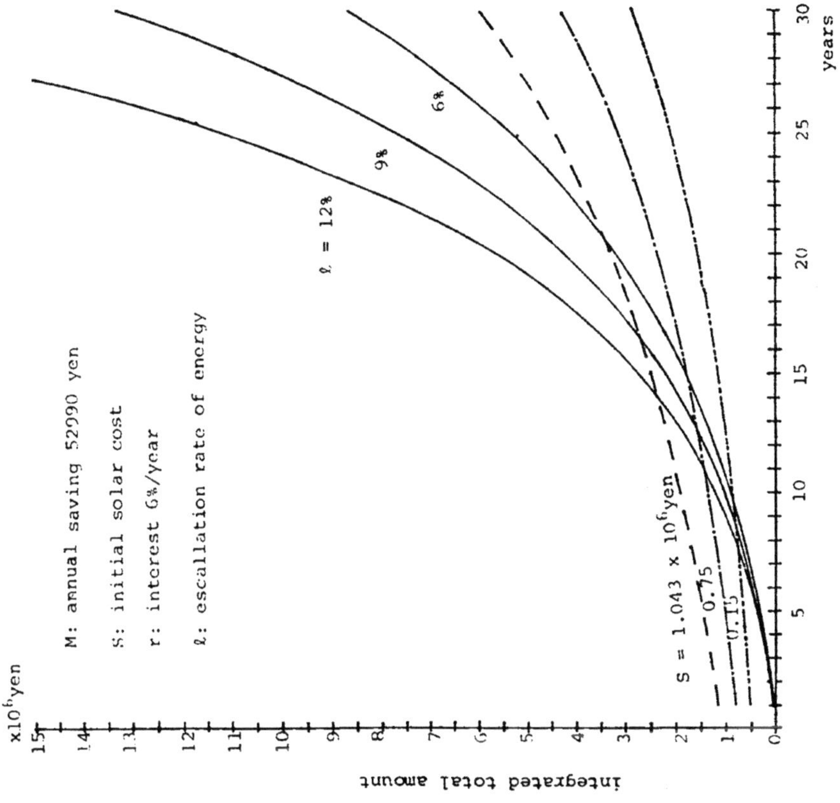

Fig. 10 - Pay Back Period of Solar Cost

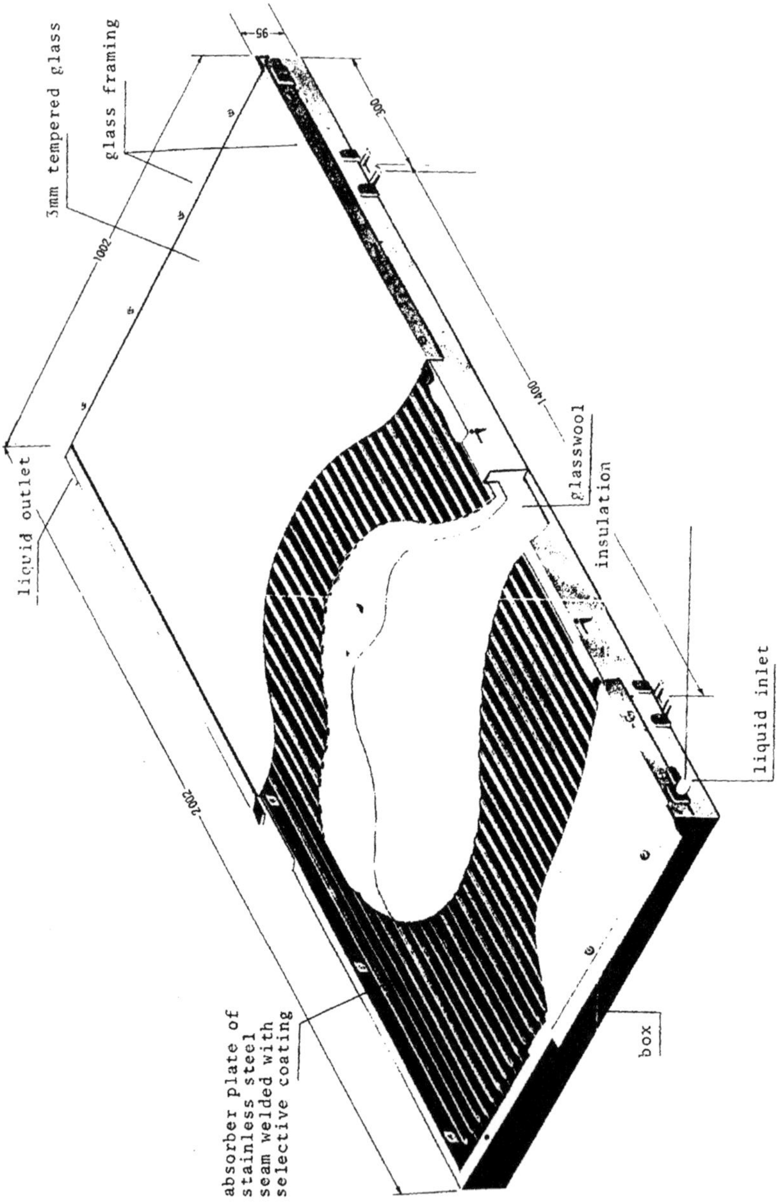

Fig. 11 - Yazaki Collector

Fig. 12 shows Yazaki's standard solar cooling, heating and hot water supply system, and Fig. 13 shows the monthly performance summarized from the operating data taken at Yazaki Solar House One and Ishibashi Solar House(10).

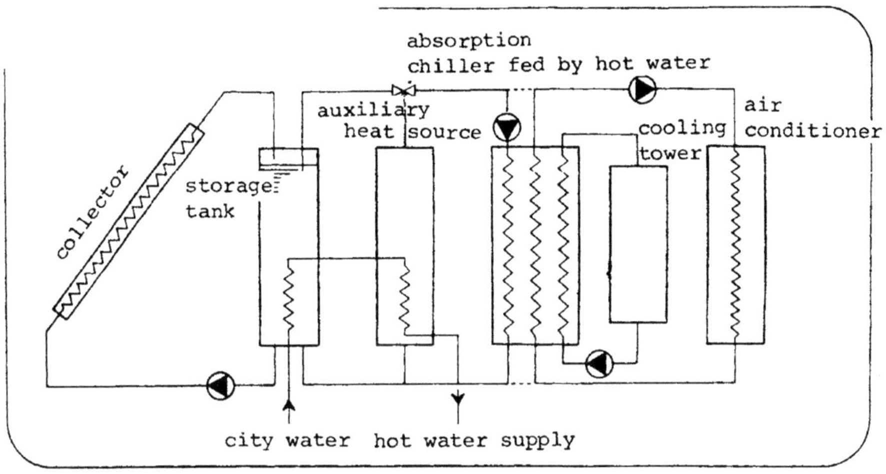

Fig. 12 - Standard Yazaki Solar Cooling, Heating and Hot Water Supply System

There are several manufacturers of flat-plate collectors with selective surfaces in Japan. The Sekisui Company installed their collectors of copper tubing in sheets for more than 2000 prefabricated homes. The Furukawa Denko Company developed a good collector of copper tubing and copper sheets. The Matsushita Company and the Hitachi Company also have their own collectors. These collectors are now commercially available in a standard size of approximately 1m x 2m.

According to recent statistics from the Solar System Development Association, 4648 solar buildings have been built in Japan out of which 125 are equipped with solar cooling system as shown in Table V. The total collector area installed by 1979 is reported as 112769 m^2.

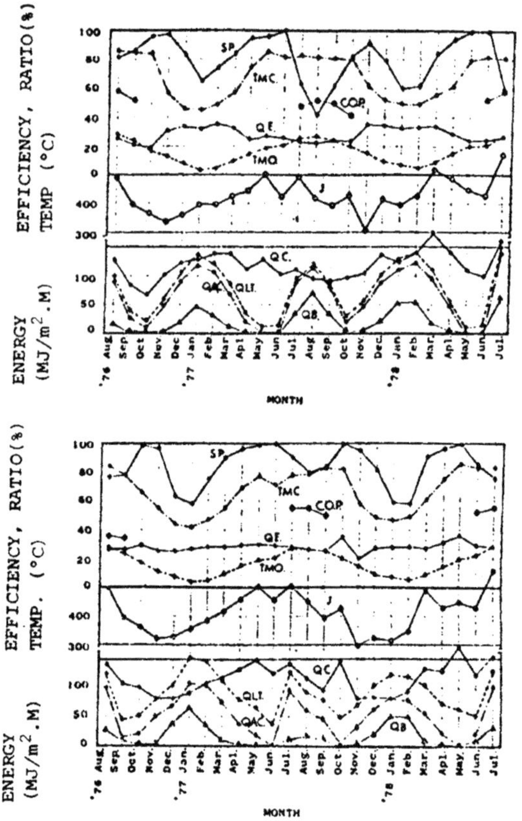

J: Global solar radiation on a collector surface ($MJ/m^2.M$)

QC: Energy collected ($MJ/m^2.M$)

QE: Efficiency of solar collector

TMC: Mean heat collecting tem. (°C)

TMO: Mean outdoor air temp. (°C)

QLT: Total required energy for heating, cooling & hot water ($MJ/m^2.M$)
Cooling (generator input) + (hot water load)
Heating (heating load + (hot water load)

QAC: Cooling & heating load ($MJ/m^2.M$)

COP: Coeffieient of performance for absorption chiller

QB: Auxiliary energy required ($MJ/m^2.M$)
(Auxiliary fuel) x (boiler efficiency)

SP: Percentage solar
* Energy saved/energy required
* Energy saved= energy required-Aux. energy

Fig. 13 - Monthly Performance of Yazaki Solar House One (top) and Ishibashi Solar House (Bottom)

Table V - Statistics of Solar Systems and Collector (Investigated by the Solar System Development Association)

(1) Installation of Solar System

(Number)

Use	System	1975-77	1978	1979	Subtotal	Total
single family house	HW	746	833	2,177	3,756	4,146
	HW . H	157	75	50	342	
	HW.H.(C)			60		
	HW.H.C.	11	18	15	44	
	Other	-	1	3	4	
Sub-total		914	927	2,305	4,416	
collective housing	HW	6	9	11	26	30
	HW . H	-	-	-	-	
	HW. H. (C)			-		
	HW.H.C	-	4	-	4	
	Other	-	-	-	-	
Sub-total		6	13	11	30	
commercial buildings	HW	40	95	192	327	443
	HW . H	3	7	17	28	
	HW.H.(C)			1		
	HW.H.C	24	14	37	75	
	Other	3	6	4	13	
Sub-total		70	122	251	443	
industrial applica-tionn	HW	3	3	6	12	29
	HW . H	-	1	-	1	
	HW.H.(C)			-		
	HW.H.C	-	1	1	2	
	Other	1	7	6	14	
Sub-total		4	12	13	29	
Sub-total	HW	795	940	2,386	4,121	4,648
	HW . H	160	83	67	371	
	HW.H.(C)			61		
	HW.H.C.	35	37	53	125	
	Other	4	14	13	31	
Total		994	1,074	2,580	4,648	
(public buildings)	HW	10	37	35	82	136
	HW . H	-	7	8	15	
	HW.H.(C)			-		
	HW.H.C.	6	8	18	32	
	Other	2	5	-	7	
Subtotal		18	57	61	136	

Note 1. Number of public buildings is included in total
2. HW: Solar hot water supply, H: solar space heating,
C: solar space cooling, (C): conventional space cooling

(2) Sales of Collector

Year	1975-77	1978	1979	Total
Effective Collector area	22,858	28,635	61,276	112,769

5. CAUTIONS WITH SOLAR COOLING SYSTEMS

It is not certain whether or not solar cooling is being successfully realised in all buildings with solar cooling systems. In most cases, the performance has been found inferior to that expected. The cost of modification may sometimes turn out economically unfeasible when compared to conventional systems. Once the system is completed, however, energy saving is realized and this may be considered more important than economics.

There are quite often several weaknesses as in the following, and care must be taken from the design stage to operation.

Long piping between the collector and the storage or between the storage and the absorption machine give rise to considerable heat loss as the temperature difference between the liquid and the ambient air is large. Imperfect insulation of piping and fittings also results in a large amount of heat loss. It is particularly important to avoid the layout of insulated pipes outside, because the covering of insulation can never be perfect and, naturally, insulation gets wet in the course of time. Care must be taken over the thermal bridge through insulation of storage tanks.

A common failing is the heat loss through the thermosyphon action during the night from the storage to the collector. Sometimes nearly half of the collected solar energy is lost to the atmosphere on account of this. Of course this would not occur with a drain-down system for collector protection from freezing.

In hot climate zones the collector might boil during the period of vacancy or power failure. If the liquid stays in the collector, air vents must be examined periodically so as to function whenever boiling is about to occur.

6. STUDIES ON OPEN CYCLE COOLING

The solar cooling system with flat-plate collectord and an absorption chiller have been promoted considerably over the last few years and is regarded as an effective energy saver by many experts. It would not be economically feasible, however, to install this system in small houses because of the recent development of small low-cost air conditioners with a high level of performance. Moreover, the absorption type of solar cooling system cannot be applicable to very small rooms. It would be desirable, therefore, to develop simple and small systems of solar cooling and/or dehumidification. Open cycle cooling may be one answer and studies have been made in the author's laboratory over the last few years.

The principle of the open cycle system has been presented by Close and Dunkle and several attempts have also been made by many scientists. The system could work as designed, but the real problem is to reduce the amount of power required to operate the pumps and the fans. a poorly designed system sometimes requires about the same amount of energy for energy transport as conventional room air-conditioners. The purpose of developing an open-cycle cooling system is to make it possible for any person to fabricate itusing easily available materials at low cost.

Experiences at the author's laboratory started with open-cycle cooling by using LiCl solution as an absorber in the summer of 1976. The regeneration process was successful with a black towel bed and a fibre-reinforced plastic sheet cover, but the cooling process of the absorber did not function well with such simple equipment. Inadequate precautions were taken against corrosion by LiCl, and the infiltration of Licl vapour into the air supplied to the room was a matter of concern. The project was in operation for a while until 1980.

Dehumidification would only be effective in the hot and humid climate of Japan, and a solar dehumidification system was studied in experiments using silica gel as an absorber. The system tested with a silica gel belt of 70kg as shown in Fig. 14 works as follows(11). A portion of the silica gel belt, after passing through the regenerative area of $2.42m^2$, is cooled down naturally and enters into the absorber box where the moist room air is dehumidified and cooled down through rock bed. The belt should be moved by handle intermittently. On September 6, 1977, the results of lhe experiment showed lower absolute humidity inside than outside by 0.003kg/kg with a slightly higher temperature as shown in Fig. 15(11). Some air leakage was found along the belt where natural cooling takes place after regeneration.

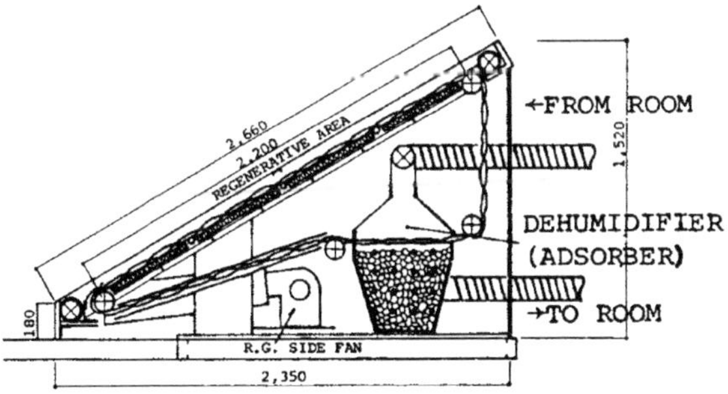

Fig. 14 - Silica Gel Belt Type of Solar Dehumidification system of Waseda University

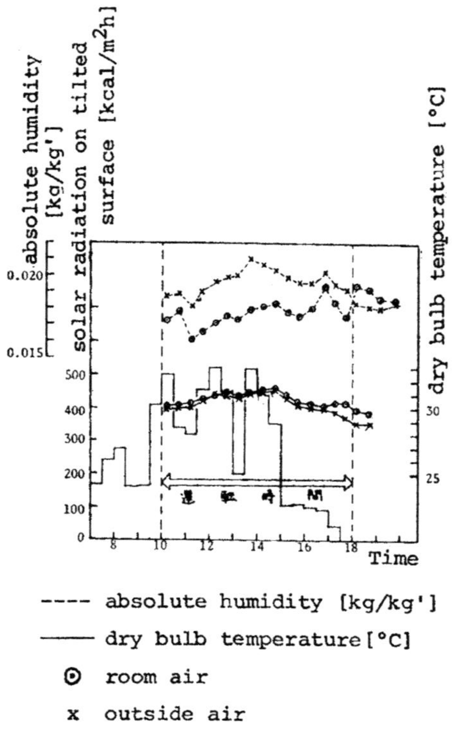

Fig. 15 - Experimental Results of Silica Gel Type of Solar Dehumification System

The second test facility was constructed by students using six drawers with silica gel instead of the belt conveyer. As shown in Fig. 16(12), three drawers were placed in the solar regenerator, one in the natural cooling box and two in the absorption box. The cooling process was planned to take place in the rock-bed storage where chilled water from the nocturnal radiation cooler made of a water tank with an area of 5 sq. metres and 10 cm deep was led through pipes.

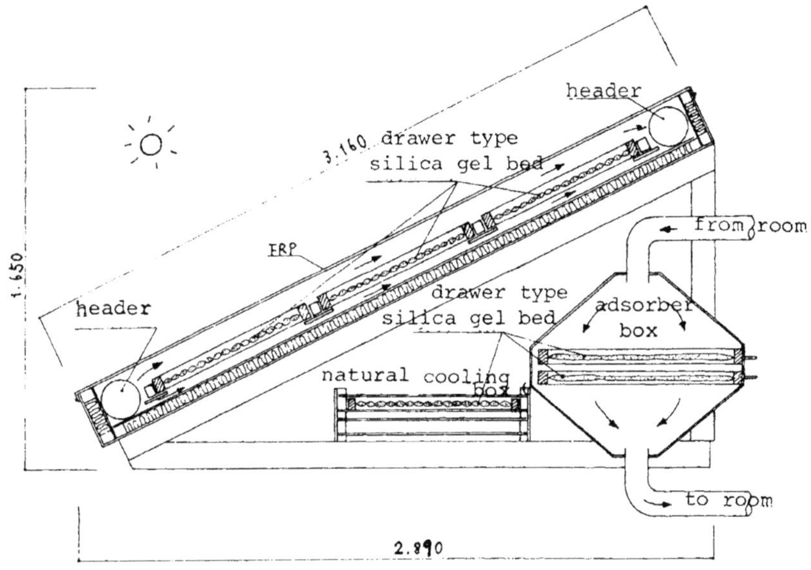

Fig. 16 - Open Cycle Solar Dehumidification System of Waseda University

Test results on August 31, 1978 are shown in Fig. 17(12). The temperature of silica gel got to 90°C when solar radiation was around 600kcal/m^2h, whereas when the radiation was less intense, the silica gel tended to absorb moisture from incoming air. The natural cooling box brought the silica gel temperature down from 60-80°C to 37°C in one hour, which seemed satisfactory. The temperature of silica gel in the absorber box was 40-45°C which was 5°C higher than expected, the effective absorption rate being 3% inferior to the design rate. Heat exchange in the rock-bed with a nocturnal radiation cooler was not as

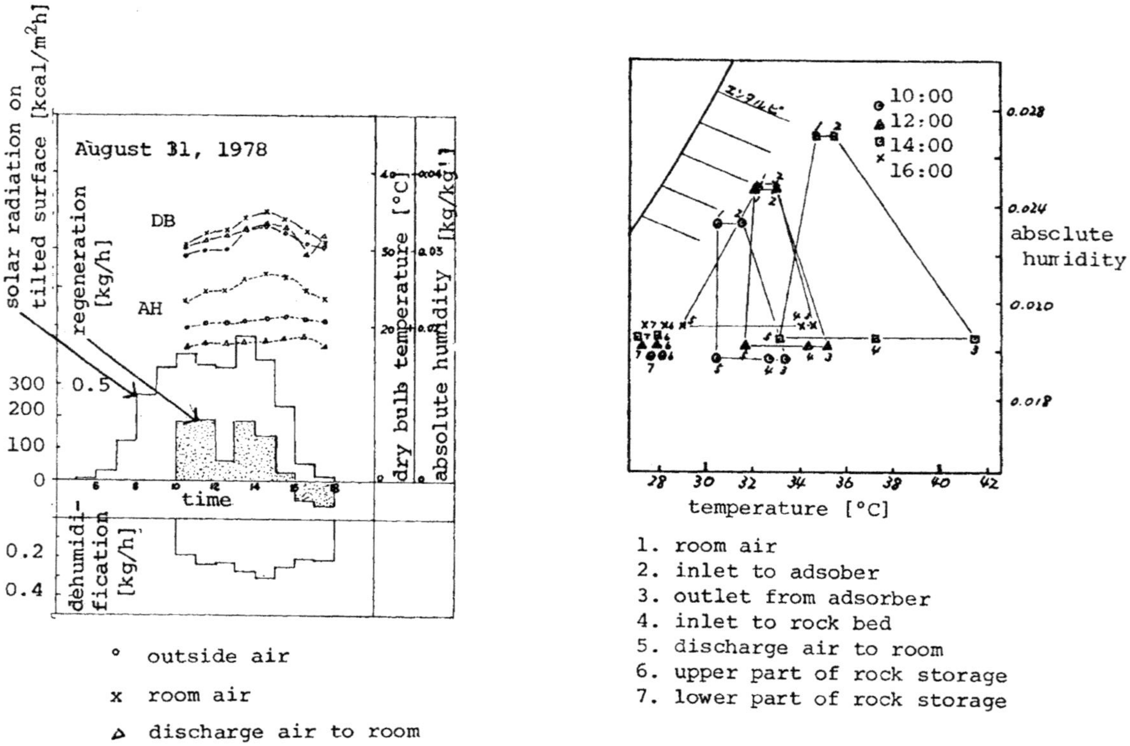

Fig. 17 - Experimental results of Drawer-Type Solar Dehumidification System effective as expected and the discharge air temperature to the room was only 1-2°C lower than the ambient air. This defect could be overcome by improvements.

The cooling process with the solar dehumidification system is one of the most difficult sections of the system. An underground duct has been proposed in order to bring the dehumidified air into the room. The evaporative cooler was not used, as it seemed unsuitable for the dehumidification process, but a small amount of evaporative cooling might be worthwhile attempting.

Another attempt at open-cycle cooling is underway. Using LiCl solution again as absorber, the system allows for the absorption heat to be used for the hot water supply.

The COP value of the open-cycle system may be evaluated by cooling output over energy input of the pumps and the fans. In the above systems tested, the COP's were found to be around 6-8, and the COP target was 20 including the energy supply for hot water.

7. PASSIVE COOLING EXPERIENCED IN TRADITIONAL HOUSES

Various possible ways and means of cooling have been devised in traditional houses to obtain thermal comfort using locally available material. There is a great deal to learn from these houses with regard to modern energy conservation practices.

Natural ventilation is a prime concern when building a house with suitable orientation to the prevailing wind. To enhance the cooling effect, people have used natural evaporative cooling by providing ponds in the garden and spreading water on the ground. A thatched roof for effective insulation and long projecting, deep eaves protect the house from intense solar radiation during summer. A lot of space between the roof and the ceiling help to reduce the solar radiation that penetrates into the living space.

After rain, the moisture absorbed by the thatch or the porous tiles on the roof may dissipate the absorbed solar heat by evaporation.

The lower temperature during the night is a cooling source. Thick interior walls made of clay and large soil floors in kitchens and working areas can keep the indoor temperature low.

Most of the traditional Japanese houses have been built with the principal room facing south, as this is the easiest way to protect the room from the sun which is at a high altitude in the summer and to invite the winter sun in. Sun-screens made of thin bamboo rods and fine wooden lattices have often been attached to the windows facing east or west to protect the room from the sun which makes a low angle with the window in summer.

Tatami mat floor made of straw naturally absorbs moisture. On rainy days in June and July, these mats tend to get very moist, and so, on a fine day after the rainy season, these mats can be dried under the sun in the garden. For this purpose, tatami mats have been developed as simple removable modular units.

These features mentioned above may be regarded as means of passive cooling. People in the past lived without active cooling systems.

It must be questioned whether perfect comfort is really needed in this modern age. Imperfect cooling and dehumidification by simple means could be acceptable even in severe conditions, when taking into account the energy problems of the future. Even if conventional or solar cooling systems are applied, houses should be designed to suit the climatic conditions of each locality. For this reason, the quality of traditional architecture must be studied further from the thermal point of view instead of being admired only for its emotional or aesthetic values.

8. CONCLUSION

In this paper, the Japanese experience in solar utilization has been summarised with an emphasis on solar cooling. Economic justification of solar cooling, heating and hot water supply systems varies according to local circumstances. It also depends on the future escalation of oil prices. In any case, the amount of energy saved by solar hot water and solar heating is considerable, and a little saving can be realized by solar cooling in Japan. Using solar cooling only would not be economically justifiable in Japan and has not been analyzed for tropical countries where cooling is needed throughout the year.

It is the author's opinion that if active solar cooling systems were applied, designing a solar system without any auxiliary heat sources would be regarded as a sound approach in view of the total energy balance. People could withstand less severe conditions without cooling when it is not too hot.

Open-cycle cooling shows some of the characteristics of passive cooling. More studies are needed to improve the system with less power consumption for pumps and fans.

A combination of the advantages of traditional architecture with modern solar technology could lead to better energy utilisation and improve living conditions in the future.

REFERENCES

1. The Handbook of the Nation's Life, Resources and Energy, Japanese Government, 1979 (Japanese)

2. The Showa Solar Collector, Catalogue of the Showa Aluminium Company (Japanese)

3. The Sanyo Solar System for Building, Industry and Agriculture, Catalogue of the Sanyo Electric Air-Conditioning Equipment Company (Japanese)

4. Ohara, S, et al., Application of Solar Cooling, Heating and Hot Water Supply System to a Large Scale Building (I) - (VI). Proceedings of Society of Heating, Air-Conditioning and Sanitary Engineers of Japan, Oct, 1977 (Japanese)

5. Summary of Reports of Project Sunshine FY 1978 (Solar Energy), Promotion Headquarters of Project Sunshine, Agency of Industrial Science and Technology, MITI, April 1979 (Japanese)

6. Kimura, K., Solar House Development, Interregional Symposium on Solar Energy for Development, Tokyo, Feb. 1979

7. Tanaka, S., Solar Cooling and Heating Systems, the Ohm Publishing Company, 1977 (Japanese)

8. Report of Investigation on the Effective Conversion Technology to Household Energy from Solar Heat (4th Report), Research Centre for Land Development Technology, March 1979 (Japanese)

9. Solar Heat Collector Blue Panel, Catalogue of the Yazaki Company (Japanese)

10. Ishibashi, T., The Operation Results of the Yazaki Experimental Solar House One and the Ishibashi Solar House, Silver Jubilee Congress of International Solar Energy Society, Atlanta, May 1979

11. Kimura, K., Matsuura, H. & Komatsu, M., Experiment of an Open-Cycle Solar Dehumidification System Using Silica Gel, Conference of the Architectural Institute of Japan, Sept. 1978 (Japanese)

12. Study of the Drawer Type of Open Cycle of Solar Dehumidification System Using Silica Gel, 4th meeting of the Japan Solar Energy Society, Nov. 1978 (Japanese)

DISCUSSION

Question:

It is sometimes suggested that photovoltaic cells are not effective for hot water heating. Could Prof. Chuah please comment?

Assoc. Prof. D. Chuah:

We have done work on pumping water to cool the cells, but I am afraid the water temperature is not high enough. We have also tried to use the thermosiphon method during the pumping process, but it has an adverse effect on the cell temperature because the thermosiphon method is a slow process depending on the combination of hydrostatic pressure plus changes in density. By obtaining a good flow-rate, heat may be transferred from the cells to the water at about 45°C.

What we hope is, with high concentration, sufficient current can be generated for heating water instead of using solar thermal. If the price is right, I think that should not be a problem.

Question:

I would like to ask Prof. Chuah about the power-conditioner. I suppose that is to prevent over-charging of the battery. Am I correct?

Assoc. Prof. D. Chuah:

Not quite. You can prevent over-charging very simply without the power-conditioner. Just put a diode in series with the panel. What the power-conditioner does in addition is to stabilise the operation at the maximum power point, which is very important to avoid small fluctuations that arise from the variation in the intensity. It is possible to incorporate an DC-AC invertor, if necessary. Basically, it is to stabilise and to track the maximum power point, which is the point of operation for a solar cell.

Comment:

There are at least a couple of us here who may disagree with you a little. I am with one of the photovoltaic companies in Thailand. In our experience, at least, the diode is not always effective. I believe that it depends on how many cells there are in the module. Some modules for a 12 volt system will have 32 cells, others have 33, 36 or 40. I am not convinced, I should say, that just putting in a blocking diode would drop the voltage enough, for instance in a 40 cell module, to bring it down without having to add some type of regulator on the battery.

Assoc. Prof. D. Chuah:

I agree with you. That is why I said that over-charging depends on the number of cells on a diode. But one may incorporate this in the power-conditioner for regulation. The power-conditioner is not necessary if only lower voltage is required.

Question:

I understand that besides the silicon type of solar cells, there are also other types. Can the panel elaborate on other types of solar cells, and their advantages and disadvantages.

Assoc. Prof. D. Chuah:

The most stabilised cell is, at the moment, the single crystal silicon, in terms of application. There are other cells such as the amorphous type, Si structure, metal-insulator cum other semi-conductor types such as GaAs alloys. Generally, all these are acceptable. But they are all at the moment still in the laboratory stage. They are not viable yet for field application. An efficiency of about 10% is normally an acceptable percentage for it to be commercially viable. There are photovoltaic types of cells I know of that are being experimented on. Then you can use materials like tin oxide, indium oxide, cadmium sulphide and cupric sulphide, a long list in fact; they are all in the laboratory stage too. Some do not even last for two hours. After exposure to light, it is finished. So commercially, the most viable material on the market is the single crystal. I do not know whether I am right or not. Would the people in the industry please comment?

Comment:

The poly-crystalline-silicon is coming in now. There are several of these, of course, that are being used in the power industry, and are fairly successful.

Assoc. Prof. D. Chuah:

I do not know whether they are really poly-crystalline or mono-crystalline.

Comment:

Some of them have large single crystals.

Assoc. Prof. D. Chuah:

We are monitoring some of these poly-crystalline cells.

Question:

From the practical application point of view, when do you think the technology will be developed enough to run motors with fewer collectors?

Chairman:

Until more efficient motors are available, we cannot answer the question yet. One is not able to get the same energy from a smaller area.

Comment:

Apart from concentrated cells, we do have 50 amp cells which are now at the production stage, so we will be able to get 50 amps out of the same area. Perhaps this is going to be the answer with regard to reducing actual collection areas. Theoretically, cell efficiency is only about 20 or 21% and in practice the maximum efficiency is somewhere around 18 or 18.5%. This is simply because of the transmission back to the atmosphere. Also 10% of the cell is taken to accommodate the grid. In general, 10% of the surface is devoted to the grid. Concentrators are certainly a means of getting more power in a smaller area.

Assoc. Prof. D. Chuah:

There is a type of cell with double-sided walls which can be placed vertically with reflectors on both sides. It has had three or four years of test-runs now in Spain, using a Winston mirror and that could reduce space further. The conversion efficiency is about 5%.

Comment:

In order to get the cost down, we should not be worrying about the increase in efficiency. Instead, we should work on covering large areas with inexpensive materials. This may be the direction of future development. One may accept 1% efficiency, if it is really economical to cover a large area, perhaps by spraying.

Comment:

There was an article in Business Week a few months ago. They talked about the different types of cells made of a very inexpensive substance. Some very big companies are doing research on inexpensive substances to be sprayed on large areas. Eventually, they will have sprayed areas that look like sandpaper. The efficiency is always rather low, however, if it is sprayed on the roof of a house or on large panels. This may be the direction of future developments, at present, the process is only in the early stages.

Chairman:

The manufacturing of photocells is no longer a secret. One can even purchase a kit. They tell you everything, including how to build a small furnace. One can make the plating and get a cell of 5% efficiency. The question is what criteria, reliability and efficiency are needed to breakeven, and those working in the industry know that they have set 10% as the minimum.

The idea of spraying came because of cadmium sulphide, which could be sprayed onto a plastic base. But work has shown that spraying may not solve the problem of stability in cadmium sulphide. Now they have gone over the sputtering or evacuation in a vacuum chamber and the spraying method is no longer used. With cadmium sulphide at 8%, I think the breakeven is there, and we can have sheets and sheets of plastic cadmium sulphide solar cells. But it does not work that way. Two companies gave up after some time of experimentation.

So it is a question of whether reliability is desirable, or low-cost production without high efficiency or reliability.

Prof. K. Kimura:

Would Prof. Chuah comment on concentration of radiation on the photo cells?

Assoc. Prof. D. Chuah:

Cooling is essential for high concentration type cells, and passive cooling has mainly been used. However, I remember a report about a bank in the United States using a photovoltaic system which concentrates four times more than ordinarly cells. This system is used for active cooling and the hot water which is used to heat the bank. I think it is a good system, but we have not made a study yet.

When normal cells are used for concentration, the magnitude of the series resistance is of importance. Separate measurements up to 6 times must be made to see if there is any shift as a result of the concentration which affects the series resistance. The concentration ratio will then indicate the maximum radiation receivable.

As for reflector systems, first we tried glass, like a green-house. But it was very fragile, and we had a big storm one day and the glass housing was badly damaged. Subsequently, we did away with the green-house effect, because our main aim was not to obtain hot water. But if hot water is necessary, the side glass may be omitted and only the top is used. Now we will go on to study the angles of incidence to obtain the optimum position and the angle of inclination of the mirror so that we do not have to do tracking which will be costly. This is still in a preliminary stage. We will have to get enough data at least for a year to find out the optimum position and efficiency. Also if there is glass on both sides, there will be more shadow.

Question:

Solar silicon cells appear to respond only to the visible wave-lengths of radiation. Is it possible to increase the efficiency by putting some materials on top of the cells?

Assoc. Prof. D. Chuah:

The response depends on the energy band-gap. If you can select materials with the band-gap responding to the short wave-length like germanium, and gallium arsenide to the long wave length, and silicon to the visible wave-length, the cells will probably end up with 35% efficiency, theoretically. Of course, these are all theoretical calculations. But whether the combination is really technologically useful or feasible is another problem. This is all by calculation. Through using the band-gap transition, the silicon efficiency is 25% maximum without any other losses. But under conditions of 25°C, 100 milliwatts per square centimeters, the efficiency is about 10% and the actual field efficiency is only about 7%.

Comment:

Concerning hot water heaters using photovoltaics, one of the problems is heat transfer. Water cannot just flow directly over the cells because that defeats the purpose of insulation. Also water may deposit minerals as it flows across the modules, depending on the type of water. In some areas, the water is particularly full of minerals. We have seen some systems where water flows continuously over the modules. After a period of about 4 months ,there is quite a build-up on the top surface of the modules.

In addition, the harder the silicon cell gets, the less current it produces. So, as the efficiency curves start coming down, the voltage decreases and the hotter the cell becomes. All these should be taken into consideration in the sizing of the system. If it gets too hot they are the wrong type of cell, and the grid might break down.

Assoc. Prof. D. Chuah:

In addition to the reasons stated by the speaker, uniform cooling is necessary, otherwise the cells will crack up.

Prof. K. Kimura:

I had some experience in fabricating the collectors. There are several manufacturers of solar cells in Japan and I asked if they could supply me with net units of the cells. But they were not available. All cells were capsulated in modules. They like to sell modules. Finally I was supplied with cells about the same price as the module. We expected that the solar cell would be cheaper. The next problem was how to stick the cells to the collector plates. And I asked them and they told me some sort of silicon sealing could be used. They say they have many secret ways of making capsulation. But anyway, for research purposes, I was able to buy the paste but it was very expensive. Finally we bought the paste and I asked the students to make it.

The next problem was to produce collector plates with the same flow rate so that we could compare the solar cells. Another difficulty was to maintain the same efficiency in collecting heat, because the solar cells were not uniform. This had to be solved by analysis later. So there are a lot of difficulties in the fabrication of experimental panels.

Assoc. Prof. D. Chuah:

In one industrial process, glass plates are clamped together and silicon fluid is injected into the space between. The spacing of the cells may be done by clear transparent plastic sheets shaped to fit the cells. The sealing still has to be very good. And that is the way they do it.

Comment:

We do ours with auto-claving. We complete our unit with layers of material into a vacuum situation using an auto-clave. We find that this limits the entry of the atmosphere, humidity and whatever. It is working very well for us. The auto-clave is, we feel, the best way for us.

Question:

The Public Utilities Board here are experimenting with photo-cells to control street-lighting, traffic signs and one could do the same thing for building as well. The idea is to save energy. And one might ask why this is necessary on the Equator for outside public lighting, when we have constant daylight. In fact, the times of sunset here vary by as much as 30 minutes. For road safety and for economy, the switch-on time should vary with the weather, as and when required. Because the main supply is on the street lighting pole itself, photovoltaic or photoconductive cell could be used. But what particular material or semi-conductor would stand up to the tropical conditions?

First of all, there is direct and diffuse solar radiation. Second, the temperature at the top of the street lighting pole may be very high. Indeed when the street lights are required to come on, the pole itself may still be hot. Other problems are humidity, insets and lightning strikes. Other countries like Brazil and Australia with somewhat similar climates to ours seem to be able to do it.

Assoc. Prof. D. Chuah:

The main problems will be the infra-red radiation and temperature. For very high temperature, gallium arsenide which is commercially available, may be used. Encapsulation against heat may also be possible.

Comment:

A PE cell or a photo electric cell does nothing more than either open or close a circuit. We normally try not to put it towards some dark object or in shadow, but normally towards an open area to get good natural lighting. Several years ago, if we got three or four years out of them, we thought we were fairly lucky. But now they are going for considerably longer than that. There are several variations, but in general the cell does measure the light intensity. It only either opens or closes the circuit and that would then turn street lamps on or off.

As for the application of photovoltaic cells, we should get the customer to think about a complete system and to calculate more accurately how much power may be drawn from the battery bank each day. All representatives of solar companies want to help put up systems that will work in the areas in which they are being placed. It is not advisable just to buy modules, which must be matched by size and capacity.

Assoc. Prof. D. Chuah:

I would like to comment on buying cells without encapsulation. We find that we pay more. "Cosmetic" charts are required to decide whether the changes in colour are below the tolerance level. Physical and visible inspection of the cells is also important.

Question:

I would like to ask people from the industry what the cost of photovoltaic systems will be five years from now.

Comment:

The cost of systems is coming down rather than the actual modules of the cells. For example, in water pumping, we are asked to supply so many horsepowers of pump requirement, and not the number of cells. But the costs are not dropping too rapidly. We will not expect a reduction of about 50% over the next five years. To my knowledge, in the industry we have reached a situation in photovoltaics where we are producing fairly reasonably priced cells. We have the cost of some fairly expensive equipment in auto-claving etc. So all these are part and parcel of the end result in terms of cost. But with regard to the overall cost of the system coming down, it is the components that make up the cost. In lighting systems, the cost may be reduced by the omission of an invertor in the line, and DC current from a 12 volt battery is directly supplied to a light without a starter or a ballister, except that an invertor is used in the fixture itself. Lower cost wire is used to deliver the power. For pumping, we are searching for more efficient pumps that run on DC power rather than use invertors. The answer is in the technology that applies to systems to solve the problem by means other than photovoltaics alone.

Comment:

There is probably one other thing to be considered, and that is production and cost. If we can double the production, the price will drop in the region of 20 - 30%. Now that holds good for most industrial goods. There are also some more development regarding materials used for the cells which will help reduce production cost of the cells. If we look at what has been done in the last ten years, the reduction in cost has not been appreciative. In five years' time we may see a greater reduction. I fully agree with the statement regarding the system as such, but, in practical terms, we cannot expect overall cost to be reduced by perhaps 40 - 50% .

Assoc. Prof. D. Chuah:

The bulk of the cost goes on from the materials. Unless one can find alternative materials it is difficult to reduce the cost very much. As always, labour is not going to be cheap in this region. It is catching up with the developed countries at about 10% per year.

Comment:

We would all like to get the price down to about 50%. We have recently had a meeting in California, particularly looking at the products being used now, and I think that the only answer is the one that Prof Chuah has just suggested. Unless an alternative basic material is found, this business of being able to halve the cost by doubling the production is not a reality. It cannot happen along that scale because in the meantime we also have to pay people to press the buttons on these sophisticated pieces of machinery and wages are escalating at a ridiculous rate. So a 50% reduction is unlikely unless new materials are being used and certainly not the ones we are currently using today.

Comment:

The technique of using titanium silver has been around for a while, and very good cells are made from it. But there is a process now being tested which involves spreading aluminium directly on to the cells. This process seems to have promise. A black surface field is also used and the efficiency of the cell is improved somewhat. But there are still labour and basic silicon costs involved. In addition to the basic costs, inflation is going up at 12% to 15% at least a year almost world-wide. It is going to be really tough bringing the prices down much more.

Assoc. Prof. D. Chuah:

I do agree with you. Using this titanium silver is not going to bring the cost down very much because I think the real cost is in the process. An electron gun is needed to deposit the materials in a vacuum. Aluminium evaporation also needs to be done in a vacuum. But what about the latest process, just dipping into soft solder? What is the result?

Comment:

I have not seen this myself, but apparently our parent firm has a joint-venture in Saudi Arabia on a concentrated system which generates 350 kilowatts.

Chairman:

I understand that the Italian Government is asking the EEC to finance a 1 megawatt system in Sicily. Would this make a very big impact on the industry, such as reducing the price to a considerable extent.

Comment:

That particular installation is unlikely to do so. But to create that type of precedent is to open up new markets. People will be made aware of the fact that that sort of system is available through photovoltaic application and if that increases the market, then the price might be reduced possibly by doubling up on production. More orders means we are going to be able to cover the overhead that affect the end price of the module.

THE FUTURE OF SOLAR ARCHITECTURE: A SYNTHESIS FOR THE URBAN TROPICS

Jeffrey Cook,
Professor, Department of Planning,
College of Architecture, A.S.U.,
Tempe, A.Z. U.S.A. 85281

ABSTRACT

Solar Architecture is described as building in tune with indigenous resources as symbolized and exemplified by the sun: autonomy and surplus production of services are goals; philosophical or ethical commitments are measures. Active solar systems driven by pumps and fans, and passive systems using directed natural energy flows are discussed as parts of a design strategy to minimize parasitic dependence on off-site resources. The climate and comfort needs of the tropics are outlined, together with several solar mechanical/electrical solutions. The interaction of bioclimatic parameters with solar design requirements is reviewed including site design, building siting and massing, envelope features, thermal mass, materials, old fashioned cooling, new fashioned cooling and daylighting. Examples are shown from warm and humid climates of common dwellings in rural areas, architects' residential designs, together with commercial/industrial and religious ceremonial buildings and studies. This broad definition of solar architecture as an extended bioclimatic machine requires new levels of performance, excellence and design skills seldom approached in history.

1. INTRODUCTION

1.1. Definition of Solar Architecture

Solar architecture can be defined simply as building with the sun. The sun as the omnipresent generator of the earth's weather systems, is the most obvious of the forces of nature that influence human activity, including the shelter of buildings. Thus the sun is both the real provider of climatic assets and liabilities, as well as the symbolic image and exemplar of the nature forces that shape man's earth. Such a definition of solar architecture is an extremely broad and exclusive one.

Bill B. P. Lim (ed.), Solar Energy Applications in the Tropics, 227–256.

Building in tune with nature once had the title "bioclimatic design" and the written works of the Olgyay twins are again central to many of the concerns of solar architects. However, the recent development of both analytic techniques and hardware components has meant that buildings can now be much more productively responsive to the sun than suggested by the pioneering solar buildings, bioclimatic designs and related energy investigations of the 1940's and 1950's. Within the U.S. even the narrow definitions of legislation have charted a broad definition of solar applications. The Solar Heating & Cooling Act of 1974 includes evaporation and sky radiative thermal processes as part of 'solar'.

The promises of economic direct solar energy conversion to electricity by photovoltaic cells as well as other innovative solar applications encourages the concept that buildings of the future will no longer be regarded as consumptive supports of socio-economic development. Rather, there is the real possibility that solar architecture of the future will go beyond degrees of autonomous internal services and become in many cases productive plants - producing surplus heating, cooling, electrical and other resources to supply adjacent old-fashioned buildings, or servicing industrial processes-urban infrastructure concepts well suited to tropical cities.

Such a broad potential implies that solar architecture cannot be measured by some formula of the percentage of solar. While statistical analyses and economic justification may necessitate estimates of the 'solar fraction' or 'annual solar saving' contributed to a building's performance, such mathematical exercises cannot account for the resolution of all the forces of nature and men that are ultimately accommodated in a sound and productive building. Worth must include social responsibility in addition to resource accountability.

A more formidable measure of solar architecture is the philosophical or ethical commitments about the energy and resources - commitments that can be evidenced in many ways.

1.2. Active Solar

Active solar systems are those mechanical systems for buildings and industrial processes in which major thermal inputs are derived from the sun. But while the sun's heat may be substituted for an oil fired boiler the differences of source magnitude and frequency require an active solar system to be designed quite differently from a conventional mechanical thermal system. Such a major technical re-alignment thus brings its own particular restraint and discipline.

In spite of the fuel independence of active solar systems, they are typically driven by pumps and fans and controlled by electronic switching so that there is a modest but vital level of support from external energy supplies. This dependence on off-site energy represents a parasitic reliance on electrical service and fuel supplies - services and supplies that in the future will increase dramatically in cost and decrease dramatically in dependability. Thus, solar buildings with active systems may increasingly risk becoming a kind of cripple, even if operating costs are not so high, because of the irregularity of off-site resources.

1.3. Passive Solar

Passive solar design is the name now often given to the architectural means of constructing comfortable interior environments by building with nature, without the use of mechanical equipment. Thus, by design, natural energy flows are selectively directed without using off-site energy. In contrast to active systems where external means are used to force energy flows, passive design often requires a more sophisticated understanding of energy exchanges than active systems. Since it uses thermal intelligence rather than physical muscle, it can be called a soft technology, even if the calculation is often hard.

A favourite example of both active and passive systems is an automobile. When it is in motion, an automobile is very much an active system. But when a car is stopped and sitting in the sun and getting steamy hot inside, it is an ideal example of a passive system. It is a passive solar air collector.

All buildings are passive - to the extent that without the use of mechanical systems driven by remote energy sources all buildings will produce an interior climate different from outdoors. In some cases these floating interior conditions will be an improvement over the outdoor climate. In other cases the passive interior performance will be considerably less comfortable than outdoors, sometimes exaggerating the worst aspect of local conditions - such as a glass box in the desert.

Obviously, only a century ago, before dependable forced heating and cooling systems, all buildings deserved their right to existence by providing shelter conditions that improved on nature But within the last 100 years, and especially since the mid-20th century, the heavy dependence on packaged environmental control systems driven by packaged energy such as electricity, oil and natural gas has allowed architectural design to diverge from immediate response to local and natural climate conditions.

1.4. Goals

The goal for all buildings should be design to minimize dependence on off-site energy services and supplies. As the supply of cheap liquid hydrocarbons is exhausted, it is imperative to move to renewable indigenous energy sources. We need to build for the future, not steal from it. Any building designed today without considering the broadest solar potentials is already obsolete.

1.5. Strategies

Building solar does not mean doing without or lowering standards. From a comfort point of view passive solar buildings already built have demonstrated improved interior conditions when compared with more conventionally conceived buildings. No one proposes we design buildings without the incredible advantages of electricity. But total dependence of new buildings on off-site electricity would be foolhardy. And the potential for electricity to be produced on-site economically by photovoltaic generation promises to become more and more viable within a decade in some locations. The recent romantic concept of the autonomous house may be a more appropriate approach for larger buildings where scale makes such investments more viable. The greatest benefit would be in rural or under developed locations where there is no existing energy support infrastructure. However, even urbal locations should benefit by allowing a shift of resource support.

Autonomous buildings that thrive on indigenous resources are not one year bargains but are long term investments in continuity of standards and in capital enhancement. Experiences with the costs of active solar systems as well as with photovoltaic electric generation underline the high capital investment necessary to construct these systems. An obvious strategy in considering those high cost systems is to reduce their proposed size as much as possible by minimizing the delivery needs within a building, thereby reducing their costs.

The appropriate design strategy and sequence then, as expressed in a lineal fashion, is:

(a) Site Design - Optimizing the relationship of the building functions to the assets and liabilities of the microclimates of the building site.

(b) Conservation - Consideration of minimizing energy use, both by reducing thermal transfer through the building envelop and by minimizing the energy needs of all the systems and processes within the project. In a wasteful country such as the United States, one of the greatest potentials is conservation.

However, conservation does not have such a great potential in countries such as Sweden, where prudent practices provide an even higher standard of living, but using only one half the energy used by Americans.

(c) Passive Solar - Attempting to maximize the environmental and mechanical advantages of integrated systems that are not dependent on off-site energy sources.

(d) Hybrid Solar - Considering passive systems whose performance can be optimized by the use of small motorized assistance. Such modest mechanical assistance as small motors, fans or pumps, can make cooling run uphill.

(e) Active Solar - Especially where daily or seasonal conditions cannot be satisfied by Passive or Hybrid systems the storage and control advantages of Active soar systems can be considerable.

(f) Auxiliary - As an operation of last resort, the use of off-site energy sources to fine tune conditions and to manage during exceptional climatic conditions puts present day practice into a different priority.

In general, if the above sequence of considerations is followed, the sizes of both Active Solar and Auxiliary systems can be kept small, thus reducing both capital and operating costs, while increasing dependability and the quality of performance.

2. CLIMATE

The tropics are generally distinguished by a warm and steady climate and dependable high radiation. For instance, on the average annual clear day, solar radiation at the equator is 1,880 Btu/sq.ft. and at 15° is almost identical - 1899 Btu/sq.ft. Throughout the tropics, the average number of hours of sun per day does not vary much according to season so that for both crops and for building design there are not significant differences between the seasons.

For tropical coastal locations such as Singapore, the average daily air temperature each month does not vary more than 3°F. Vapour pressure is high. But it is also extremely stable, averaging at just over 20 millimeters of mercury. Obviously, large bodico of water nearby account for the typical high humidities. Generally, higher air temperatures are only associated with lower humidities.

The effect of elevated land within the tropics provides significantly different climates. Since average temperature drops at the rate of 3.6° per 1000 ft. (0.65° per 100m) there are many variant regional climates that differ from the coastal conditions typical of Singapore. Many of the world's highest elevations occur within or close to tropic latitudes. These high locations will experience a variety of strong local microclimatic variations because of cold air drainage, cloud patterns, local winds and night sky radiation. Thus, the equatorial bank of lands may include warm humid, warm dry, hot dry, temperate and even continental alpine climates, as well as composite climates, each of which requires a particular bio-climatic response.

From a solar point of view as well as a bioclimatic basis of achieving comfort, we should identify two different tropical climates. Island tropical zones, such as the Philippines, usually have clear skies of low luminance. This means high direct beam solar potential with a small diffuse component. Equatorial tropical zones, such as Singapore and Jakarta have a higher vapour pressure and a more cloudy sky conditions - cloud cover varying between 60% and 90%. This produces a strongly luminous sky dome, lower annual solar radiation with a large portion of diffuse solar radiation. Because of the turbidity of the sky there are low night sky radiation losses and small diurnal temperature changes.

2.1. Comfort and Needs

Since the natural climatic conditions in the tropics are within or near the top of the human comfort zone the architectural goal should be to reinforce natural cooling potentials, and to avoid heat build-up. Conventionally, we usually think of producing cooling by lowering the air temperature, or by lowering the mean radiant temperature. However, by reducing humidity the effective temperature can be reduced. There is a similar result from ventilation which is usually much easier to achieve. Typically, these techniques are applied in some combination, but human heat exchange in the warm humid environment of the tropics can benefit most by ventilation coupled with overhead shading to control radiant gains.

Generally, in the tropics the air temperatures are continuously below or near to human skin temperature. This means that human metabolic heat losses to the air by convection or conduction is very low. Slight breezes and high humidity reduce the possibility that evaporative body heat losses will be large. Similarly, radiation requires control by design - to avoid exposure to direct solar radiation and to capitalize on the presence of radiative enclosures whose mean radiant temperature may be lower than body skin temperature. But with small diurnal temperature swings, most shaded radiant surfaces will be close to ambient air temperature.

Other human needs, aside from space comfort, in the tropics however, have much in common with more northern and southern locations. The provision of domestic and service hot water is universal. Commercial and industrial heat processes are also the same the world over. Electricity is a similar common need although the lack of the variation in day length and seasonal temperature should reduce the provision for large margins of peak capacity.

2.2. Current Architectural Solutions

Among the current architectural techniques of rejecting solar radiation is the popularity of buildings clad in mirror glass. This unique spectral method of rejecting heat simultaneously lays the directed load on someone else. Such adjacent buildings have had to substantially increase their mechanical air conditioning plants and, in addition to that, have looked forward to a lifetime of operating a much larger air conditioning system. One legal solution to the reflection problem is to buy up all adjacent buildings such as has happened in at least one instance. Thus, for overheated climates, the architectural solutions to radiation control probably have fewer disadvantages.

2.3. Solar Mechanical/Electrical Solutions

Common mechanical and electrical needs for commerce and industry will undoubtedly benefit from the same solar solutions that are becoming significant elsewhere. Solar domestic and service hot water systems have standard solutions. In the tropics where source water temperatures are already warm, the job of heating water is easier.

One recent innovation is the use of photovoltaic cells to drive the pumps of such a system. When the sun disappears behind a cloud or at night the pump will stop, and variable speed direct current motors can respond to any variation of solar intensity. Such systems in the U.S. now cost little more than $100 extra because of the advantages of tax breaks. Such a "hybrid solar thermal/photovoltaic system" then becomes independent of urban electrical supplies and provides an autonomy similar to passive solar applications. Especially in the tropics where freeze protection is not a problem, the use of such direct solar heating systems have great promise.

Others have spoken about the potential of photovoltaics in more general application to urban needs. Probably, urban locations will benefit least from larger photovoltaic installations, because more conventional electric generation and distribution is in place. Although the U.S. industry average costs for photovoltaics are dramatically declining, they are still some distance from the 1984 D.O.E. goal of U.S. 70¢/watt.

Present and projected average industry costs =

1980	US$11.00
1981	8.80
1982	7.45
1983	6.00
1984	5.00

These costs for modules or arrays need to be multiplied by some factor (150% to 170%) to obtain total installed costs that would include storage and conversion. And of course, the availability of high quality silica or world events may change these cost projections.

Nevertheless, a tract house in Phoenix, Arizona, by John F. Long Homes to whom I am a consultant, has already demonstrated a house shingled with photovoltaic cells - 7,200 three inch cells. The roof system was designed by ARCO SOLAR. It is the first installation to be permanently connected to the electric grid and since May of 1980 has been producing about twice the electricity that the model home presently is using. John F. Long Homes is already proposing a 100 unit photovoltaic subdivision which would be heavily subsidized by government support. Thus, although the fantasy of electricity autonomy for buildings is still 10 to 20 years away for non remote locations, it is still within our own lifetimes.

From a popular point of view, the possibility of a photovoltaic wall paper and veneer for the walls and roofs of buildings is easy to understand. More complex but possibly closer at hand are small and medium sized solar/power plants that could supply electricity to smaller communities, or to neighbourhood segments or functional complexes within larger cities. Among prototypes are the central receiver/power tower system, the parabolic dish solar power system and the parabolic trough solar power system. Both simulation studies and field experiments have indicated all these systems offer economic potentials with good interface characteristics with existing electric power systems. Indeed, hybrid electric systems have potentially lower busbar costs and lower risks of interrupted electric generation and, thus, are likely to precede stand-alone solar plants.

At present the largest operating solar-powered facility is the 150 KW ACUREX parabolic trough system that operates deep well irrigation pumps on the Dalton Cole cotton farm near Coolidge, Arizona, U.S.A. This system has had a good operating record since starting on September 21, 1979, with collectors in service for more than 80% of the available time. It operates by producing electricity which then drives conventional pumps. Surplus electricity is dumped into the grid and is borrowed back from the grid as needed. Thus, the power grid serves as a storage system to balance loads.

Generally, systems like the ACUREX system support and supply needs of an industrial process. They demonstrate design for urban planning and resource management but are not directly applicable to most architectural solutions. However, they do demonstrate the large fields of solar collection hardware necessary to support such applications.

Other technologies may make significant architectural contributions within a much shorter period. They may also have a more immediate impact on building design and site planning. For instance, the development of low cost air-to-air heat exchangers or of passive/hybrid desiccant systems could dramatically change mechanical cooling practices in the tropics.

3. BIOCLIMATIC DESIGN

Bioclimatic design approaches continue to provide an ideal basis for solar responsive buildings regardless of new technologies. For instance, access to the overhead sun of the tropics for photovoltaic roofs and for active solar systems is completely serendipitous with the needs for roofs and shade in the tropics to reject heat. One way to think about active solar collector arrays is that they are cooling devices for the solar surfaces on which they are mounted. And even photovoltaic modules provide both shade and convective heat which can thermally shield activities underneath.

3.1. Building Siting

The ideal bioclimatic disposition of buildings in the tropics is low density in dispersed patterns with access to ventilation on all sides by verandahs and porches. This ideal is usually modified in an urban context where density must increase and the distance between buildings is often minimized because of the cost of land.

Lower densities for urban tropical locations have three particular advantages:

(a) maximum free flow air movement around all sides of all buildings, avoiding hot spots and solar heated coves and recesses;

(b) providing privacy by distance so that air movement is not impeded by walls or privacy screens;

(c) allowing outdoor use spaces as an alternative or complement to indoor functions

The concept of sealed buildings and forced mechanical systems implies a conflict with the bioclimatic approach. In fact, the typical high cost of infrastructure and services in conurbations is associated with the higher density and tighter packing of building masses/design strategies that need to be reconciled with passive or bioclimatic approaches. The potential of solar mechanical and electrical services for autonomous buildings implies a lower density, determined in part by the quantity of solar flux available on the site.

Placing tropical buildings on pilings or stilts and exaggerating their vertical massing are techniques of maximizing access to cooling breezes. Unlike cold climates the ideal sites for buildings in the tropics are exposed and unprotected locations such as promontories, hilltops and ridges. The spaces under raised tropical buildings also provide more comfortable outdoor areas as well as maximizing air flow on all sides of a building.

Orientation is another design strategy in site planning. Obviously solar gain surfaces must face the sun to maximize irradiation. All other surfaces should be shaded. Sometimes this advice is in conflict with local conditions where for instance, cooling breezes may blow on an East-West axis - an orientation that should not be open to the sun. Thus, properly shaded structures should be wind oriented.

At higher latitudes where heating is a problem, the choices of orientation and solar access geometry are within a narrow range. Within the tropics where solar collection surfaces are likely to be for active systems, there is a broad range of orientation and tilt angle choices, with only a small penalty for deviation from the ideal angles.

3.2. Site Design

Because vegetation is so prolific in the tropics there may not be so much concern about removing trees to avoid the shading of solar collector surfaces. Because the sun is so consistently overhead, trees do not offer the potential obstacles for solar that can happen with the low sun angles at high latitudes. Preservation of legal access to the sun, and solar rights should also be less of a problem because the sun's altitude is so high! On the other hand, in those locations where vegetative shade is desirable there is no difficulty in encouraging growth.

The general desirability of vegetation to lower ambient temperatures implies both high canopy and ground cover. Vegetation should allow breezes from prevailing directions while maintaining shaded conditions. Dense hedges, bushes and screening growth should be avoided to facilitate shaded breezes. The general goal of encouraging vegetation is to achieve favourable microclimate modification - both to improve conditions for outdoor activities, but also to minimize thermal and moisture impacts on building envelopes. Well developed landscaping should be an important part of the site strategy for all solar architecture.

Large masonry paved surfaces are not thermally recommended unless completely shaded - then they will conduct the same temperature as the moist ground beneath. But paved areas close to buildings may increase both thermal radiation and reflection and thus need special consideration.

3.3. Building Massing

In urban situations where detached one family houses are not the predominant building type it is difficult to prescribe ideal building size or surface/volume ratios. In general, where the building envelope or skin is the passive source of comfort or climatic relief such as is typical in many parts of the tropics, building designs should provide every room or space with access to an outside wall. Indeed, in many tropical locations the exterior skin may be more of a screen than a weather wall. Similarly, the design of office buildings, schools and public structures to use interior daylight would restrict the internal distance to an exterior wall. Larger markets and exhibition spaces can be daylighted and ventilated from above.

Typically, the urban massing of buildings, all of which are competing for the same breezes, can produce a 'no win' situation. Generally any major structure will significantly alter the adjacent air movement patterns. Planning to allow 'natural ventilation rights' may become an important urban requirement. The higher the building, the better the access to any breeze. For ventilation reasons as well as to avoid large hot rays, there is good bioclimatic sense in considering higher blocks, even point blocks, rather than low compact developments.

Only when buildings are completely mechanically conditioned and lighted do they allow complete freedom in massing. Generally in those cases the larger the building the less influence the skin has on the comfort loads.

3.4. Envelope Features

In the tropics, roofs have the strongest thermal impacts. Since this potential for heat gain is primarily by radiation, the use of ventilated double roofs and wide overhangs is most appropriate. Roofs can benefit most by insulation, reflective surfaces or by self-shading construction. Generally parapets are to be avoided. But roof ponds or roof sprays may provide a major active or passive cooling source.

In the tropics, walls generally have very modest requirements as thermal barriers. Thus good insulation or thermal resistance qualities are not necessary if solar radiation can be kept off the skin. Multiple glazing should not be necessary and indeed there may be instances when no glazing is needed at all - only louvred openings for privacy, ventilation and shutters for security. Solar architecture for the tropics would have shaded walls and would be aesthetically distinguished by screens, shades and overhangs whose open pattern would allow views and breezes but not direct sun.

Buildings in the tropics must occasionally respond to intense storms. The disastrous impact of spasmodic tropical storms is well known and all buildings should have emergency response modes to those irregular events. Primary effects to be resisted include high winds: hurricanes by definition have winds of 34 meters per second or 120 km per hour, and peak gusts exceeding 67 meters per second or 240 km per hour have been recorded. The other primary effect of tropical storms is torrential rains and up to 20 cm of precipitation in 1 to 3 days is not uncommon. Extreme rain conditions can be extraordinary and most world rain records come from the tropics.

Secondary effects of tropical storms are flooding, landslides, storm surges of the ocean and storm generated ocean waves. Severe local storms include thunder and lightning, storms and tornadoes. All of these hazards of weather make special demands in terms of protecting windows, tie downs and anchoring, as well as storm water management.

Urbanized areas have particular problems in relation to extreme weather events. Typically, the natural stream beds and drainage have been built over and result in paved areas with fast run-off flash flooding. Similarly, high buildings could accelerate the speed of extreme wind gusts.

3.5. Thermal Mass

Great attention is now being paid to the design strategy of thermal mass in passively heated buildings in colder climates. It is usually assumed that buildings in warm, moist climates should be without thermal mass to avoid interior heat sources. Lightweight construction is the usual standard for tropical locations - a standard somewhat at odds with high density urban development. However, depending on the local climate, there can be contrary recommendations.

Abundant examples exist of successful indigenous practices in the warm, damp climates of the lower Mississippi Valley. Here one finds the traditional porch/veranda type of plantation house with shaded large windows facing all orientations and often a ventilation monitor, like a belvedere, at the top of the house for exhaust air. What goes unnoticed is the use of masonry for the core walls in such 18th and 19th century buildings. This masonry is rooted in cool, moist ground and acts as a 'thermal wick' or 'cool conductor', providing a cool vertical radiant heat sink within the building.

Thus there may be places in tropical applications for thermal capacity materials - either as a passive thermal sink, or, alternatively, as passive thermal storage that is charged by active cooling systems during off-peak periods.

3.6. Materials

Because of the high humidity in the tropics there are special problems of materials choice and assembly details. Aside from the exclusion of rain, the control of insects and pests must be included in wall tasks.

The old necessity in damp and sticky climates to use floors that are easily cleaned, durable and discourage pests and insects continues to be ideal. On the other hand, highly acoustically absorbent surfaces are desirable to provide noise controls and privacy where large openings are exposed to wind and air movement. Choice of building materials should recognize that high humidity accelerates mould and aids algae growth, rusting, rotting and organic decay. Along coastal areas high salt content of the atmosphere encourages corrosion.

3.7. Old-Fashioned Cooling

All the traditional low technology solutions for keeping cool in the tropics are at a stage of rediscovery. Alternate outdoor use areas from work and play would return flexibility to building design functions. Sleeping outdoors may require minimal architectural design - but it is a consideration often missing in modern buildings. In general, windows in the tropics should reach to the ceiling line so that hot, humid air can be swept outside. Cross ventilation continues to be the most dependable natural method of air change. And air movement without air change produces physiological cooling by lowering the perceived temperature. Fortunately, loose, light clothing has never been discouraged in the tropics.

The use of large low-speed overhead fans or punkahs are again becoming popular - not just as decorator novelties. The elimination of heat sources from human environments could affect cuisine, family and social arrangements, for instance, by reinforcing joint food preparation or encouraging restaurants. In the work world, the tendency of business machines toward miniaturization has greatly reduced the need for electricity use and reduced the production of incidental heat in office environments.

3.8. New-Fashioned Cooling

Dehumidification is a mechanical process that can bring warm temperature and high humidity conditions within the human comfort zone. Normally, dehumidification is achieved as part of the air conditioning function of air handling equipment. The concept of an effectively dehumidified building implies controlling the moisture content of the interior air which can be done most efficiently if outside air is not allowed into the building. Such an approach of sealed buildings, while currently the practice, is at odds with the intent of minimizing dependence of buildings on forced systems. The interface of such an approach with bioclimatic strategies implies a dynamic building mechanically - sometimes it would be a weather selective pavilion, at other times it would be a vapour resistant container.

According to Report #448, published by the World Meteorological Organisation on Weather Climate and Human Settlements, by H.E. Landsberg:

> "... roughly four per cent of the total energy used in the U.S.A. (is for cooling). This type of climatization, even though it has been occasionally designated as luxury, is readily compensated by the savings in food spoilage, improved efficiency of workers, and protection of the health of the very young and the very old fraction of the population. The latter two groups have undeveloped or impaired heat-regulation mechanisms respectively. Living and comfort conditions in vast areas of the globe could be vastly improved if the power demands could be met. An assessment of global cooling degree-days shows that really about one-half of the land surface of the Earth, with 1.5 thousand million inhabitants, requires some cooling in dwellings and work places to afford an optimum environment."

Thus energy effective architectural cooling must be seen as a major challenge for the near term in the very heated parts of the world. The potentials of active solar cooling systems in concert wth bioclimatic passive and hybrid cooling strategies including dehumidification offer great promise in reducing the electric power requirements that are associated with conventional compression refrigeration.

3.9. Daylighting

Daylighting is being rediscovered as a viable low energy strategy for larger buildings in the U.S. Fundamentally, it is seen as a passive cooling approach because it substantially reduces the heat produced inside the building envelope. It saves energy by reducing electrical use first for lighting and second because air conditioning is not necessary to remove the excess heat that is an unwanted by-product of electric lighting. Generally, the effectiveness of daylighting designs is dependent upon electronically controlled auxiliary lighting which is automatically dimmed or brightened in response to changing natural conditions. In tropical locations where sky luminance may be much more dependable than in higher latitudes, daylighting is a solar energy application with immediate viability. Some concept of the solar resource of daylighting in the tropics is revealed in typical lux illumination values of design skies:

London	lat 52°N	5,000
Hobart	lat 43°S	5,500
Sydney	lat 33°S	8,000
Brisbane	lat 27°S	10,000
Darwin	lat 12°S	15,000
Nairobi	lat 1°S	18,000

Indeed, the problem with daylighting in the tropics is that it is too easy. The general luminous quality of the sky invites daylight design. In the tropics, the physical and the psychological results of overlighting could be disastrous. The design goal then in such overheated regions is to keep daylighting adequate, but at the minimum edge of the scale. For the building designer there is the particular challenge to avoid "gloominess" and to avoid excessive contrast while providing adequate brightness and interest.

Typical architectural solutions include avoiding views of the bright sky or of brightly lighted scenes outside the window, often by using louvres or other shading devices such as plants to control glare.

3.10. Examples

From this series of strategies or principles from a theoretical point of view, we may review examples that may stimulate their imaginative application to more immediate building designs. Certainly, where I come from, there has been a re-discovered interest in indigenous buildings and those successful examples that have responded to the intimacies of local climates. Models built by some students of mine at the Arizona State University in the early 1960's demonstrate indigenous buildings from tropical regions.

First of all, examples from Sumatra in Indonesia, and from the Philippines, share a common conceptual point of view: they are all very similar. They are raised buildings to get above the flooding conditions and up into the breezes. They have wooden sides which in some cases can be hinged open, but in any case, sufficiently open so that the breezes can blow through. They have high roofs which throw off the rain but which also provide a vertical dimension to the interior. The high interior volume will develop a venturi effect. The stacking of hot air can be relieved by vents which are hooded against rain penetration. Thus, natural convection cooling is built into the interior. Since the building is raised and the floor boards do not fit tightly, not by reason of poor craftmanship but deliberately for thermal reasons, it is like living in a kind of ventilated sphere. The breeze from any direction can enter the building and will rise up underneath the thatch and be exhausted in the triangular gable openings near the ridge.

Those presumably primitive examples have parallels in historical and Western buildings. In the southern states of the U.S. where there are overheated, warm, damp climates, one finds similar approaches, although perhaps not quite so thorough. Certainly, the idea is present of building with volumes that have access to breezes in all directions. Buildings are only one room deep so that they can be ventilated regardless of the orientation of the breeze. These are the so called "I" designs; "I" shaped plans from South Carolina in the United States from the 19th century.

One 20th century example of a bio-climatic building designed by an architect is a fairly old one from Pondicherry - GOLGUND was designed in 1938 by a Japanese architect, Antonin Raymond. He designed this multi-storey dormatory for the Sri Aurobindo Ashram in Pondicherry which is around 12 or 13 degrees North latitude in South India. Everyone who has had anything to do with Pondicherry always claims that this is by far the most comfortable building in the city. Within the enclosure of the garden wall, there is landscaping. The surfaces are all designed to keep it cool. The sun shines on both the north and the south elevations of the building. There are 2 wings or dormitory rooms which are slightly split off-centre. A single access corridor runs along one side of the building. This allows for thorough ventilation as well as preserving some sense of privacy. This building was finished in the early 1940's and when one realises how remote its location, it is truly impressive - the degree of refinement, finish and maintenance. It looks as pristine today as it did when it was finished.

The roof is a double roof. In other words, the concrete roof vaults are serving as solar collectors simply to collect the heat which is then disbursed convectively through the openings at the ends. The walls have no glass but rather are pre-formed louvres made of asbestos cement which are controlled in gangs. The wooden doors are panelled in such a way that they are ventilated by means of woven panelling which allows the doors to be transparent to air movement while providing privacy. The original furniture which was also designed by Raymond is very appropriate to a tropical climate being lightweight and airy. It is memorable that the two people who manage this building today were part of the construction crew when it was built in the late 30s and early 40s.

Another important project is the design for a large office building - a million square foot office building for the Southern United States which was designed in 1979 for the Tennessee Valley Authority. The Tennessese Valley Authority is considered to be the world's largest electric utility. It has also been named by some people as the world's largest environmental pollutor. It also has the largest number of nuclear electrical generation plants either in operation or under construction. It is, therefore, perhaps rather appropriate that they should take some leadership in looking at a new generation of office design and, indeed, that was their goal - this was not to be a state-of-the-art building, but to demonstrate an advance in the state-of-the-art. To do this, a team of consultants was assembled which includes some of the most prominent names of energy designers in the United States, both from an architectural point of view such as The Architects Collaborative (TAC) and Caudill, Rowlett & Scott (CRS) as well as ome of the most distinguished engineering consultants in the U.S. The site is one at around 33°N. at Chattanooga which is just at the border between Tennessee and Georgia on the Tennessee river. It is in a valley and it is in a warm, humid location, very unpleasant during the summer. It has a climate very similar to many tropical locations although it also has a rather damp, cold winter. So, from that point of view, it is a more complex situation to deal with than more tropical locations. Because of atmospheric inversions and the high humidity, as well as the air pollution, this can be a very uncomfortable city.

The parameters of the design were extensively analysed and discussed by the team before arriving at a solution. An understanding of the climate was a rather important place to begin, not just from a solar point of view. An innovative analytic technique was an examination of the sequence of weather events. By using computer search techniques, characteristic sequences of days that are repeated, maybe 3 or 6 or 9 times, during the season were identified. In winter, there were two frequent winter climatic events, one cold front and one warm front event. During the summer, there was an oveheated climatic event that one could track through 7 or 10 days. Thus, there are sequences of winds blowing in storms, of overcast days, of temperatures rising, of radiation going down, and so on. It is the kind of sequence that one needs to understand so that the transients of the building can respond.

Then there is a whole urban design aspect which has important energy implications, both for the city and for TVA. What should a building be like if it assembles all of the employees of the largest employer in the region in a small city for the first time? This can make an enormous impact because now the day-to-day activities are decentralized and, thus, are not obvious. One of the interesting aspects is that TVA, the Tennesse Valley Authority, has a company transit program that has reduced people driving their personal car to work. Instead, there are company buses which take people to and from work. But another design alternative is that everybody should drive to work in their own car and that people should not just live in the down town area but should commute from a large territory. Thirdly, there is the mass transit idea where you deliberately develop higher density with arteries and some kind of communal transit to take people from work to home. Lastly, there is, hopefully, the kind of ideal which includes a mix of those previous stylizations of what the relationship of workers to their home locations would be.

These are some design parameters of this particular building. But its design goals were very ambitious. First of all, it should impact all levels of the energy world so that this would not just be a TVA electric energy efficient building, but it would affect people from an individual basis right through to, potentially, its asset as a world model for this kind of activity. Of course, as an electric utility, there was a strong desire to reduce both peak power use and total energy used as well as to reduce the energy content within the building materials of the structure. There was also the desire to have a mix of users within a building that was originally intended to be only an office building, so there could be a better balance of services.

Another issue is the problem of solar zoning. Solar zoning ordinance is a 2-edged sword. On the one hand, if one is going to build a solar building, one wants to be assured of perpetual solar access. On the other hand, one ought to demonstrate a good neighbour policy that one is not depriving potential neighbours of their possible solar rights. As a result of the design of this building, a model Solar Ordinance was proposed for the city of Chattanooga which would achieve this kind of mutuality.

From the very beginning, the directions in which the architectural solution finally emerged could be detected. A kind of low profile building which would not disrupt the scale of the city would also provide the easiest access. Such a design would not be so highly dependent on escalators and elevators, and would integrate with the existing circulation system of the city, the street pattern, as well as with the arteries that lead out of the city. Such a low profile also allows a variety of daylighting design options. Early sketches of such a building were made without really understanding totally what might happen. To reduce the thermal impact of the envelope, one might enclose the series of building wings by some kind of green house construction. In effect, this is the germ of the solution.

Within a large office building, typically space heating is not the problem, even in a colder climate. Cooling is the problem and one of the principle sources of the excess of heat is lighting. Therefore, if one wants to be simple minded about it, the single most important thing to design for an efficient office building is the lighting system. That is the point of departure for this design: There are series of office floors which are open on both sides. Where the sides face each other, there is a landscaped courtyard in-between which has a glass roof - the courtyard in effect serves as an enormous luminaire. How do you design a glass roof in an over-heated climate? One must have some kind of valve or screening that will keep the heat out but will let the light in. On a winter day, the louvres of the screen which are on the outside of the glazing should be totally open because on those days, one really may need some heat in the interior. The light that comes into the courtyard is bounced off by reflectors across the high ceiling and becomes a light source extending half way across the interior of the workspace. At night, this green house louvre cover is tightly closed to keep the heat in when the building is not in operation. This kind of concept requires not just lighting design but an integration of all services. The structural supports have got to run the right way so that the beams do not project down and cast shadows. The integration of auxiliary lighting has to be such that the lighting is coming from apparently the same source. One also has to integrate the usual services of electrical and telephone connections and so on. On the exterior wall, the integration of light source is somewhat easier because of immediate accessibility to the outdoors.

The secret of any design has to do with the architectural detailing of certain very critical points of transfer, in this case, light and heat. As you may be aware, daylighting can be calculated. It can be pretty troublesome in terms of the convolutions of calculation procedure. But lighting can also be modelled 3-dimensionally. For this project, a large size model was built and instrumented and was then tested outdoors in a variety of natural lighting conditions. Literally, one can directly scale what will happen in the real buildings from how the model performed. One of the absolute necessities for the success of such an approach is that the auxiliary lighting must be automatic. In other words, if the sun goes behind the clouds or if a storm blows over and the lighting levels drop below a certain desirable point, the auxiliary lighting has got to come on by itself. Similarly, when the sky gets brighter, the artificial lighting has got to turn itself off. One cannot depend on human beings to do that because human beings only have one way switches. They only know how to turn things on. So, this is a fairly sophisticated control problem. The fluorescent auxiliary lighting system is based on automatic dimming to provide a balance light for the interior office environment.

Incidentally, this building has a miserable site because the ground water is almost at the surface. But this is a case where a liability has been turned into an asset. The ground water is at 59°F, and it can become a cooling source. It then becomes a major resource in designing the mechanical cooling system for the building. In terms of energy, there is a 70% total reduction in terms of the annual energy consumption or annual energy budget, counting the net results of heating, cooling and lighting design strategies. In appearance, the exterior of this new architecture is not the flashy kind of thing one often expects from a prototype that is intended to redirect the future of a building type.

The other large project, I would like to share with you is much more theoretical. It is the work of some of my students in response to a problem in South India - in particular a large ceremonial structure located in the new city of Auroville, which is being built near Pondicherry by followers of Sri Aurobindo. The climate is always humid with tropical temperature ranging from warm to hot, and a severe rainy monsoon season. This is one of those dream projects which has had a certain amount of world exposure since its initiation in 1968. The founders and settlers of this new city are very aware of the problems of building a city. Fortunately their idea is a spiritual one - perhaps a more attainable Utopia than the kind of hard bricks and mortar that most of us think of as "city". The plan of Auroville is a spiral kind of thing, with a wide green sward of open space in the very centre which was intended to be the spiritual as well as physical focus for this new settlement. In the centre of that parkscape is a meditation structure which is called the Matrimandir. It is the heart and soul of the new community. Its construction has been a continuing activity in Auroville and construction has been underway since 1971. The original architectural design for this meditative centrepiece was in essence a sphere, or rather a spheroid - It is not absolutely circular. It was covered with gold discs and set in a highly formalized garden with enormous flanking structures called petals. Without getting too far into lessons in physics, but in the tropics, to pave a square mile with stones without any shade relief or vegetation, is one of the poorest concepts in terms of providing an advantageous micro-climate. Nevertheless, that was the original proposal and based on this design, construction began. As the project proceeded, there were thoughts about modifying the site plan and of softening the effect not only visually but also climatically. Construction has proceeded using reinforced concrete, mixed and placed by hand. The status of the project, when I visited the site in 1978, was the preparation of form work for the roof of the meditation space. This octagonal structure then would be enclosed by a spheroid which was proposed to be assembled from pre-cast concrete planks which would be formed in triangles.

One of the early fantasies was that the golden discs which would be suspended away from the structural skin would thereby provide some shading. In addition, the gold discs might be covered with photovoltaic cells to provide electricity to drive a conventional compression refrigerator to condition the interior. It turned out that this was not a very practical idea due to the enormous expense and complete lack of efficiency of such an investment. There have also been other studies to try to simplify the original design.

One of the most obvious aspects of the original design of the Matrimandir is that it is being built in the tropics with all of that sun, and yet it does not take any advantage of the natural conditions of the site. Such a structure could be built anywhere. Thus, my involvement on a voluntary basis was to use this as a sketch base for a series of studies in passive and hybrid cooling which I worked on with a number of graduate students in the United States. A selection of these concepts will show some ideas of a passively cooled architecture, and may stimulate your imagination.

In one scheme, the exterior skin is thought of as something transparent which the breezes can blow through. The balance of the design primarily provides a clustering of trees planted very tightly to the building to provide the shading of a sunken space in which whatever cooling is generated by the vegetation could collect. It would provide an antespace next to the building to give some kind of adjacent micro-climatic advantage. The use of trees near the building and also for the gardens seems to be a highly desirable alternative to paving all of the gardens around the structure. The development of symbolic gardens around the perimeter of the building with flowers seems to be even better than paving as proposed in the original part; thus, the shape of the building and its appearance in the landscape could take on different character - one which is softer and less architectural, but more related to the natural world.

The next scheme simply sees the problem as one of making a fresh start and tries to shadow all of the interior with selective sunshades so that the direct radiation will never hit any of the interior space, and yet the entire enclosure would be open to the breezes. Double layers are provided for the skin of the building so that natural convection could be encouraged to cool off the shaded interior surfaces. A pleasant aspect of this approach is the potential of the skin. It is proposed to be made of folded golden metal with a faceted appearance. The skin is designed to accommodate the prevailing winds with the facets forming scoops which would introduce breezes into the interior. However, the hot and wet winds which are the monsoon winds and come from a different direction, would roll over the top of those scoops and, therefore, the interior would not be soaked or wind-driven.

Another conceptual design proposes to use a major undergound water reservoir as a cool source, with great earth embarkments to trap air breezes which could then be directed across those underground cisterns.

This cooled air would be led up through the interior of the spaces and exhausted at the top. To get cooled air to rise up out of the cistern, one must generate the equivalent of solar chimneys at the top of the structure which would heat air and thus draw up air from down below.

The last solution illustrated perhaps best represents the synthesis of several of the ideas that have already been seen. There is a reservoir and a movement of air vertically from that up into the meditation space which is lighted by a single light beam. There are also double skins and black exterior sufaces to encourage natural convection by a kind of solar chimney. Obviously, this design is not totally passive, it requires pumps at least to pump the water into the reservoir. In effect the design almost runs the whole building into a kind of water machine. It also really articulates the difference between the top half of the outside of the spheroid building and the bottom half. The top half obviously should reject sun, and keep radiation out, whereas the bottom, which is easier to shade, could be more open to encourage breezes to come in. Therefore, what is developed is a series of louvres. Air is brought into the underground cistern. Water falls behind the outer skin which presumably will keep it cool, and as the water falls in a continuous screen, it produces a kind of evaporative cooler. The interior again is almost totally enclosed by wet surfaces, to capitalize on the stable temperature of ground water and thereby to keep the mean radiant temperature of all of the surrounding concrete surfaces as low as possible.

4. CONCLUSION

With this broad definition of Solar Architecture, responsibilities extend far beyond the successful attachment of solar hardware to the building to the point that the building itself becomes a piece of solar hardware. Since the sun is symbolic of climatic forces, the solar conception of a new architecture is a bioclimatic machine tuned with passive, hybrid and active systems to the particulars of a local climate and indigenous resources. The goal is a building fully harnessed with the forces of its site, and as independent as possible from off-site resources.

The professional responsibilities of such an ambitious Solar Architecture require buildings to perform environmentally on a level of technical competence and operational excellence seldom approached in history. Thus, the professional skills necessary to design solar architecture are demanding of new levels of understanding and wisdom. I wish us all well in that undertaking.

There are perhaps all kinds of conclusions for this presentation.

Obviously, there are many variations of technical and architectural solutions for tropical buildings. But one observation is that this paper was assembled half the world away and was from more of a theoretical point of view than an experiential one. It has been very reinforcing to come to Singapore and to see that there is, in the making, a kind of city that begins to follow fairly closely sets of parameters that seemed to me, from a distance, to be most appropriate for a solar architecture in the tropics. I am tremendously optimistic about the possibilities of what could happen in a place that is building in this way, this city of Singapore. I thank you for the opportunity to contribute.

Batak Dwellings, Sumatra - a model prepared by seconc year students of architecture, Arizona State University, 1963-6

Typical plantation house, Lower Mississippi Valley U.S.A. (Houmas, 1840)

Typical Kampong dwelling, Malaysia

Typical pre-war bungalow, Singapore

Golgund, Pondicherry by Antonin Raymond - view of dormitory

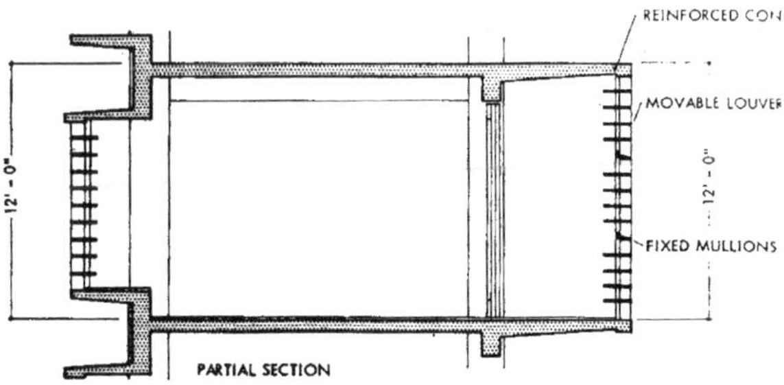

Golgund, Pondicherry by Antonin Raymond - cross section

'Fiesta', John F. Long, Phoenix, Arizona, U.S.A. - front view of house

'Fiesta', John F. Long, Phoenix, Arizona - photovoltaic on roof

Dalton Cole Cotton Farm, near Coolidge, Arizona, U.S.A. - a 150 kW ACUREX parabolic trough system generates electricity for deep well irrigation pumping

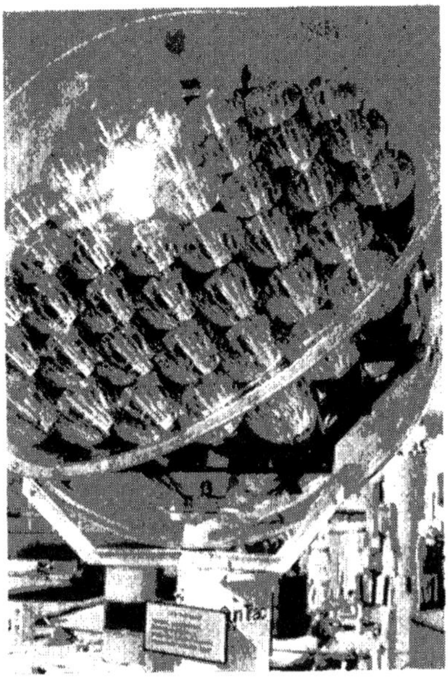

SANDIA SUNTRAC - a photovoltaic system tracking the sun

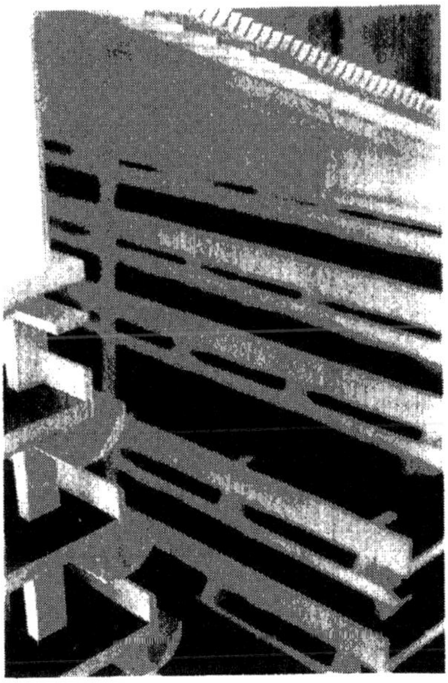

Tennesse Valley Authority, U.S.A. - model of proposed office building showing roof louvres and daylighting reflectors

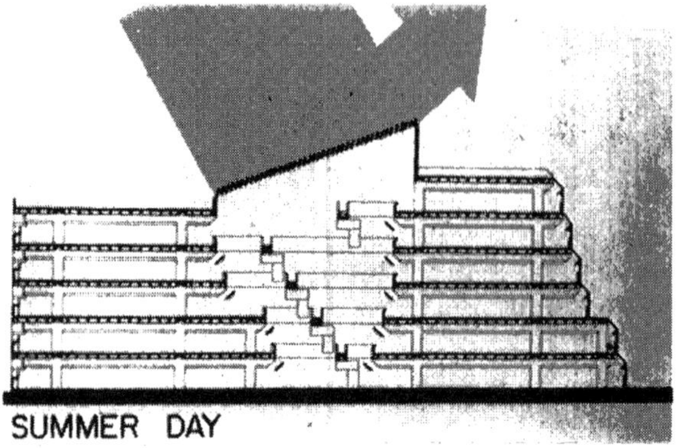

Tennesse Valley Authority, U.S.A. - cross section of proposed office building - roof louvres covering internal court

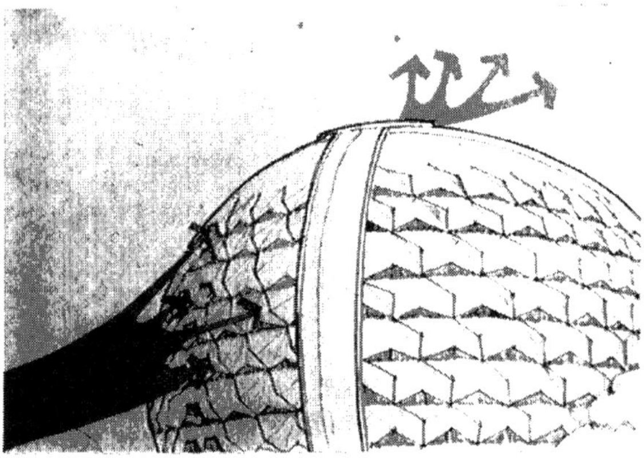

Matrimandir scheme 1 - perforated external skin for ventilation (Katz)

Matrimandir scheme 1 - details of external skin (Katz)

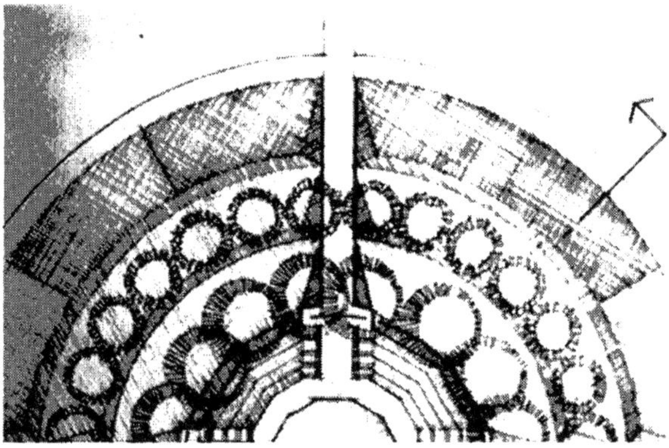

Matrimandir scheme 2 - plan showing shading trees over sunken garden (Cooper)

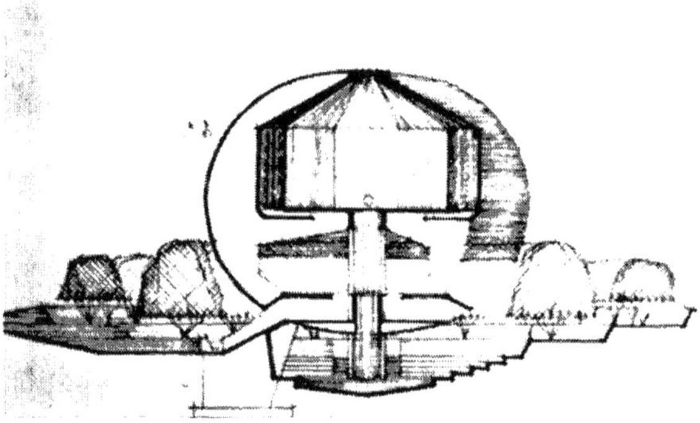

Matrimandir scheme 2 - cross section showing double skin concept with external skin of folded golden metal (Cooper)

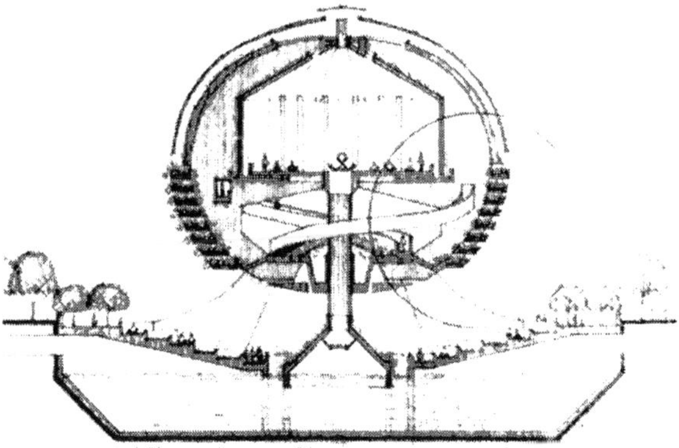

Matrimandir scheme 3 - cross section showing underground reservoir which cools incoming breeze (Finsen)

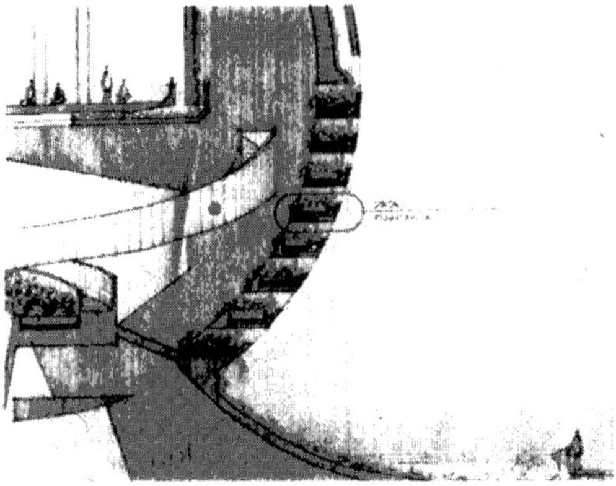

Matrimandir scheme 3 - details of planting boxes acting as louvres (Finsen)

Matrimandir scheme 3 - details of reservoir and sunken garden (Finsen)

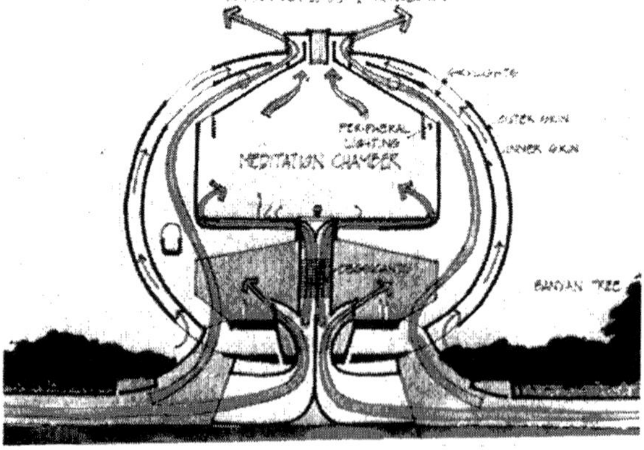

Matrimandir scheme 4 - cross section showing double skin arrangement and ventilation system (Finsen)

WORKSHOP 1: SOLAR ENERGY SYSTEMS

SOLAR HEATING AND COOLING SYSTEMS AND THE OPERATION RESULTS

Toshihiro Ishibashi,
Airconditioning Research & Development Laboratories,
Yazaki Shigen Co., Ltd.,
1370 Koyasu-cho, Hamamatsu-shi,
Shizuoka-ken, Japan.
(presented by Mr. Hiramatsu)

The basic systems for heating, cooling and hot water which are at present used in Japan are introduced in Table I.

Although water and air can be used as the heat collecting medium, water is employed in almost all cases since air has poor heat exchanging performance and is not suitable for cooling application. When using water as a heat collecting medium, caution must be exercised especially for preventing heat loss and corrosion. Every system requires the use of corrosion inhibitors. In Table I, the System D is widely used at present, and the features are described as follows:

(a) 'Drain-down' system: water in the collectors is returned to the heat storage tank through piping whenever heat collection is finished.

(b) The auxiliary boiler is located parallel to the heat storage tank and the water-fired absorption chiller.

(c) A motorized three-way valve is used to bypass the boiler or directly connect the boiler to the absorption chiller.

The operation results of the Yazaki Experimental Solar House and the Ishibashi Solar House which adopted System D during the two-year period from August 1976 to July 1978 are introduced herein.

The following are the summarized results for both houses:

(i) There was almost no difference regarding the solar energy received in the cooling season. However, in the heating season the energy received by the ISHIBASHI Solar House was 15% less than that of YAZAKI Solar House because the solar collectors were angled 50 degrees south-west.

(ii) The heat collection temperatures for both houses were between 78° and 86° for cooling and between 44° and 60°C for heating.

Bill B. P. Lim (ed.), Solar Energy Applications in the Tropics, 257–261.

Table I - Basic Systems for Heating, Cooling and Hot Water Supply

Illustration of system

SC:	SOLAR COLLECTOR
ST:	HEAT STORAGE TANK
AH:	AUXILIARY HEAT SOURCE
CT:	COOLING TOWER
CH:	ABSORPTION CHILLER
RC:	RANKIN CYCLE ENGINE
CR:	COMPRESSION TYPE CHILLER
HP:	HEAT PUMP UNIT
CHST:	COOL OR HOT STORAGE TANK
HST:	HIGH TEMP.STORAGE TANK
FC:	FAN COIL UNIT
EW:	ELECTRIC WATER HEATER
WS:	WATER SUPPLY
HW:	HOT WATER SUPPLY

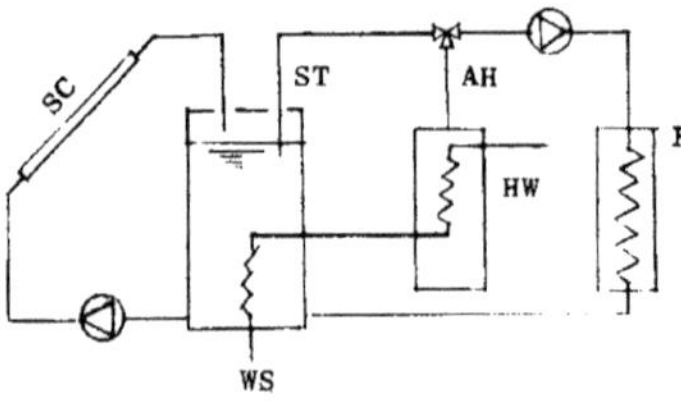

A. SOLAR HEATING/HOT WATER SUPPLY (DIRECT SYSTEM)

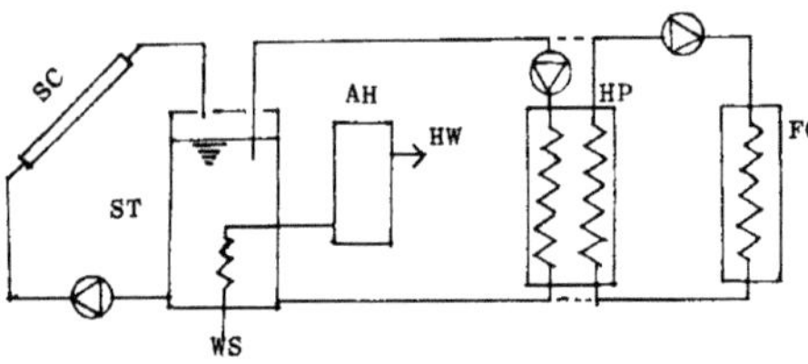

B. SOLAR HEATING/COOLING/HOT WATER SUPPLY (HEAT PUMP SYSTEM)

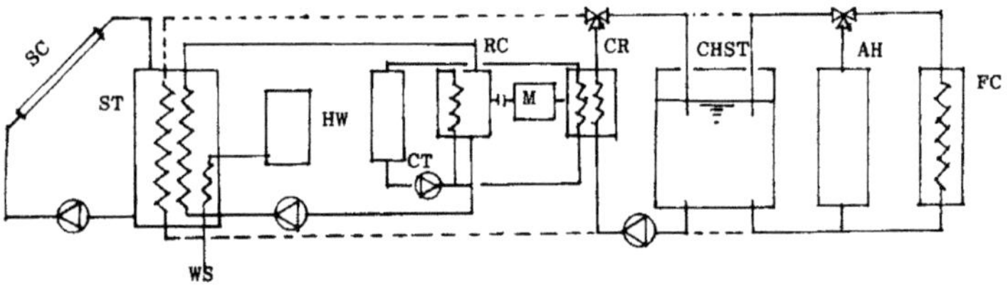

C. SOLAR HEATING/COOLING/WATER SUPPLY (RANKIN CYCLE SYSTEM)

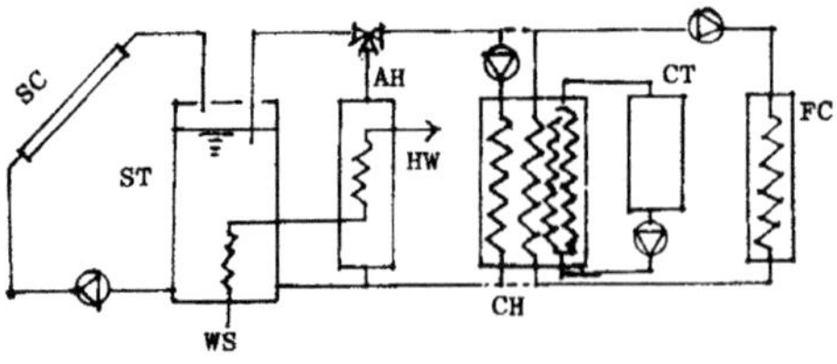

D. SOLAR HEATING/COOLING/HOT WATER SUPPLY (ABSORPTION CHILLDER SYSTEM)

(iii) The heat collection efficiency ranged from 0.19 to 0.36, the annual average being from 0.28 to 0.29 for both houses. During the cooling operation, the YAZAKI Solar House had a collection efficiency slightly lower than the ISHIBASHI Solar House and slightly higher during the heating operation.

(iv) Energy saving per unit area of the heat collection surface for the two-year operation was 1366 MJ/m^2 for the YAZAKI Solar House but 1797 MJ/m^2 for ISHIBASHI Solar House. This means that the ISHIBASHI Solar House was 1.32 times as efficient as the YAZAKI Solar House.

(v) The COP of the absorption chiller ranged from 0.42 to 0.58 and the average was 0.51.

(vi) The year round solar-dependency of the YAZAKI Solar House including the hot water supply was 73%. This consisted of 60% for cooling, 76% for heating and 90% for the hot water supply. On the other hand, the solar dependency for the ISHIBASHI Solar House was 81%. This consisted of 72% for cooling and 79% for heating and 86% for the hot water supply.

Outline of System	Present Situation
In the case of heating, this system collects heat 40 - 60°C, stores the heat in the heat storage tank, and circulates the hot water through floor panels or fan coil units for heating. For hot water supply, water is heated by means of the heating coil in the heat storage tank or by the hot water reservoir. If the temperature of the hot water is low, the hot water is further heated by the auxiliary heat source unit before being supplied. This type of system includes the water heat-collection and water heat-storage type as well as the air heat-collection and gravel heat-storage type.	Commercial stage
In the case of heating, heat is collected at 10 - 30°C, and the temperature of the collected heat is raised by a heat pump. Heat is taken up from the ambient air when the ambient temperature is high, or from the heat storage tank when the ambient temperature is low by switching the water heat source system and the air heat source system. If the temperature in the heat storage tank is higher than that required for heating, the heat is used directly for heating without using the heat pump. For cooling, the normal compression chilling operation is performed without utilizing solar heat.	Trial selling stage
In the case of cooling, heat is collected at 90 - 150°C and is used to evaporate the heat transfer medium by the vaporizing coil in the high-temperature heat storage tank. The expansion of the generated vapor drives the compression chiller through the Rankine cycle engine. If the temperature in the high-temperature heat storage tank is lower than the required value, a motor is operated to drive the compression chiller. Heating is performed directly without using the Rankine cycle engine or the compression chiller. An auxiliary operation is conducted when the temperature in the heat storage tank is lower than that required for heating.	Development stage
In the case of cooling, heat is collected at 75 - 100°C, and the water heated absorption chiller is operated. Heating is effected through the direct heating method. An auxiliary operation is performed when the temperature in the heat storage tank is lower than that required for the cooling or heating operation.	Commercial stage

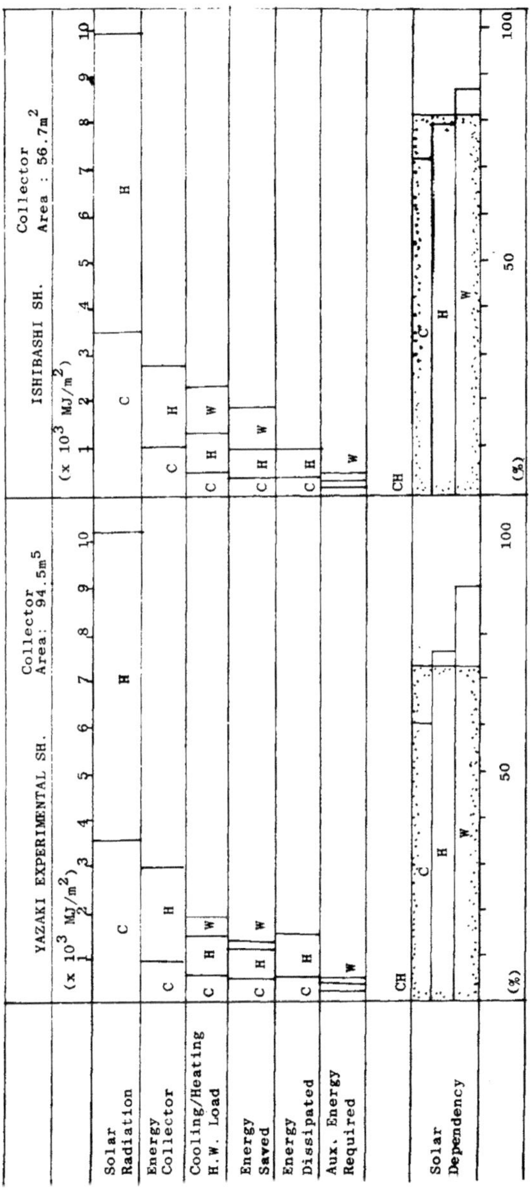

Fig. 1 - The Operation Results of Yazaki SH. & Ishibashi SH. (from Aug. 1976 to Jul. 1978)

DISCUSSION

Chairman : Mr. S. Sujarittanonta

Panel Members : Dr. N. R. Sheridan
Assoc. Prof. H. B. R. Exell
Assoc. Prof. D. Chuah
Assoc. Prof. T. Y. Bong
Mr. Hiramatsu

Question:

Dr. Sheridan, in your estimates you gave on the construction cost, did you have any way of accounting for the actual cost of installing the solar collectors or is that just included in your plumbing cost?

Dr. N. Sheridan:

In the list of costs that I gave for those Western Australian projects, certain installation costs were laid out. If you want full details of them, I have got a further breakdown on actual installation costs. From memory, I did not think the installation cost of the collectors was particularly significant. Perhaps some of the cost was absorbed in the building cost, as the buildings were specially prepared for solar collectors; the actual installation of the collectors on the roof was about 5% of the collectors' cost. The plumbing, of course, includes the connecting of piping, etc. A large proportion of the installation cost could be related to the water system.

Question:

How reliable are solar installations and what are some of the operating problems?

Dr. N. Sheridan:

One of the first solar collectors I ever saw was built in 1935 and was still in operation in 1955. That is reliability. But in Queensland recently, there was a solar air-conditioning project installed to a very exacting specification or so people thought. Yet the collectors failed after less than 12 months in service. By failure, I mean these collectors are completely unserviceable. I think the problem is that collectors used for solar air-conditioning are operating at much higher temperatures than those used for hot water, and the manufacturers have not much experience operating at such temperature levels. A water heating collector, which might operate for most of its life around 50°C is not necessarily suitable for solar air-conditioning which does not start operating until about 75°C, and will run much of its effective working life over 80°C. There is a high risk of material failure especially if the collector runs short of water or circulating fluid at these high temperatures. The temperature can go to relatively high levels of 200°C or so, and this can cause damage to the collector. Another comment I would like to make is that we imported American collectors for a solar thermal engine running at 100°C. These are probably the most efficient flat-plate collectors that I know of. We have had them in operation for about 12 months and there are some problems with out-gassing of the insulation which is made of fibre-glass. It must have some resin in it. Some of that resin must have evaporated and deposited on the inside of the glass. That means that the glass is not as effective or as transparent as it used to be. This problem arose within 12 months of going into service. So I would say that, while collectors should be able to last 15 - 20 years, I think that the current type of collector, particularly for the higher temperatures, does not really give that level of performance.

Question:

Mr. Hiramatsu, have you solved the problem of crystallization of lithium bromide in your absorption system? If so how have you solved it? Also, I understand that the water system in the absorption air-conditioning system is most energy-consuming. Is it possible that the heat rejection in the cooling tower can be re-circulated so that the total heat energy can be saved in some way?

Mr. Hiramatsu:

With regard to the first question about crystallization, in the Yasaki absorption machine, there are two main devices used to avoid crystallization which are not applied in ordinary gas-fired absorption machines. One is the solution adjustment chamber, and the other is the refrigerant by-pass tube. By means of these devices the lithium bromide solution is always maintained constant without crystallization. At the lowest input, crystallization may occur in the absorption machine. In this situation, we use a by-pass tube from the source to the lower part of the generators, and if it boils in the lower generator portion, then we can avoid crystallization. As for lithium bromide solution, the concentration is lower than that used in ordinary gas-fire absorption machines to overcome the crystallization problem. However, there is another possible crystallization, that is when the cooling tower temperature drops to less than 21°C. We are now trying to avoid this point.

As for the total energy from the cooling tower, the heat rejection temperature is limited and is not high. I think it is impossible to retain the heat rejection temperature from the cooling tower.

Question:

Mr. Chairman, I wonder if I may raise a question here. I was talking with one of the Yasaki representatives in Bangkok and he was telling me about the case in which they had crystallization in the Yazaki machine which was designed actually for a customer in the Middle East. What they found was that the ambient condition under which this air-conditioning system was working was different from the condition which had been specified by the customer. It was working outside the range for which it was originally designed. My question would be, to what extent or how close must one be in specifying this precise ambient condition so that the machine can be designed for the particular application?

Mr. Hiramatsu:

Under high bulb temperature the cooling capacity will be reduced. However, the standard maximum wet-bulb temperature is 28°C. If wet-bulb temperature is increased more than 28°C, the capacity of the standard-type absorption machine will also be reduced. Therefore, for the Middle-East, the ambient temperature and the wet-bulb temperature are very high, for instance, 29, 30, sometimes 31°C. We have another type of condenser to meet such a requirement.

Dr. N. Sheridan:

Mr. Chairman, I have some written questions here. The first is on an ejector type system. In my paper, I mentioned that there were three possible energy convertors that one could use for heat-operated refrigerators: absorption, Rankine and ejector systems. The ejector system is sometimes called a jet refrigerator. It's been around for many years, using steam as the motive fluid. The general idea is that the motive steam is expanded in a nozzle to a high velocity. It picks up the refrigerant which has to be water, and compresses it in a diffuser to the high pressure. For solar operation, it would be better to use Freon as the working fluid; such a system has been described in the literature. The Russians have particularly worked on the system and they claim that the co-efficient of performance of the system is the same as one can achieve with the absorption and Rankine Cycle systems at temperatures up to 100°C. The future advantages of such a system would depend on whether it could be developed to be competitive with the other units.

The second question concerns cooling by dehumidification, and the economics of such a system. Personally, the only extensive work I know has developed a device using rotary heat exchangers and rotary dehumidifiers. In my opinion, the performance that they have obtained with the device is more or less unattractive. It does not give as good a performance as other competing types of equipment. It is after all an open cycle machine and this means it uses air as the fluid for energy transport within the system. The disadvantages of air are that it does not have much heat capacity and is not as attractive for heat transfer, since it cannot carry as much energy per unit mass of fluid moved. Nevertheless, I note that recently the CSIRO in Australia and the University in Wisconsin in the United States, have a joint project from the Department of Energy to develop further such a system. Dehumidifi-
cation cooling is not yet economical at this time and it is still under development.

The third question is what type of absorption system we have used. The only absorption system that we have used is lithium bromide and water. The original machine was an ARKLA which is an American machine and this proved very satisfactory. It is similar in principle to the Yasaki machine, and I think that for solar work there is no reason at this stage to go beyond the lithium bromide machine. However for high temperature, if collectors become available that can operate satisfactorily at, say, 150 to 200°C, the lithium bromide machine is not very suitable for working at these temperatures. Ammonia water may be a better substitute at these temperatures; and this machine could be developed to have a much higher coefficient of performance. The reason why the lithium bromide water machine has not improved in coefficient of performance lies in the limited regenerative heat exchange. As the temperature is raised, an improved performance is generally expected from thermodynamic equipment. This does not happen with the lithium bromide-water absorption machine; but by improving the regenerative heat exchange, which might be more difficult than it sounds, at least in principle the co-efficient performance of the machines may be raised.

The last question concerns the cost of water systems. A major component of cost is due to the fact that the collectors are put on the roof individually as panels and the plumber is responsible for connecting these together. Now these collectors are operating at 100°C and allowance must be made for expansion and contraction of the water pipes as they will operate over a considerable range of temperature. Since these pipes have water in them, they normally have to be built of copper or some non-corrosive material. Another problem is that these collectors usually have to be arranged to operate in parallel rather than series circuits. The collector temperature rise must be the same as the chiller temperature fall which is of the order of 4 - 5°C, a value that can be achieved by a single pass of the collector. Thus, the panels of an array are usually placed in parallel to give an array rise of 4 - 5 °C. But the parallel circuit requires more header pipes than a series circuit. The other implication in this question is the use of rejected heat from the water system. Putting in energy at one temperature means rejecting heat at a lower temperature, in order to make the system run. That certainly is the law of thermo-dynamics. There is little possibility of recovering heat which is rejected to the environment, especially at a temperature close to the environmental temperature to make the original system efficient.

Question:

Mr. Hiramatsu. In both your solar cooling systems, the solar contribution is around 70%. What are the considerations when you design for such a percentage?

Mr. Hiramatsu:

From the economic point of view, our conclusion for solar dependency of homes should be more or less 60%. And if high solar dependency is required, the initial cost will be high. However, in the case of the tropics, because a cooling system is required throughout the year, a high solar contribution ratio is required to compete with conventional electricity. So I think the solar dependency mainly depends on the location.

Assoc. Prof. T. Y. Bong:

If I may ask Prof Exell, how much did you spend on your solar refrigeration system in AIT? What is the capacity?

Assoc. Prof. R. Exell

The capacity of our refrigeration system which could be measured by the amount of ice produced in one day is actually very small, in terms of tonnes. The system produces about 25 - 30 kg of ice per day from a collector which is 5 sq. metres in area. This system is an intermittent ammonia water system. It is very inefficient compared to the continuous system. The continuous system produces about probably 2 or 3 times that amount, but our system has no auxilliary energy at all. The cost of that particular system is about US$2000/-.

If one assumes that the system lasts about 10 years, and we spread the cost of building it over that period, the cost of the ice will come out to approximately twice the factory price of ice in Bangkok, and about 4 times the retail price of ice in Bangkok at that time.

Assoc. Prof. D. Chuah:

My main interest being in photovoltaic, and not being an economist, I wonder if my colleagues could enlighten us on what the cost would be if the present thermal machine is replaced by a photovoltaic cum electrical machine, and what do they expect the cost to come down to in order to be reasonably economical?

Dr. N. Sheridan:

We have recently done a study looking at photovoltaic systems for remote area power generation. The objective of this study was to forecast the cost of photovoltaic systems around 3 kilowatt capacity, which could be used for electricity generation in areas remote from distributed electrical power. We indeed looked at six different types of system using flat-plate solar cells; concentrating solar cells, at two levels of concentration, viz., concentration ratios of 10 and 40; and systems which produced both electricity and hot water. These latter were called total energy systems since remote area residents required both electricity and hot water. One could imagine that such systems could be optimised to produce more electricity or more hot water depending on demand, with the option of running air-conditioning either from the electrical energy or thermal energy. In looking at these systems, we decided that it would be after 1985 before these systems could be competitive, remembering that the United States has got a programme under which the cost of solar cells should come down to 70 cents per peak watt in 1985. With these systems, it does not matter much whether a flat-plate system or concentrating system is used, they will both cost about the same. By the end of the decade, electricity produced by the systems should be competitive with diesel-generated power. That means it will be costing about A 12¢ per kilowatt hour. It is interesting to note that in this system the price of the cell becomes more or less immaterial. In the concentration system, if I remember, the price of the cell represented about 1% of the total cost of the system, and in the flat-plate system, this is certainly less than 10% of the cost of the system. Remaining costs include optics, structure, power conditioning, cooling and installation. If an air-conditioner is connected to the system, it does not seem to matter much if it runs from the electric end or the thermal end. They are going to cost much the same. If it runs from the electrical end, a few more cells are necessary; but they do not constitute the major cost anyway, and there is the advantage of the higher coefficient of performance obtainable from the vapour compressor machine. From the thermal end, it can be argued that the system will probably produce more thermal energy than necessary and, therefore, it is better to have a look at pushing up the thermal demand. The coefficient of performance of the thermal machine, however, is less than the electrically driven vapour compression machines. Solar photovoltaic programmes in the US are very much on target, and the US government has guaranteed the purchase of a minimum amount of solar photovoltaic equipment per year. If they can guarantee to lower the price as well, it is my opinion that solar photovoltaics is certainly the area to be looking at.

Assoc. Prof. D. Chuah:

I agree that the normal flat-plate cells and concentrated cells would not make any difference when costs do come down, but would you consider the type of concentration which is 10 times to 40 times more than the normal flat-plate type? Do you think the nature of radiation would make a difference, i.e. could you use medium and high concentrations in the climatic conditions like here, or would you prefer the flat-plate type?

Dr. N. Sheridan:

The point is that the 10-time concentrator is of the CPC or the compound parabolic concentrated type, which has a fairly wide acceptance angle, and, therefore, it does take advantage of diffuse radiation at least, the diffuse radiation that emanates from around the direction of the sun. Up to 10 to 1 concentration is a possibility for this sort of climate. I agree perhaps that the 40 -1 concentrator might be suspect in this region. We had a look in Australia - the comparison between a concentrator and a flat-plate collector for water-heating purposes. This study would generally tend to show that perhaps the concentrating type is of more interest in the drier climate where humidity is not a problem. A rough estimate is that in a less humid area like Central Australia, such a collector will be able to produce something like 30% more collection, whereas in the coastal areas its performance will be somewhat similar to a flat-plate. So I suspect that for the Singapore area, in the humid tropics, there are some advantages in going to the flat plate. We were unable in our photovoltaic study to show very much difference in performance between the flat-plate and the concentration devices. But, of course, the reason for staying with the concentrated devices is that if the US cost programme (i.e. the cost of photovoltaic cells) does not meet its target, one is better off with concentrated devices because they are not so sensitive to the photovoltaic cell cost. One should not say that we should stop working with concentrated cells. They are more efficient than flat-plate cells, a significant output from them is possible. The best way we have had from one of our cells operating at 40° at 40 to 1 concentration, is about 20 amps current drawn from a single cell. So a special kind of technology is necessary to deal with by-current flows at such low voltages and currents. I would like to see some continued work on concentrating cells, although, as I said, there is no indication that they will, in the long run, be better than the flat-plate system.

Another question is cooling of the cells, the concentration of radiation means concentration of heat.

Prof. J. Cook:

With the photovoltaic cells, higher temperature of the cells tends to decrease the electrical output, and it is important to control the temperature. There is one American system that you can buy which is the Martin-Marietta system. It is available with passive cooling of the concentrator cell with 40 to 1 concentration. I think the operating temperature of the cells is around about 30 - 60°C. These units are scheduled for operation in the Arabian desert and I think the idea is that they prefer not to go to water cooling of those units. We have only tested cells and we tested them in the range of ambient temperature to 100°C. They are water cooled so that we can indeed control the temperature. I suspect that probably the passively cooled cells have long aluminium fins at the back of the cell which are used for cooling. These would be simpler for installation in remote areas and when the thermal energy is not needed. But we have some interest in total energy systems where we would like to extract that thermal energy from the cells and so we have been thinking more of water cool systems. But there is not much of a margin in either system and so from a reliability point of view, the passive cooling in relatively high ambient temperatures can control the cell within operating temperature of about 50 - 60°C, and this does not deteriorate the concentrating cell's performance significantly and maintains an efficiency of 12 - 13%.

Assoc. Prof. T. Y. Bong:

Regarding photovoltaics, we bought a 20 ampere-hour system. We were only thinking of running a colour television set or switching on a 20 watt light at night. The cost came to about S$4 - 5,000, of which one-third went on the storage system; so it means that even though the price of the cell itself will go down to practically zero, the price of the system will not go down to one-third of it. Regarding the use of compound parabolic mirrors to capture some of the diffuse radiation, according to our experiment, it really did not capture much of the diffuse radiation because the acceptance angle is too small, just a few degrees. So, even though we can capture some of the diffuse radiation for this purpose, there is not very much of it.

Question:

How does solar photovoltaic compare with the wind machine?

Dr. N. Sheridan:

That's a good question. We have not done very much on the wind machine ourselves but there was one Australian study done by the Philips Company and funded by the Government. It produced a wind generator backed up by some solar photovoltaic cells. It was made in an ordinary shipping container and run in Adelaide for about 12 months. It had batteries for energy storage and also a diesel engine for back-up. It is claimed that the diesel did not have to be used during the twelve month test period. In other words, there is a good match between the availability of wind and the unavailability of solar energy and vice versa. The Company decided that the system was not marketable because it would have to sell for about A$20,000 which was higher than the market could bear. I think the wind generator in that system actually cost about A$4 to A$5,000. But one must remember that Australia is the continent with the least wind in the world. Our only possibilities for wind usage are in the South where, at 40° latitude, there is a prevailing 40° westerly wind; but most of the rest of the continent is more or less unsuitable. As a result, there has not been much incentive in Australia to look at wind as an alternative.

I remember about three to four years ago, we had a workshop in Bangkok, the objective of which was to give some priorities to different aspects of solar energy and their applicability in the South-East Asian area. I think the top priority was for solar crop drying; solar electricity was pretty well up in the list, and solar air cooling was pretty well down on the list. Could anyone here say what they think the market for solar air cooling or solar air-conditioning would be in this area, and what application you would see it being used for, and why indeed you would give it a reasonable priority against other applications of solar energy.

Prof. R. Exell:

About 40% of electricity used in Bangkok is for air-conditioning, especially in the hot season, and I know that the electricity generating authority of Thailand would like to see solar air-conditioners all over Bangkok. Of course, it is difficult in the city where you have tall buildings to make solar air-conditioning work. Though it had a very low priority in that meeting, one feels that if it is economically viable to have solar air-conditioning in buildings in South-East Asia, it is a good proposition. Most of these cities generate electricity from oil, and it certainly is an attractive use of solar energy.

Dr. N. Sheridan:

I suspect rather that one would be better off thinking about generating electricity from solar energy and using that electricity to operate the air-conditioning system. For large buildings, there would be insufficient area on the building to operate the collectors required to produce the building energy. As a result, solar generation would be located at a site remote from the building.

Assoc. Prof. T. Y. Bong:

Our priority is also solar cooling: solar air-conditioning. We are not thinking of using a solar concentrator, even though it has obvious advantages because under our conditions, it is cloudy practically every day. Unlike in Australia where there can be a few days of clear blue sky followed by cloudy days, in our case, very seldom will one see any clear blue sky here. On average, our percentage of diffuse radiation is about 40 - 50%, and if you want to use a concentrator, a large percentage will be lost. At the moment, we are still thinking of using an absorption chiller instead of an electricity operated Rankine cycle.

Prof. R. Exell:

I could raise one more point about the utilisation of solar energy in a place like Bangkok. What is actually happening is that the solar water heating business is gradually increasing and it is being used by hospitals and hotels and so on, and possibly will be used by industries. However, looking from the point of view of national development at the grassroot level, solar water heating is actually unnecessary - it is a luxury, it will add to the quality of life of the people of the richer section of the community.

Assoc. Prof. D. Chuah:

Before I make a comment on your enquiry, Dr Sheridan, could you clarify which meeting at Bangkok, which one in particular are you mentioning where decisions were made as you mentioned giving priority to crop drying with solar air-conditioning at the bottom?

Dr. N. Sheridan:

This was a meeting organised by ESCAP in 1976. It was mainly aimed at rural areas, not the city development. It had representatives from about thirteen or so countries from the South-East Asian and Pacific areas.

Assoc. Prof. D. Chuah:

There was another meeting early this year, I think it was the Asean working group committee on non-conventional energy sources, and they came up with the same conclusion. But the order of priority was fixed according to the national need rather than from the point of view of commercialisation, and from the point of view of seeking funds from external bodies such as the E.E.C. and American funding research in the institutions in each country. I think Singapore gives top priority to air-conditioning. I happened to be at the meeting and Malaysia was crop growing and Indonesia, I think, was bio-mass going along with the Philippines. This is purely to get funds for R and D, not for commercialisation. It is entirely a different aspect. I think in Malaysia, we would see if solar water heating could be brought up to a level of producing boiling water which I think is essential and not a luxury. If thermal processes can be improved to produce boiling water, I think that would be a good commercial proposition.

Dr. Sheridan:

There is really no difficulty in producing boiling water from a flat plate collector. We use some American collectors which have an efficiency of 50%. When the top temperature is at 100°C, one has to design the collector with the proper materials. The main advantage of this collector over others is the use of low iron, non-reflecting glass. This means that the glass does not absorb 15% of the energy before it gets to the plate and also of course, it has a very good selective surface which means that there is little loss. But at this temperature, it has a significantly higher efficiency than most other types. So there is no problem in running flat plate collectors at 100°C.

It is not obvious, of course, that the tropics is really the place for solar energy. There is no doubt in my mind that the place or solar energy is where heating is needed, and this, of course is not related to the availability of solar energy. It is related to the needs of buildings, such as in the mid-west of the United States where the winter temperature is minus 15 - 20°. One cannot live in this area unless there is heating. Solar energy is not plentiful but sufficient to make a significant contribution to the energy needs.

Morever, one does not need very efficient collectors since the temperature is not high. One of the successful collectors for building heating is an air collector. This is not very efficient but it is cheap to make. It uses simple materials. It can be put on the building fairly simply, and it has a big job to do because there is a lot of energy needed for heating buildings. Space cooling by solar energy is not so attractive. It is just too complicated. One can do space heating with today's technology quite simply. There is a demand for it and it is already economical against higher priced electricity.

Comment:

One place where solar energy would make a significant contribution is in crop drying, which was one of the conclusions reached at the ESCAP meeting. The point is that the traditional method of drying crops is simply to spread them out on the ground in the sun. Then the crops dry very quickly and very easily.

There is certainly a problem with drying in the wet season, especially when the new varieties are harvested during the wet season. Like a factory which boils shrimps and then has to dry them, it operates in the wet season. They have problems simply because it rains on the produce when it is being spread out. Really all you need to do is to shield the produce from the sun by means of some kind of cover which lets the sunlight through. It is coming into use rather slowly. Now this is useful in South-East Asia in general.

Assoc. Prof. T. Y. Bong:

Going back to the issue of hot water heating, we mentioned that it is possible to produce hot water at 100°C using flat plates. However, we also need to know at what radiation intensity and ambient temperature this 100°C is reached, because sometimes our radiation intensity can be quite low. We do feel the need, as Dr Chuah said, to produce hot water around 100°C boiling for industrial processes. Actually we are now looking into the matter as to how much one may substitute oil with solar energy.

Assoc. Prof. D. Chuah:

In addition to industrial hot water, we also have old peoples' homes. They are complaining that the electricity bills are mounting. For these old people who depend on contributions from the public, they want a system that can produce boiling water and cut down the electricity bill. I am sure not only in old peoples' homes but also in our own homes throughout the country. If one can produce a system that can boil water at low cost, I am sure that will be very saleable.

Assoc. Prof. T. Y. Bong:

Regarding the production of boiling water or steam from solar collections, it is possible. At one time, I almost got burnt by steam coming out from the flat-plate and evacuated tube collectors. There was an endless stream of invisible heat coming out from the collector.

Comment:

I would like to comment on solar crop drying. I am from the Philippines and we have developed a solar dryer. We are also developing a solar lumber dryer to replace the kiln dryer. If we make use of this solar dryer which we have designed, the cell-life of the product is longer and is free from diesel contamination. I think those who attended the 1978 workshop on Solar Drying have seen this solar dryer we have designed.

I would like to ask Dr Sheridan whether in Australia right now, there is any programme for solar thermal electric generation like what has been done in New Mexico in the United States.

Dr. N. Sheridan:

In Australia, this sort of programme is not very seriously taken by our Government. The problem for solar energy in Australia is that our electricity is not too expensive. We have a lot of coal and practically all of our electricity is produced from coal. We sell the coal to our electricity generating authorities for A$9 a tonne or so. The cost of exporting coal to Japan is from A$30-40 a tonne. So you can see that we are really under-selling this for our own use. This cost makes it very hard to justify solar projects but the situation is not the same in Singapore. In Singapore, electricity is produced from oil. Therefore, you can say that whatever you do to save electricity you save oil. In Australia, we do not have this position. To save oil, we have to save it from the motor-cars. As a result, the emphasis on energy research is to make the motor-cars more efficient, an area in which solar energy cannot contribute in the short term.

One story I must tell you is about our University. One of the oldest residences or colleges had their oil-fired water heater blown up and they decided that, before replacement, alternative systems should be considered. So they asked me how much a solar installation would cost. I said that hot water from a solar installation would cost about half that of water from a new oil heater. But, while I am interested in solar energy, I have to be practical about it. I said that the electrical authorities will sell electricity at a price that will produce hot water at half the cost of solar. The price for electricity for off-peak hot water in Brisbane is A$0.017 kwh. There is no solar system in the world that can produce hot water to compete with that sort of price. I am not saying that the competition is unfair when the authorities offer an off-peak rate. It is a give-away rate in a way, but they are making some profit out of it, and so it is a business proposition.

The problem in Australia for solar energy is that it is very hard to find an area where solar energy is really going to compete against the existing energy system, and at the same time save oil, which is the energy problem in Australia. Even then there is not much pressure, because at the present time we are 70% self-sufficient in oil; and I think our Government really has to face up to the problem that there is an energy crisis around the world.

Of energy research funds, the Government of Australia is distributing per year a very significant proportion to coal, because this is funded directly from the coal industry. I think it may be as much as A$10 million. Then a very significant amount goes to transport research which would save oil, and something of the order of A$1 million or about 4% of the total budget goes to solar energy. So you can see that the Government really does not consider solar energy to have any major impact. In fact, in a recent report by a Senate Select Committee, we made a submission that solar energy could be considered as an alternative to petroleum, if one considers it far enough along the line to the solar hydrogen system, where hydrogen could be used for fueling cars. This system did not even get mentioned in the report. This omission means that we have not got through to the Government that solar energy is something which can be looked on for future usage, even to the extent of replacing petroleum. Perhaps, there is still work to be done in such things as flat-plate water heaters in the short term. But we ought to be making some noise about solar hydrogen or solar electricity as something for the longer term.

Question:

First of all, was there any noise about the renewable resources of Australia during the last election, and really what pressure are you bringing to bear on the Government and the various parties to make sure that Solar is there or the renewable resources are there?

Dr. N. Sheridan:

In Australia, we have six States. The State Governments are responsible for the coal and oil that is mined in their state. They do not have all the responsibility for the oil that is mined offshore. The states are very keen to ensure that they will maintain a responsiblity for energy. One result is that most States in Australia have started solar energy research groups. The funding to the Western Australia group is of the order of A$800,000 and will rise to well over a million dollars in the succeeding year. So to be fair, I must say that there is some interest in solar energy from the different States.

The only problem with this arrangement is a lack of coordination, and that means there is overlaping and not very much discussion between the state groups. West Australia has achieved significant success, particularly with demonstration programmes. But I think that in a federated country like Australia, the states can do just so much. One really has to convince the Federal Government that there is a need to be concerned about many aspects of energy. As far as I can see, they are not convinced. There is hardly anything constructive in the energy policy. The Government is really unconvinced about the fact that there is any energy crisis at all.

I can say that, for Singapore, solar energy is something that can contribute very much, because no matter what you do with solar energy you are going to save some oil and that makes a really good deal for the country. From the Australian Government point of view, to save some electricity by installing some solar collectors in houses in Australia does not really save oil. Therefore the Government does not give this the top priority. Whereas in countries like Singapore, it is going to save and have a direct impact on the country's economy and make one more aware of the advantages of solar energy.

Question:

I would like to ask Dr. Sheridan and Mr. Hiramatsu from the Yasaki Company a question. Have you ever encountered the case of the absorption system that fails to operate? Is there any technical perimeter or criterion to justify the capacity of the storage volume in an absorption system?

Mr. Hiramatsu:

In Japan we have no significant problems with absorption machines at the moment. But there are other problems. For instance, failure occurs due to inexperience in installation. The solar industry is fairly new and the installation people are unfamiliar with the systems. In Japan, we try to educate these people from time to time to avoid system failure.

The second question has to do with the volume of storage tanks. In a solar system the most important point is to optimize the design. If the collector areas are too large and the storage tank too small, there will be too much heat loss. On the other hand, if the storage tank is too large, the temperature inside the tanks is not high enough, and the absorption machine cannot operate. A computer programme is used to obtain the optimum design for the correct storage tank volume and adequate collector area. It is very difficult to obtain such an optimum design by hand calculation because the climatic conditions are always changing every day and every month.

Dr. N. Sheridan:

We must remember that absorption machines have been commercially available probably for 50 years. There is sufficient experience with the machine; there are machines in operation without maintenance for a period of 20 years. In general, I would say the absorption machine is really quite reliable. It requires minimum maintenance and, in general, has a very small incidence of failure. However, the problem is that if there is a major failure, like some inadvertently bad weld or some fault in the manufacturers' inspection, it is very difficult to repair the machine on the job. The working unit is taken out and returned to the factory, becase the lithium bromide machine works under vacuum, and with the highly corrosive lithium salt involved, it is relatively difficult to undergo any major repairs in the field. Unavoidably with manufactured goods, one gets the occasional failure that would escape any examination system of the equipment. But from what I can see with absorption machines , generally, you can expect them to be reliable. There will be a major failure in a small percentage of cases.

In discussing the size of the storage tank - in several American cooling installations that I saw in 1976 when I was there, the biggest problem was the over-sizing of the storage tanks with the result that they are unable to get up to a satisfactory operating temperature. I think it is wrong to size a storage tank in relation to the load to provide for cloudiness or the amount of building load. It is better to determine the collector and size the storage tank in relation to the size of the collector, at so many cubic metres of storage per square meter of collector; I think that is the best way of going about sizing the storage, remembering to approximate on the smallest size tank to ensure that the machine is brought up to operating temperature. These absorption machines will not start to operate until the temperature is around 80°C with the Yazaki machine. If the minimal running temperature is 75°C, one has to raise the temperature up to 80°C before the machine will start, and it will continue to run until the temperature falls back to 75°C. If the storage tank is too big, it is difficult to get the temperature up high enough to cause the machine to start.

The other problem is with cold storage. I think Yazaki's opinion is that they do not encourage cold storage. A problem with cold storage is insulating the cold tank. Water vapour tends to penetrate the insulation and condenses. It destroys the insulating property and increases the heat losses from the tank. The ordinary insulating contractors are not able to insulate cold tanks to satisfactory standards. As regards the vapour penetration, there have been probably more failures of cold side storage tanks than any other part of an absorption air conditioning system.

When storage is being designed, it is important to consider the interrelation of hot and cold storage with the machine capacity and the need for auxiliary energy. Hot storage is needed if the collector capacity is larger than the maximum machine capacity, while cold storage can be utilised to improve the system control and ensure that the machine does not stop and start under the influence of the room thermostat. Generally, the storage provided should not be more than will carry the load through the night. Back-up or auxiliary energy will be required if the system must produce cooling under all weather conditions. But auxiliary heat is likely to be expensive with absorption machines due to the poor coefficient of performance. System simulation is desirable to understand the detailed operation.

Question:

It is true that the countries which are not so blessed like Australia can look forward to the photovoltaic cells to give an economic breakthrough. The flat plates which have been tested and proven viable - the collector area that is required is rather large and the cost per unit area is also high. Now what research has been done or can be done to improve on these?

Assoc. Prof. T. Y. Bong:

The efficiency is limited by the transmissivity of the glass cover, and the absorptivity of the absorber plate. The glass transmissivity is about 87%, absorptivity 90, 92 or 93%. When these two factors aremultiplied together you get the maximum possible efficiency without heat loss, about 85%. Of course, besides the absorber, there are other things that take up the space. This will drop the efficiency further, if the efficiency is based on the gross area occupied.

Now, regarding the heat loss from the collector, the higher the temperature of the operating fluid, the higher will be the heat loss. One way to overcome this problem is to use evacuated tubes. This can drop the convective heat loss practically to zero, but the radiative heat loss is still there. We are getting better and better collectors with advancement in research. However, the efficiency will always be tied to the kind of temperature required, radiation available, and ambient temperature prevalent at the time.

WORKSHOP 2: SOLAR ARCHITECTURE AND TECHNOLOGY

RESEARCH DEVELOPMENT AND DEMONSTRATION EFFORTS IN FRANCE ON CLIMATIC CONTROL OF BUILDINGS USING SOLAR ENERGY

Louis Drouot,
M.S., University of California, Berkeley,
Manager, Solar Techniques Department,
Commissariat a l'Energie Solaire (COMES)

In 1979, the energy consumption in France was 180 million tons of oil equivalent, of which two thirds was imported. Since 1973, the country has been facing a sharp increase in its expenditure on importing oil, and more recently on importing natural gas. This fact eventually led our government to set up a very ambitious energy programme whose three main components are as follows:

(a) diversification of oil and natural gas imports;

(b) enactment of a very determined energy conservation policy in industry, housing and transportation;

(c) development of national energy resources, among them, nuclear power and renewable energies.

Thus, it is expected that by 1990, 30% of our energy will come from oil, coal and natural gas will account for another 30%, nuclear power for 30% and renewable energy for the remaining 10%. The Ministry of Industry has been made responsible for the implementation of the energy programme. It was decided to set up special government agencies to co-ordinate these programmes. Therefore the National Energy Savings Agency was set up in 1975 to implement the energy conservation policy. More recently, the French Solar Energy Authority was established in March 1978. Known as COMES (Commissariat a l'Energie Solaire), it is a government agency with full authority over the national energy programme. Broadly speaking, it is responsible for the 1990 objectives, whereby the renewable energy should contribute some 5% to the total energy budget, that is to say, 10 - 12 million tons of oil equivalent. COMES co-ordinates the activities of the many national laboratories engaged in solar research, development or demonstration such as the National Centre for Scientific Research, The Atomic Energy Commission and The Scientific and Technical Centre for Building.

These teams represent a community of over 500 scientists and technicians. COMES received budgetary appropriations of about

Bill B. P. Lim (ed.), Solar Energy Applications in the Tropics, 281–282.

US$45 million in 1980, and some US$55 million is foreseen in 1981. These are mainly for incentive plans, private or public institutions or industries or even individuals who are eligible for such subsidies. The wide financial autonomy which has been granted to COMES makes it possible to offer the most appropriate solution for each programme. For instance, COMES subsidises local authorities or individuals for demonstration purposes, purchases solar equipment or components, supports the solar industry through grants or group orders, enters into joint ventures with private or public corporations. The activity of COMES covers the fields of application of solar energy in the broadest sense - solar heating and cooling, solar electricity, energy from bio-mass, wind energy. However, the sectors of hydro-electricity and ocean thermal energy are not operated by COMES.

COMES is also in charge of international co-operation, either as the representative of the French government in the case of formal internal agreements, or for direct relationships with formal organisations. It is an active participant in the policy of the French government towards developing countries. It also has the responsibility of promoting solar energy application in French overseas territories.

DISCUSSION

Question:

What does the French government provide in terms of incentives to French citizens to install domestic hot water systems in their houses?

Mr. L. Drouot:

There are a number of incentives for solar heater programmes. As far as COMES is concerned, we deal with the industry to get quality programmes, equipment and standards. When it comes to specific financial incentives, the French Government has provided specific plans on income tax. The people can deduct part of the investment in solar heaters. Maximum deduction per family is 7,000 francs and 1,000 francs for each child. Not only can solar heaters be deducted but also all kinds of investments for energy savings. Insulation for the roof is also deductable. As a matter of fact, when people are spending such a large amount, it is better for them to insulate their roofs rather than spend the money on solar heaters. One can also obtain medium-size loans with reduced rates of interest for the sum needed to buy solar heaters, about 10,000 francs. Lately we have had specific financial incentives for low cost housing programmes at a low interest rate of 9%. Then one can get 3 - 4,000 francs more for loans if a solar heater is used.

Prof. J. Cook:

I would like to relate the experience of France to the U.S. where there has been a tax incentive for people to install solar hot water heaters. The incentive is about several hundred dollars up to a thousand dollars. It is my estimation that this was a very poor incentive because it took the government up to 15 months to make up their minds. During these 15 months, there were no solar hot water heaters sold at all. So the industry came to a total halt waiting for the government to make up its mind to initiate the programme. When the government finally moved, then sales began, with the price going up almost to the same degree of the payback. So the home owner really never benefitted at all. It all went into the pockets of the industry. And so it seems that, at least in that case, meddling with commerce has not been very helpful.

Prof. K. R. Rao:

The COMES organisation has an important job in co-ordinating solar energy in France and outside French territories as well. I would like to know whether you have any collaborative research programme with countries which are not French territories.

Mr. L. Drouot:

We have been developing governmental relations with Mexico, for example, and we are also trying to develop joint ventures with other countries.

Question:

Most governments, about 20 years ago, encouraged the population to use as much energy as possible. At the same time, the government centralised the energy supply in the form of gas or electricity. Now the government encourages the population to decentralise, and to go to the smallest unit possible, and to install solar collectors and solar cells on roofs of houses. The equipment needs a lot of space and there is not so much space left to supply the whole world with all the energy they need. If we refer back to crude oil, we face a political problem because the suppliers of oil are based only in certain parts of the world. For solar energy, some parts of the world have a lot of sun, other places do not. What is solar energy contributing to the total energy needs, both now and in the future?

Mr. L. Drouot:

First, solar energy collectors are going to use a lot of space and there is not very much space left on earth. For space and water heating of French homes, the area for solar collectors is about the same as the floor space. In other words, by filling the roof with collectors, one gets enough space for heating, so there is no need for additional space. Additional space is needed to get the electricity generated by a conventional or nuclear power plant to the homes that are far away from it. There is also the need to balance the physical space required and the economic returns of using solar energy, in view of the increasing cost of conventional energy.

As to the proportion of solar to conventional energy, we do think that 5% of the total energy target by the year 1990 is a feasible objective.

Prof. J. Cook:

Would Mr Drouot like to comment on the political aspect of centralisation versus decentralisation? Is it an issue in France?

Mr. L. Drouot:

Yes. France is administratively divided into 22 regions, which are further divided into many cities and towns. We initiate energy programmes whereby the people take the responsibility for decision-making locally. Our work is to ensure that the concept is applicable to local conditions and that the equipment supplied functions properly. We also provide technical assistance to the local authorities so that there is some assurance of the success of the operation. The rest of the responsibility is locally taken. We need more emphasis on education of engineers to implement energy programmes.

Question:

Could Mr. Drouot supply the figure for low temperature solar panels in Mexico? Is there a figure for the overall efficiency of the system?

Mr. L. Drouot:

The question is about the thermal efficiency of low temperature operating cycles. In Mexico, there is a 20-kilowatt solar generator. The heat source is from regular flat-plate collectors. The overall efficiency is 1 - 2% at the most. Hot water is heated through the solar collector which, in turn, transfers its heat to boil an organic fluid which drives a small turbine. The fluid is then condensed by water and is re-cycled. This is how the thermo-dynamic cycle works.

The rise in temperature in the flat-plate collectors is 50 - 70°C, and the cold water temperature must be some 10 - 15°C because it comes from underground. Th real efficiency is about 1%.

Prof. K. Kimura:

Would you mention more about solar cooling?

Mr. L. Drouot:

Generally speaking, there is no air-conditioning of homes in France. The only air-conditioning we need in France is for high-rise builings in Paris and other cities of France. For compulsory air-conditioning such as computer rooms, food containers, and some private homes, passive devices are used by means of appropriate architectural treatment. Air-conditioning systems such as vacuum tube collectors and absorption pumps have internal and overseas marketing potential and they are being developed.

Prof. K. Kimura:

It was mentioned that 5% of national energy in France was to be covered by renewable sources, excluding hydro-energy. The energy required for air-conditioning must be very small. Is there any attempt to recover some electrical power in air-conditioning by solar power, thermal power or photovoltaics, rather than absorption type air-conditioners?

Mr. L. Drouot:

The question is about the proportion of air-conditioning in the total energy budget. Heating of homes in France, in terms of oil consumption, is one-third the equivalent of the total consumption of the country. That is the problem of heating space in relation to the problem of air-conditioning in the country. If we can achieve the saving of 5%, we will be very lucky and very proud. We estimate two thirds of this will be bio-mass from wood and straw.

Prof. K. Kimura:

Since nuclear power may not be increased in the future, we have to limit the requirement of electricity demand. This is the reason why photovoltaic and thermal power are suggested by many people. Would you comment on that?

Mr. L. Drouot:

The energy structure in France is very different from that in Singapore. In Singapore, emphasis is placed on limiting electricity demand, but in France, it is on energy conservation in homes. We really do need to have better insulation in our homes.

DISCUSSION

Chairman : Assoc. Prof. P. S. Kooi

Panel Members : Prof. Jeffrey Cook
Prof. Ken-ichi Kimura
Prof. K. R. Rao
Dr. Chiam Joon Tong
Mr. Louis Drouot

Comment:

Although solar energy is free, the hardware required to harness the solar energy can be rather costly. Another problem is to incorporate this hardware into the building design so that it will give maximum efficiency, and look attractive as well, especially for passive cooling systems where the solar storage is to funtion as part of the building itself. Most of the architects in practice nowadays were trained during the time when energy was abundant, so not very many of them are really familiar with solar energy. One should look into courses of solar architecture we should offer in the school of architecture.

Prof. J. Cook:

This is a subject dear to my heart, since what I do is teach Solar Architecture. The solution within my School is that we have established the graduate solar programmes as distinct from undergraduate studies. It works very well because our graduate students come from all over the world. When they come to us, they already know how to read, to write and to draw. They all know how to design a building and a bathroom, and they are now ready to begin their real education, and we start all over again. It is like kindergarten except it is now solar and thermal kindergarten. They know what buildings are made of and how to design buildings, and now we start again thinking about the new kind of architecture.

Within the United States, our programme is unique, and I personally do not think it is the best way. In the more distant future, solar architecture should be taught as architecture that is responsive to not just human needs, but also to the relationship of buildings with all natural resources. Therefore, it should not be a special course. It should not be a graduate curriculum, but it should be integrated in all the courses that architects take as part of their education. If one considers the Parthenon, the Pantheon or the Pyramids, each of these has an energy component which was a determinant in how these buildings were designed that way. This is how one may regard the cultural values of the time they were built as well as weigh those values in measurable terms familiar to our own time.

Prof. K. Kimura:

Passive heating in architecture is easier than passive cooling, because cooling has to be related to air flow, which contains not only heat but also humidity, and the dehumidification process is necessary for comfort. Passive heating is mostly done by radiation, conduction and natural convection to retain sensible heat. I have tried, in Japan, several passive cooling systems by convection and conduction. Radiation may also be used, as shown by Prof. K. R. Rao. It is not so effective as in dry climate conditions. I am also trying for passive cooling, to use underground heat 2 or 3m deep, which is more or less stable throughout the year. I put a duct about 1.5 metres deep and drew air above the ground on the northern side of the building, and brought it up again in the middle of it. I expected to have a chimney effect that could take this cooling air to the attic space. The cool air is now coming out successfully. I have not measured the amount of heat removed but the occupants are very much enjoying it! Towards the end of summer, the ground being warmer naturally warms up the air drawn from the outside.

In conclusion, I feel one must provide a much longer duct in the ground, and try to spray water in front of the duct inlet so that a more evaporative cooling process can be provided for the air stream. If one provides some small fans, it may give more effective flows of air. In general, I feel that in this kind of hot and humid climate, to prevent heat gain is very important. Shading devices or insulated roofs as in traditional Japanese architecture are some of the solutions. There are many devices of natural means for hot and humid conditions.

Chairman:

May I ask whether in the Department of Architecture in Waseda University, there is a course in Solar Architecture?

Prof. K. Kimura:

Not a course in Solar Architecture, but we have a somewhat different type of education system. In the Department of Architecture where I am teaching, the environmental aspects in architecture, such as climatology, insulation, shading devices, solar radiation, and lighting are taught. This is so-called Building Physics and is one of the undergraduate courses. The mechanical portion of air-conditioning and heating is taught by other professors. In the graduate school, I have in my laboratory 5 to 10 students working on Solar Architecture. But there is no regular course on solar energy. The courses I am teaching at the School of Architecture are environmental theory and engineering physics. In that context, I teach estimates of the heating/cooling loads and other aspects of indoor architectural environments.

Chairman:

I understand that the French have many solar houses. I wonder whether Mr. Drouot knows if they have any courses in solar building in France.

Mr. L. Drouot:

We have many solar houses. I would say a few experimental or demonstrative houses, but all of these houses are being lived in by families. Quite recently, we had a competition involving 5,000 solar houses which were opened by our Ministry of Construction, and special financial incentives were given to the winners so that these 5,000 solar houses could be constructed in the coming 2 years.

Now you asked me whether we had training courses in solar architecture and techniques. The answer is yes, and we have had courses on education in the solar field, especially in architecture, for a number of years. The new generation of young architects are really pushing forward and coming up with new designs, new concepts and many different styles of houses in France.

Question:

Mr. Chairman, let us come back to Singapore from the many examples we saw this morning in U.S.A., in Australia nd now in France. These examples really reinforce my fear that solar energy applications may not be so applicable in the Singapore situation. Let me explain. Within the Singapore context, for development in the Central area, the sky is the limit. Now if I were a developer really wanting to install solar energy, say, in a 30-storey high-rise office building, I would be very worried because my neighbour might put up a 60-storey block and all the money spent on solar application would go down the drain. Do you have any comments on this?

Chairman:

Yes, this involves the problem of "solar rights" which was discussed by Prof. Jeffrey Cook. We may have to do the same in Singapore if solar energy becomes more popular and if everybody is trying to use it.

Prof. J. Cook:

The law I was referring to has been enacted in the USA. It is simply a prior right law. Whoever gets there first can have it, so if you build your 30-storeys first, then you have already taken the sun that may be falling on your neighbour's land, but you are entitled to it because you were there first. This is an area for the lawyers to work out and, obviously, if you want to encourage solar building, solar rights have to be anticipated.

Mr. L. Drouot:

Yes, I would like to comment on the question which was asked earlier. I see that this problem of right to the sun is not only a problem for 30 or 40 storey buildings. The problem is the same regardless of the height of the building. The climatic conditions in France are very different from what we have here in Singapore. There is a very great contrast between our winter and our summer, which means that every single dollar we spend on solar energy is only used in a fraction of the year. This is the greatest difficulty because the use of the solar energy must be financially sound for the people who spend the money. The climatic conditions in Singapore mean that this contrast is not very ponounced throughout the year, and that every square metre of collector you spend money on is for all the year round.

Question:

I am an architect by training, but we have a company that is specialising in solar energy in East Malaysia. Up to now we have been discussing the application of solar energy which is more for the temperate countries, its application in the tropics is the reverse in the passive form. I think we will be thinking of keeping out the heat, rather than keeping it in as in a green house.

The other application, the active part, would be to use it to heat up water and perhaps for air-conditioning and generation of electricity by photovoltaic cells. Now I have been trying to convince a couple of clients to look into the application of the construction of solar houses.

My main problem at the moment is to look into the reduction of solar heat gain of buildings and perhaps, utilisation of collectors on the roof top for hot water and air-conditioning if it is viable. The main element we can look into perhaps, is the roof. That is where most of our solar energy is received. For example, we could use the flat plate collectors as roofing itself, and one would save cost by taking away the normal roofing.

Prof. K. Kimura:

In solar cooling, the aim is to minimise the heat gain and to maximise the heat loss from the building, if possible. Efficient solar collectors may be used as a roof covering as well, because, for solar cooling, you need good quality solar collectors to produce a high enough temperature to run the absorption machine. Another problem we found is that the area of the solar collectors is quite big in our climatic conditions. In an experimental house which we intend to air-condition, the room has a floor area of only 28 m^2. According to computer calculations, we require 28.6 m^2 of the solar collector area which is more than 1:1. The house is designed to eliminate or minimise heat gain as much as possible. We also require a storage tank of about 1 m^3 plus an auxiliary boiler, if the air-conditioner is to run for 24 hours.

Comment:

I am an architect and my higher degree is in Psychophysics. 36 years ago, I lived in an attap grass-roofed hut on the borders of Thailand. I wore a sarong most of the time. My metabolic rate was fairly low during the day because I did very little work. Most of my activities were in the early morning or late at night. But I felt fairly cool, and when I went out in the sun I felt warmer; but when I came back into the shade, especially if there was a little breeze to cool down under the attap which was good natural insulation, I felt cool. My comfort condition was fairly tolerable.

Twenty years ago, I worked at the PWD Headquarters in Kuala Lumpur and there we had an old little colonial building with a verandah, red-tiled floor and ceiling fans. They moved very slowly but the air movement was fairly reasonable. I did not wear a sarong but I wore shorts with a fairly loose shirt and sandals. During the day I did not work very hard. In the early morning, I did most of my creative work out on the site and when I came back I found that the coolness of the building plus a small amount of the electricity we used for the fan gave me cooling comfort by contrast with the heat of the building site. But my tolerance was not quite as much as when I lived in the attap and wore a sarong.

Now I live in an air-conditioned flat. I get into an air-conditioned car to go to the University. I have an air-conditioned office there. When I come to the Conference, I wear a suit which I feel comfortable in. But when I go outside I feel most uncomfortable because my tolerance has come down to a very narrow limit. (The tolerance here is about 24.7°C, effective temperature). The more technological applications we have the less tolerant I am.

It is difficult and expensive energy-wise, and this is our big problem: to boost technological advances and to make everyone much more tolerant than our forefathers were. In fact, in Singapore, the only practical application which I have heard of so far is Prof. Rao's thermally induced chimney effect to draw air into buildings. I lived in Hong Kong for 16 years and there we reduced the size of our buildings to smaller perimeters so that the energy input expenses were as little as possible. The temperature differentials in Singapore are very small and, therefore, we cannot gain by exposure to an Arabian sky, as I have done in Saudi Arabia, using thermal capacity at a bigger temperature differential.

The only possible cooling source we do have is the sea nearby; and the sea has a slight differential by day and night between the air temperatures and the water temperatures, and, therefore, we have some sort of thermal reservoir. This is the only means. I suggest we concentrate on it for the time being until we overcome the development problems by lowering production costs or by making the solar technology market more popular. In Singapore, I would recommend that we try and concentrate on the simple passive achievement, which we can do. Instead of trying too hard to run, I suggest we walk gently before we go too far.

Prof. K. R. Rao:

We have seen a number of applications in the temperate zone. Some of those principles can easily be applied to the tropics. For instance, movable insulation insulates the building by the day and is removed at night to permit the structure to cool more rapidly. Similarly, we may have thick walls with a large heating capacity during the day, and convert them into light-weight thin walls during the night. This may be achieved by using water walls, pumping water in during the day time and out at night time.

Now as far as solar architecture and solar control are concerned, even in Singapore we are teaching our undergraduates a part of the solar architecture content, that is, sun control. To eliminate direct solar heat gains of buildings as much as possible, students study solar geometry for designing effective shading devices, shading coefficients of different types of glass, etc. Colonial houses or, even earlier, the traditional Malay houses, were very well suited to the local climate. We got into the age of glass houses because this was the fashion of the day. It is true that we have to go back to climate-conscious designs. We can easily adapt our modern architecture to our climate by the use of new materials and a better understanding of the sun and building relationship and of principles of natural ventilation.

Prof. J. Cook:

I would like to reinforce that even though I live at 33°N, I live in an overheated climate. I live in a desert and I do not like it during the 4 months of summer. I cannot stand it at all. Depending on how one measures it, we have at least twice as much cooling required as heating. So I am very sympathetic towards the problem here in the tropics. One way of thinking about cooling is the absence of heat. So an elementary concept is to see what people are doing in cold latitudes and avoid it. If one really understands how glass houses work, do not build glass houses in the tropics. This is a very heavy responsibility for architectural schools. We can no longer assign comfort problems to the physicists or to the engineers.

Dr. J. T. Chiam:

I tend to agree with the sentiments that have just been expressed. Now perhaps I could quote from my own experience. Eleven years ago, I built myself a house and this was at the time when energy was cheap. My young architect influenced by modern European architecture decided that he would build a house similar to those in Europe and, as a result, I have a house that looks like a green house. So, in the morning when I would like to breakfast in comfort, I have the sun streaming into my glass dining room and it is extremenly hot. In the evening, my bedroom with glass windows facing West is extremely warm and I have to put on my air-conditioner just to cool it down. Now this underlines the fact that many of the modern houses built in Singapore during the time when energy was cheap are not really geared for the tropics.

We have come to a situation where we now need to conserve energy in buildings. I think we are all aware that the Singapore government is very concerned about what we can do to reduce the heat influx into buildings built about 10 years ago in Singapore. We know that the Government would enforce on existing buildings the OTTV of 45 watts per square metre beginning 1st January 1982. Within the next few years the task for architects and engineers who are involved in energy conservation will be to try to solve this problem of how to retrofit buildings which have been built with high energy influx. This is one of the major tasks because it is certainly the government's policy to tackle this problem seriously by penalties and incentives. The task ahead of us in the next few years is really to come up with ideas and innovations to reduce the heat influx in existing buildings and thereby reduce the cost of air-conditioning. This is an urgent problem and I think architects and engineers in this region should put their heads together to resolve this, especially in Singapore and other cities in this region.

Mr. L. Drouot:

Mr Chairman, I shall make one comment on retrofitting of buildings. This idea does apply in France, and I do not know if it would in Singapore. Our figures are the following: We have about 20 million individual houses in France, and the rate of construction is about 2% a year, which means that it will take about 50 years to convert the whole lot to solar architecture. We are already worried about retrofitting, and our architects are really working on that subject.

Prof. J. Cook:

It is interesting that when a new technology is proposed, one finds all kinds of reasons why one should not use it. "It will be too expensive", "It will be ugly", one would have to wait until next year so that the "technology could be improved". There are all kinds of reasons for not doing it. Considering retrofitting, the real problem is to rehabilitate the existing capital represented in buildings that are already built. We can learn most effectively by building new buildings in the correct way. In doing so we can learn to retrofit. I see new buildings as being like pieces of clay with which we can sharpen our tools to make them more effective, and thereby develop the means of tackling all building stocks for all climates.

Prof. K. Kimura:

I think the earlier speaker made an interesting point concerning the degrees of tolerance in temperature and humidity. In Japan, we have discussed this problem with many scientists, and the conclusion is that we cannot go back into ancient times where we did not have any sophisticated machinery. So what should we do? We know there is so much energy wasted in our modern community, and we must learn from ancient devices. We should aim somewhere between these two means, which is called intermediate technology. This means we are to learn the real ancient ways of response to the climate and environment and use the idea of modern technology.

Mankind has had some bad experiences because we have had too much oil for the last twenty years. Now we are approaching the era of oil shortage. The consolation is that the environmental disruption by spending more on oil will be avoided. We always talk to students about this kind of problem. I realise that your city planning is better than ours in Osaka, because we have been destroying more green areas, and the buildings are very close together. There is very little green left in the metropolitan area. But here in Singapore, I see many, many trees among buildings and they will help make the climate stable in the city area. So, I feel the temperature and the humidity less than what

I had expected. It also happened in ancient times in Japan when there was a certain amount of land between small individual houses. This is now done in the city of Singapore.

Chairman:

I think the situation in Singapore is a little different because most of our housing is high-rise and high density. Prof. Rao has also looked into whether we actually need air-conditioning in Singapore. Perhaps Prof. Rao wishes to comment.

Prof. K. R. Rao:

This depends on individual taste. Physiologically, we are on a borderline and, under the local conditions, a slight variation in clothing, activity and air movement would make one change from feeling comfortable to feeling uncomfortable. For example, if one is required to come to the office or a meeting in a suit, it is necessary to have air-conditioning; probably no air-conditioning is required if light clothing is worn. Again, if there is good natural ventilation and fans are used, to a large extent, one does not need air-conditioning. But we cannot use natural ventilation in the office building for other reasons, like noise. There are so many social factors intermingled with other factors that determine the need for air-conditioning of a building. Whether we need air-conditioning or not one should question, "Have I used all my design tools? Have I exhausted all my faculties and methods to design a building with minimum energy requirements? Does it minimize the discomfort period for the occupants if it is not air-conditioned?"

One should not design the building first and then air-condition it, which is the practice now. The need to air-condition residential buildings may be very marginal. Even with office buildings, one may have air-conditioning but the energy requirements may be much less than what they are now, if all proper thermal and solar design considerations have been taken. If we know the art of controlling the sun and maximizing the air movement, I think in 80% of the situations we may not need air-conditioning.

Chairman:

The solution will be to design buildings with natural ventilation or with the passive approach, and to use active cooling systems only in extreme conditions, when one cannot tolerate the heat anymore.

Comment:

There is a very great responsibilty for architects and engineers in our society to work out cost constraints, because when we think about eliminating solar heat gain we have to think about incorporating many features such as cavity walls. This involves quite a lot of cost. Our professionals have a great responsibility to do cost studies and prove to the client that it is worthwhile investing in the additional cost, so that the capital cost may be recovered in time. The professionals will have to look into this area of budget control.

Prof. J. Cook:

I would like to ask if there are some new social responsibilities that the architect is expected to execute and whether there should not be some common standards whereby architects have to conform to a certain level from the energy and comfort points of view.

Comment:

It is the duty of the architect to be responsible, or to be energy-conscious, in the design of buildings. The client now realises that the need to conserve energy has become very obvious because of the high energy cost. I also feel that we should transmit energy consciousness to architecture students, as they will be the future architects of Singapore.

Singapore is on a path which is clearly for economic growth, and it sees itself as a part of the international financial world. In that context, when one talks about Singapore I can only think of the architecture of Singapore as city-architecture and commercial architecture.

The first question is that, in the international financial world anybody asking about renting an office building or putting up shopping centres, will want to know if the building is air-conditioned. And this, I think, is a constant problem. Once one ventures out into the market system which is then linked with the international financial world, any kind of commercial viability exercise done on commercial buildings will reveal that air-conditioning has become an important factor. The question is: how does one introduce solar architecture when the basic principles are non-existent? Unless there is a change of attitudes, as long as the market forces dictate which certain types of buildings are to go up, it puts the architect and the engineer in a very difficult situation to introduce buildings in terms of tropical architecture or even solar architecture.

Prof. J. Cook:

I feel very sympathetic to the problem that is put before us. Certainly, in the United States in the last 10 years, anyone who wanted to build a solar building has been at a great disadvantage. Mortgage lenders and bankers would not look at the solar building except as a freak. It was not considered normal or acceptable or resaleable. That attitude is changing very slowly and we now do have some constructions in which they actually give a mortgage advantage. They are allowed to give points for a lower mortgage rate or give a shorter payback period. In other words, there is a financial advantage to having a solar building. So at least in the United States the economic controlling forces are changing. But it has been slow.

I can also identify a similar change in the general popularity of the subject. We have been holding national passive solar conferences now for several years. We just finished one last week. We have had 2,500 people for the technical sessions. In terms of numbers that is a lot of people for a conference. So I think the momentum is certainly coming. One of the problems for education is not to educate for the moment but to project the future when those people who are in school now will be hitting their prime and will be the principals in architectural firms 20 years from now. So, somehow or other we have to identify the ethics that will be appropriate 20 years in front of where we are at the moment.

Question:

In Japan, there are so many solar buildings. How does the architect manage to convince the client to build them?

Prof. K. Kimura:

The case that the architect convinces the client is rare. In most cases, clients take the initiative in having solar buildings. In Japan, the government has finally decided to give one-half allowance for public installations and low-interest loans for private houses. The people who would like to build solar houses, first of all, can borrow some more money on top of the regular lending terms for normal houses. This is a low-interest loan at the rate of 5.5% annually. For large buildings, the local government or national government also has some demonstration subsidies to provide more examples of solar installations. I would like to make one comment. In design, the green house effect in the winter for colder regions insulates heat from inside to outside for the living space. But this is only one function, and people will not like it. So one will have to consider dual function devices such as solar collectors also acting as shading devices. For the tropics, when one designs a shading device, why not make water flow through the shading device to get hot water? This kind of dual function device could be more cost-effective. But one has to account for the energy used on an annual basis to work out investment returns.

Prof. K. R. Rao:

Solar devices which can improve the internal comfort condition need not necessarily be more expensive than existing systems. It is only a question of looking at alternative materials and designs for maximizing their dual functions, as mentioned by Prof Kimura. Energy conservation should not be at the expense of one's comfort nor add to the cost. It is no easy answer, but every application need not be an expensive application.

Chairman:

Prof Rao is also considering solar integration in building design, possibly in drying clothes - an example of dual function.

Prof. K. R. Rao:

If one goes round the housing estates in Singapore, it becomes apparent that the space for drying clothes is a real problem. The danger of falling objects is also there. One could have a solar room in the form of a green-house type of attachment to the balcony or verandah for this purpose. With such a provision, the clothing is also protected from rain. One could look into real practical problems like this, without asking people to change their whole pattern of living. One could also combine solar ventilation and clothes-drying. There are the possibilities of direct solar application even in our climate. If we adopt a positive attitude, we will find some real solutions.

Comment:

In time past, many architects were trained overseas in temperate countries. So one could understand their enthusiasm when they came back to the tropics. The first thing they would think of was to put up curtain wall office complexes in the days when energy was abundant, and no thought was given to energy conservation. Fortunately, the recent crises (fuel shortages) have opened our eyes, and the government in Singapore has taken the first step in implementing the OTTV regulations. Personally, I was fortunate enough to be trained locally in the University of Singapore, and I can still remember the first year I had with Prof. Bill Lim. The local context of training was more geared to tropical architecture. It seems to be more advantageous if students in architecture receive their education locally instead of abroad.

Prof. J. Cook:

If one studies the careers of great architects and one looks at where they came from and where they went, one will find that, even with the people we all recognise as geniuses, Michaelangelo or Frank Lloyd Wright, their earlier work is always an imitation of their masters, and most of us never get any further than that. We simply copy whoever our teachers were. That gives the teachers a very special responsibility, because it means that our students, the poorer ones, are never going to do any better than copy their professors. Even though we hope to teach principles to a limited extent, they are not always easily transferable. Those who have gone to another climate to study their profession may have brought back the wrong principles.

Question:

Mr Chairman, in the Singapore context, we are considering solar energy as an alternative source. But so far we have only been hearing of using solar energy as a dual purpose or for keeping houses warm in a temperate country. One cannot apply a dual purpose in the use of solar energy in this country. Also one does not need as much hot water here as in temperate countries. So we have to design buildings that will, as Dr Chiam puts it, recede or absorb the minimum amount of heat so that we can reduce air-conditioning. Prof. Rao said that we would have to design houses in the style of the good old days which cooled themselves so that air-conditioning is not required. The next point is how we may tap the abundant solar energy for uses other than just heating water, perhaps when technological advancements are such that we can convert solar energy into air-conditioning energy. But at the moment the applicability appears to be limited to hot water which is useful in hotels and hospitals. Otherwise, solar heaters are not suitable to our country and climate.

Prof. K. Kimura:

In fact, we could use solar energy for air-conditioning. In Japan, solar collectors are quite extensively used for air-conditioning. Of course, at the moment the cost would be prohibitive for any individual house unless electricity prices go higher.

I agree with you that we do not have to have hot water for low income housing but the water may be put to other uses. For example, there is a glass building in Denver - a five or six storey building, with internal movable louvres of aluminium extrusion sections, and water runs through the centre of the sections longitudinally. The water absorbs solar heat, with the result that the building requires smaller refrigeration plants. This is one of the ways of reducing heat gain.

Prof. K. R. Rao:

We are exploring what is available today, and how much we can make use of it. Solar water heating is the simplest and easiest - that is why most of the people go for it. In Singapore, PUB once published that more than 20% of the homes use hot water. If that can be made solar - not electrical - one could save some energy. We should not think in terms of 100% of the population using solar energy. The conversion of prime energy to electricity is about 3 times more costly than oil fuel. This is because the efficiency of electricity generation is 30%. Therefore, if electricity is used for hot water purposes, we are utilising very many more energy resources which can be replaced by solar energy with present day solar technology.

Dr. J. T. Chiam:

It is true that, at this point in time in Singapore, solar energy is used mainly for the generation of hot water for hotels, restaurants, hostels and so on. There are, of course, wider uses of solar energy. It is true, too, that at this point in time we have not fully exploited this form of energy. Industry in Singapore (and also in other industrial nations of the world) has hardly utilised solar energy. For example, in the cosmetic industry. My experience with one of the firms near Jurong is that wax is melted by electrical power at 60°C. Now this can easily be replaced by solar energy. A study has also been done in Australia where it is found that in the food industry one could easily replace 50% of low heat energy, below 80°C generated by fossil fuels, by solar energy. In Malaysia, one could think of solar heat drying timber and crops. Now these forms of exploiting solar energy are largely under-utilised, and I feel that the time will come when it will be possible to use solar energy not only in urban areas, in homes and hotels, but also in industry at large. It makes economic sense, of course where possible, to replace diesel fuel or petroleum gas with solar energy, especially in the vast agricultural industry in this region. We should consider also the wider context of solar energy, particularly the technology of bio-gas applied in an agricultural environment. The energy from the sun absorbed by plants could be generated through bacteria action, from agricultural products or wastes. For example, the palm-oil industry in Malaysia produces a large amount of waste effluent which could be converted into an alternative fuel, bio-gas which is essentially 60% methane and 40% carbon dioxide. From my experience with a palm-oil mill in Malaysia, I noted that this mill with a capacity of 60 tons FFB/hr is producing biogas equivalent to 2,100 gallons of diesel per day. The potential in solar application in its wider context, especially in the ASEAN region, is tremendous. What we must do is to educate industrialists and the people in this region on the economic viability of solar technology and how it can be used to conserve conventional energy sources. The technology is there, it therefore remains for us to apply it.

Question:

I agree with the various speakers that most Singaporeans, in fact, do not use hot water and we may not need very much of it. We should think in terms of drying clothes. Why not use the hot air from solar energy in a drying room, say, in multi-storey flats to dry clothes?

Comment:

In my opinion, there is no point in spending money on building a hot room just to dry clothes when one can just hang the clothes outside to dry. I support fully Prof Rao and Dr Chiam's statement that we need in Singapore to consider solar energy in its wider sense and not only in the building or architectural sense. We have not considered the use of solar energy in places of low energy heating and low energy manufacturing where electrical energy may be replaced by solar energy. We would probably like to know more from the industralized countries about how solar energy can be utilised in the manufacturing and food industry or in the agricultural industry where low temperature heating may be used. Let us encourage our industry to use low-heating energy from the sun in place of electricity.

Question:

I'm an architect. I work with New South Wales Public Works Department in Australia. I was interested to hear what Prof. Cook said about loans through bankers of the the finance companies in the United States. In Australia, one tries to eliminate as much of the sun as possible in summer, and to use the sun as effectively as possible in winter, and this can be easily done if one can orient the building properly and work out the overhangs required. And, in this regard, we do have some components of passive solar design in buildings. (Active systems in buildings at this stage are not very common and are only being used to some extent for domestic applications such as heating of water by solar energy.) How then does one define a solar building?

Prof. J. Cook

In the US, houses are not always architect-designed. Typically, architects are not at all involved in speculative housing as done by developers. And, from that point of view, solar houses that have been built have not come from the speculative market, and therefore are quite exceptional. We do not have any professional definitions as to what solar building is, and what solar building is not.

I suppose the only way one could discover that is if the drawing says "Solar House". One of the things that we have learnt as professionals is not to label our drawings "Solar House", because we find that, if we do, the builders will double or triple their estimates of the costs of the houses. They will think that it will be something terribly difficult and that there will be lots of dangers and that they simply have to put in a fault factor or fear factor to cover all of the difficulties. So at least the drawings are not labelled "Solar House".

When it comes to looking at the sizing of the heating plant, of course, one can see that solar energy is making some contribution. I guess the need to define "solar" comes up with the regulations regarding the sizing of space heating or space cooling plants. Certainly, the provision of domestic hot water by solar is not a very troublesome issue. The costs involved are not major, therefore the definition is not a problem. I guess I am avoiding the question because it seems to me that there is not a very good norm against which to judge it. I would really be interested in hearing other people's comments on this question, "What is a Solar Building?"

Prof K Kimura:

I think "solar" is just symbolic. The sun represents nature and the natural way of thinking. Perhaps, for the last 30 years we have neglected nature and the natural way of thinking. Does that give any answer?

Prof. J. Cook:

Well this is my classic answer. But that is when I talk to other architects. But when I talk to bankers I cannot say that this is a symbolic solar house. He probably wants to know how many barrels of oil it will use, and therefore I have to come up with a standard set of numbers.

Prof. K. Kimura:

It is not just the practice of solar energy, it is a way of thinking.

Prof. J. Cook:

Yes, totally.

Comment:

In any kind of solar design input, there is some sort of passive solar design in any design situation, whether it is just a minute amount or whether one is trying to get a very beneficial effect from the sun. Either every building is a solar building or no building is a solar building at all. I do not think there is something in the middle.

Prof. J. Cook:

Perhaps, "solar" buildings come into a sharper focus in North America than they do in the rest of the world. We have wandered further in building "non-solar" buildings in the past. It may be said that American architects in particular have been and continue to be involved in architectural values with very little relationship to, what I would like to call, the real world. If that is the case, then a solar building represents the opposite set of values. A solar building represents a design which comes from a solar or nature point of view. It tries to focus on what the natural conditions are on its site and in particular the relationship of the building to natural energy forms. So from that point of view, the term "solar building" does indicate that it is now another kind of building. It is really surprising in the United States that this polarisation has taken place. For three years, I have been on a Committee on Design of the American Institute of Architects, which is a voluntary association of architects in the United States, This committee is concerned about the quality of architectural design, not just houses, but of all building designs, especially public buildings. We have had quite a number of discussions about how the criteria of energy should be identified in terms of programmes sponsored by the AIA. Many architects are of the belief that comfort and energy questions are what one gives the engineer to do. An architect considers building as a design problem, and the design has not got anything to do with comfort - that is for the engineer to solve. Only a few of us are of the view that design and comfort should be considered together. With that kind of experience, it is not hard to define a solar building.

Comment:

One important consideration we haven't taken into account is the question of cost. In this respect, I mean both capital and operating cost. Then we would have to consider how the application will impinge on the other elements of the building fabric. Only when one can sum up the total cost equation can one make the final decision. If these are not taken into account, we would be talking generally in the abstract.

Comment:

With reference to solar drying, I see great possibilities in the suggestion of making use of solar energy aesthetically. At the moment, the public housing residents dry their clothes from the 20th storey downwards. I think the size is a bit staggering. Now if we can provide a little space for people to dry their clothes, I think the public housing image would be greatly enhanced.

Question:

I think I would like to define a solar house in a tropical country as a house with the following achievements. A house that has achieved minimum heat gain in its design, a house that has managed to utilise solar energy for air-conditioning and hot water requirements, plus a house that has added the feature of photovoltaic cells to produce electricity. There is a general need in our tropical countries to educate the population on solar energy and its application. When installation of solar heaters was suggested to one housewife, she replied, "Well, will I get cancer from the solar radiation used to heat up the water?"

Another application which we haven't actually explored is its application to generate electricity. If someone could manufacture solar photovoltaic cells on a very large scale to bring down the cost, we could perhaps envisage our office complexes with photovoltaic cells on the roof to power all the electrical requirements of offices in the day-time; but and perhaps there would be objections from the PUB. I can see the potential of photovoltaic cells in our country because to generate electricity using diesel fuel is costly. In Sarawak, we are experimenting with the potential of hydro-electricity, and perhaps that eliminates the use of photovoltaic cells. Perhaps this seminar could come to some conclusion on education and exploration of the potential of solar energy in air-conditioning and photovoltaic cells. This would be its major contributions to the tropics.

Mr. Barry:

Firstly, I would like to apologise for Mr. Brown not being here. I am sure he would have stimulated the meeting. Now that the question of voltaic is mentioned, I thought I should say a few words. My name is Barry from Solar Transmissions in Australia. I've been involved in photovoltaic electricity for about 10 years now. Our main area of endeavour has been in islands such as Fiji and New Guinea mainly to generate power for beacons, and now we are getting into village lighting. There is one thing I would like to comment on: this business of hydro-power. We are in a very lucky country down there. It may be the cheapest electricity in the world, but the expense is in getting the power to where it is required. In Australia, on a per kilometre basis in New South Wales for instance the cost is about A$2,000 per kilometre just to run the line to somebody's homestead. And then, as we are dealing with 240 volts, we have to put a transformer on Queensland, stimulated initially by an inquiry from a gentleman farmer who had a homestead about 14 miles from the nearest power source. The quote from the local government to put the power in his house was A$27,000/-. Now that is only 14 miles at A$2,000 per kilometre plus the transformer. With photovoltaic cells we got that man 19 lights in his house, which were individually inverted to 240 volt flourescent tube plus his TV for 4-5 hours a night. He was not doing any cooking, but the man got his TV set and his lights on for A$6,500 of power package, including 20 year long lasting batteries. So the entire package under the circumstances worked out more economically than conventional power supply.

Chairman:

We should ask Prof. Cook about passive cooling which we have not touched upon, because I understand in Arizona there is a house using passive cooling. Do you see any application for passive cooling in Singapore?

Prof. J. Cook:

Passive cooling tends to work the way passive heating does. In other words, one tries to tap natural sources which are much less intensive than heat sources. A typical solution probably, is a hybrid approach. In other words, where one uses a small amount of parasitic power, namely, the use of electric fans or pumps, to help move the heat into a cold source or alternatively to move the coolness to the place where one would use it. Ventilation obviously is the number one potential for this tropical climate. I will comment further during this conference.

WORKSHOP 3: SOLAR DRYING AND REFRIGERATION

Chairman : Prof. K. R. Rao

Panel Members : Assoc. Prof. H. B. R. Exell
Dr. N. R. Sheridan

Assoc. Prof. R. Exell:

I was asked to say something about my working in solar drying and refrigeration at AIT. For solar drying, the basic problem is that the farmers traditionally grow rice in the wet season, and the harvest comes after the wet season: the dry season in the winter, that is December and January. The harvest is very easy to dry. If the rice is simply spread out on the ground, it dries because the weather is always fine. But with the new varieties of rice coming in and with improvement of irrigation, it is possible to plant rice in the dry season and harvest it in the wet season. Then it is difficult or almost impossible to dry the rice and to store it. The farmers have to sell it quickly and they get a lower price for it because it has a high moisture content.

So we wondered whether it would be possible to make a very simple solar dryer to dry the rice even in the wet season. What it consists of is a layer of burnt rice husks on the ground which is black, so it absorbs solar radiation. And above that there is a clear plastic sheet stretched over, and at the bottom in the front there is an air inlet about 10 cms high. The sides and the back of the unit are closed. It is a very simple air heater made of bamboo poles, wire and plastic sheets. The hot air rises through the box at the back, and the tall chimneys are put there in order to give a much taller column of hot air for better convection. The flow of air through the rice bed is sufficient to dry a rice bed 200 mm deep.

We were rather careful about the design because it would be very easy to build a dryer which did not work due to either the lack of air flow or heat supply in the air heater. The amount of heat required to remove the moisture from the rice, of course, must be provided by the solar heater. So we did our sums and that is the design that came out.

There have been a number of units like that put in the countryside in Thailand and other parts of the world. This particular unit has

Bill B. P. Lim (ed.), Solar Energy Applications in the Tropics, 307–319.

many chimneys, and some outlets are without chimneys. The units are also divided up into compartments for experimental purposes. But if a farmer builds one in the field he only needs one chimney and it is not necessary for him to divide his dryer up into many compartments. It holds one ton of rice, depending on the moisture content when it is put in. In the worst conditions when it rains, or during intermittent rain and cloudy weather, the rice will dry in about 4 days, so that for intermediate conditions typical for the wet season, it may just take two days. This is very simple technology. A farmer can build it himself. If he uses that particular model, or the simplified one, he get his money back, and with a little profit due to the extra value of the rice.

Now we come on to the refrigeration unit. In my paper there was one diagram which illustrated the circuit for the AIT refrigerator. This is also an intermittent ammonia water refrigeration system. There are two units; each one has two collector panels and two large tanks. The collectors contain ammonia-water solution. The large mirrors are used to increase the amount of direct radiation falling on the collector. By rotating these mirrors at 10 degrees per hour throughout the day, one can double the amount of direct radiation on the collectors. The tank at the back contains water and a pipe used as the condenser of ammonia. The ammonia which is distilled from the collectors passes out through a pipe and is condensed in the condenser. The receiver for the ammonia is behind the tank. It is rather important to keep the water in the tank as cool as possible. We now have insulation on that tank. At the back, there is a receiver for the ammonia, and a cooling box which contains the evaporator and a cold compartment. The evaporator goes through a small vessel of water, and on each side of the water vessel there are two other compartments for holding the food to be cooled. The evaporator produces ice which cools the food. The compartment in the centre is always filled up with ice when the system is operating.

Question:

This morning we were talking about optimum flow rates of flat-plate collectors, and the figure of 1 - 2 litres per minute per square metre of collector area was mentioned. For Melbourne, the optimum flow rate is suggested to be between .01 and .015 litres per second per square metre (i.e. 0.6 - 0.9 litres per minute per square metre). Could the panel have some discussion on this?

Dr. N. Sheridan:

I am not sure whether the figure quoted this morning was on a unit area basis or not. In my experience, the flow rate through the collectors is not of great importance. If the flow rate is too high, you pay a penalty in increased pumping losses and additional energy is required. If it is too low, the heat transfer coefficient of the water is poor and the value of the heat removal factor is decreased with a corresponding decrease in efficiency. And so a balance between these two limits is required. In addition, the flow rate depends on the intended use of the collectors and may have to produce a certain temperature rise to satisfy the equipment being heated. Pipe size also affects the choice of flow rate since less friction will be involved with larger pipes. Flow rates in the range of 0.01 - 0.015 litres per second per square metre will satisfy most purposes, but where small temperature rises of the order of 2 - 3°C are required, flows as high at 0.03 litres per second per square metre might be used without seriously increasing the parasitic pumping power.

Question:

What is the volume of the solar dryer, the collector area, capacity of the solar dryer, and the rate of the air flow in the chimney of the solar dryer?

Assoc. Prof. R. Exell:
The capacity was one ton of rice, the rate of air flow, if I can remember correctly, is such that the air change in the rice bed is about 4 to 5 times per minute. What happens is that the air flows very slowly up through the rice bed.

Question:

And when you told us that it takes 4 days to dry the rice, do you operate the solar dryer at full capacity?

Assoc. Prof. R. Exell:

Yes. The rice is fairly moist when it is put in the dryer. As a matter or fact, rice when it is harvested can have a very large range of moisture content. It is better to have rather high moisture content because there is less loss in the field. The custom is for the farmers in Thailand to harvest it to somewhat lower moisture content than perhaps they need because of the drying problem. However if you harvest the rice at a moderate moisture content and if the weather is very fine, by the time the rice has been threshed and put in the dryer to dry, it loses some of its moisture anyway, and then it is left in the dryer for several days. What happens is it would dry during the day time and at night time nothing happens. It may even pick up a little moisture, and it will dry the next day. If it is cloudy there would be some drying effects. And even if it pours with rain, the rice will be protected. If there is a thunderstorm nothing will happen to the rice, but it will continue drying when the sun comes out again afterwards. So I would say two days is a reasonable average in the wet season.

Perhaps, I should also mention that the area of the rice bed which has a depth of rice from 100 mm to 200 mm will determine the amount of rice that can be put in the system, and the area of the air heater in front of the solar collector should be from 3 to 4 times the area of the rice bed itself. Now, this means that the amount of ground which is used to dry this rice is actually about the same as it would have been if the dryer were not used at all. Of course, the rice is protected from the rain, and continued drying is possible even in bad weather.

Chairman:

Can this be done in different levels or layers with air gaps in between?

Assoc. Prof. R. Exell:

We have not tried that. But if there were several layers with air gaps in between, then presumably the layer at the top would get the worse deal because there would be moist air coming up from the layers below. I somewhat doubt whether that would very much help. The very top layer of the rice bed actually receives heat from above because the cover over the rice bed itself is transparent.

There is some drying from the top downwards. The slowest drying occurs about three quarters of the way up the rice bed. In order to get uniform drying, one should stir the rice from time to time. But I am rather doubtful whether using several different layers would be an advantage. It might be for some crops. In the case of rice, there is the problem of loading and unloading two layers one on top of the other and it is not so easy to do by hand.

Question:

Prof. Exell, have you conducted any experiment on the drying of rubber sheets, timber, and copra?

Assoc. Prof. R. Exell:

No, we have not tried any other materials. We have just concentrated on rice. We have a project for rice. But I think a solar drying system should be developed for certain other products. One which I have seen is boiled shrimps. I have visited a shrimp factory. The shrimps were obtained from the sea, and they boiled them and spread them all out on the ground to dry, and they have problems in the wet season.

Question:

In your system, you are using plastic as the covering material. Would it be difficult and inconvenient to replace?

Assoc. Prof. R. Exell:

It is not too inconvenient. The plastic itself which we use deteriorates rather rapidly and must be replaced once a year. It is not very cheap, but it is worth it. It is not much trouble.

Question:

So it is not worthwhile to use glass as a cover?

Assoc. Prof. R. Exell:

I think some people might prefer to use glass. The one I showed you was a very simple construction: bamboo poles, wire, wood, rice husks and plastic sheets. But we have had requests from villages to develop a more durable model because the plastic would deteriorate in a few years. What they want is a concrete slab painted black with wooden posts and glass cover. They want something more sturdy. That would cost more but it would be more durable.

Question:

Would it help if you added ventilation fans at the chimney to extract the air? Would it speed up the process of drying?

Assoc. Prof. R. Exell:

One of the professors at AIT in the Agriculture and Food Engineering Division has done experiments on drying rice blowing unheated air through the rice with fans and his opinion is that one can get a long way with that. The other possibility is to use a fan to draw air through the rice bed, and use large black plastic cyclinders which will heat the air. The large cylinders are used as air ducts for the heat to enter the system and to be blown by the fan through the rice. That might be an even simpler and better way of getting good rice. I think one can get better performance with fans but the dryer which you saw is intended for the farmer out in the countryside where he has no source of power at all.

Question:

What about the use of photovoltaic cells to generate power?

Assoc. Prof. R. Exell:

One has to give them to the farmer, he will not be able to buy them. He can hardly buy that dryer let alone photovoltaic cells or a fan.

Question:

I would like to discuss the application for the curing of timber and rubber because in my country there is a lot of rubber, timber, copra and so on. Now, I am thinking, if we could use flat-plate collectors to generate hot water and a system of copper pipes could be run through a chamber and hot water be continuously pumped through, the heat generated would be enough to provide a controlled environment for curing timber, rubber sheets, copra, etc., because in such a system perhaps one could incorporate a thermostat to regulate the temperature. At present, the curing of timber, I believe, is about a one week process, where the temperature should be something like 52°C reaching about 74°C. Our Forestry Department may be interested in trying to get a solar system working. So I wonder if, from your experience, you think there is a possibility of such a system?

Assoc. Prof. R. Exell:

I do not know about rubber and copra but there has been some experience in Australia on timber drying. Perhaps Dr Sheridan might talk about that.

Dr. N. Sheridan:

To my knowledge, there have been two timber drying kilns built in Australia. These have air heating collectors with a selective surface, a single glass cover, and a reasonably good efficiency. The array of collectors is mounted on the ground horizontally in tiers and is tilted at an angle. Air is forced through the collectors with a fan before passing to the kiln, which is a brick building, dimensions about 6m by 4m, and about 3m high. Below the timber stacked inside the kiln is a brick matrix through which the heated air can be re-circulated.

As the re-circulation proceeds, any excess heat that is not used by the timber is stored in the bricks; and during the night fans circulate the air, from the bricks up to the timber. One of these kilns is installed in the middle of New South Wales, in a place called

Griffith, a very dry climate. It is an irrigation farming area. Another one is installed in Townsville in North Queensland which is on the coast but not a very high rainfall area. The experience is that the time taken to dry the timber is about 50% longer than the normal process of drying with a steam-heated kiln. The longer drying time produces higher quality timber, but it is doubtful whether a higher price would result. In Australia, the situation is that these kilns have not become commercially viable, and I do not know of any commercial operator that has actually installed one. The kilns I am talking about have been installed by CISRO on an experimental basis.

Question:

Will there be any high risk factor involved if the collectors are mounted on the ground; the timber may drop on the collectors, smashing them all?

Dr. N. Sheridan:

The collectors are at the side of the timber shed. I do not think there is much risk. The timber is simply laid in the kiln. I do not think there is going to be any problem in that. The air-type heater is fairly well-designed. It has a steel plate absorber which is deeply grooved to increase the convective area, and give some geometric selectivity to the surface. The steel is treated to give a chemical selective surface. So overall it is a fairly efficient type of collector. I think it operates up to 68°C.

Question:

Could you comment on the possibility of a chamber whereby the entire roof is covered with solar collectors; where water is heated and pumped around the walls of the building and into the floor, and where the heat released will warm up the air inside. Perhaps a system of exhaust fans will be needed to remove the humidity. This might be a way to cure timber and rubber sheets.

Dr. N. Sheridan:

It may be possible. I am not sure that, if air is used as a drying medium, it is necessary to introduce water into the system, for that will complicate the system to some extent. Admittedly water heating collectors can be a little more efficient than air-heating collectors. The heat removal factor is relatively low for air-heating collectors because of the poor heat transfer properties of air. But for the water heating collectors, heat exchange between the air and water will cause an effective temperature drop. So it may not be necessary to go to water heating collectors if air is heated up within the temperature range of 50-55°C which is satisfactory for most of the timber drying operation. It is only for the finishing off that a higher temperature is required. These air collectors are totally different from the water heating variety.

Question:

Can the same system then be turned around and designed for application, say, in a sauna bath?

Dr N Sheridan:

That requires a much higher temperature of course. Steam has to be generated and the collectors need to operate at a higher temperature.

Question:

Would Prof Exell elaborate further on the refrigeration system? What is the lowest temperature you can achieve and what size of freezer may be used?

Assoc. Prof. R. Exell:

The system is designed to produce ice. So the lowest designed temperature in the evaporated coil is about minus 7°C. The ice of course is at 0°C and the temperature in the cooling compartment will be in the range of 0-5°C, or 0-1°C. Perhaps I should just elaborate. If a particular temperature is required by means of an absorption system, then the difference between the ambient temperature and the cold temperature according to the design, is always lower than the difference between the temperature at the hot end and the ambient temperature. So, if one starts say at 30°C to go down to 0°C or, say -10°C, there is 40°C drop. That means the temperature rise in the generator has to be more than 40°C, call it 50°C, which would mean 80°C or more.

This particular unit of freezer will make 25 - 30 kilograms of ice. So the actual ice box has to be 25 to 30 litres in size, and the compartments on each side of it are about 4 times that size.

Question:

I would like also to know the maximum temperature that can be achieved in that solar dryer.

Assoc. Prof. R. Exell:

The maximum temperature is probably something like 60°C maximum. Maybe not that. To dry rice, the temperature should not go above 45°C otherwise the rice will be ruined. In some experiments we have had the temperatures going up too high, between 50 and 60°C. In that case, we have to bring the temperature down. In that particular experiment there were two things wrong with it. It was built outside AIT. The chimney was too narrow and there was some resistance to the flow in the chimney. The other thing was that the collector area in front was actually rather large. So we reduced the collector area and increased the size of the chimney, and we got the temperature down to about 45°C. The highest temperature in that collector, I would guess, would be up to something like 55 or 60°C, but we do not use that.

Question:

In that solar dryer, is there any thermostat or any device to control the temperature inside the dryer?

Assoc. Prof. R. Exell:

No, there is no control. It is a very simple system, intended to be built by the farmers and put up in the fields. Any kind of control would be manual control, e.g. opening and closing the openings or the spaces. Perhaps, if the temperature is to be cut down, some kind of covering should be placed over the collector, such as coconut leaves, to prevent the collector from getting too hot. It is not such a complicated piece of engineering.

Question:

May I ask Dr Sheridan about the pros and cons of the absorption system versus the Rankine system?

Dr. N. Sheridan:

The Rankine system or the Rankine refrigeration machine consists of a turbine driven by vapour generated in a boiler. The vapour is condensed when it comes from the turbine and is pumped back to the boiler as in the Rankine cycle. The system is normally used in steam power stations. If we want to operate this system at lower temperatures, we would prefer to use an organic material as the working fluid; and, hence, it is sometimes called an organic Rankine cycle engine. The turbine then drives the compressor of a conventional vapour compression refrigerator. It would be possible to use the same fluid in the Rankine cycle and the vapour compression cycle, and such a fluid might be Refrigerant 12. In that case, there is a combined condenser for the refrigeration cycle and in the Rankine cycle.

Rankine cycle machines, as far as I know, are not commercially available. The interest in them arises because engineers have noticed that when one operates an absorption machine at higher temperature, the coefficient of performance of the machine does not tend to increase. Now this tendency is contrary to the general experience with most thermodynamic equipment in which, as one raises the temperature of the heat supplied to the machine, one tends to increase the efficiency or the coefficient of performance, as happens with the ordinary Rankine Cycle.

Accordingly, if we are going to have more effective solar cooling, we should improve the coefficient of peformance of the machines. One way to achieve this improvememt is to raise the temperature by using higher efficiency collectors such as concentrating collectors or evacuated tube collectors. High temperature fluid, obtained from these collectors, is supplied to the Rankine refrigerator to produce cooling at a better coefficient of performance than current absorption machines. It is probably realistic to think that these concentrating devices will soon be available.

Coming on the market in Australia this year is a concentrating collector designed for industrial water heating, operating up around 100°C or for low temperature steam of about 125°C. And so I think that in the future we will see high temperature collectors becoming available at reasonable prices. Then they may become cost-effective for cooling systems. They certainly are not going to be useful for absorption machines of the current type which are not able to improve their coefficient performance as the temperature is raised.

Nevertheless, in my paper on this topic, I did indicate that regenerative heating may be used in the absorption refrigerator; that means, recovering some heat from one part of the cycle and returning it to another part. This is commonly done in steam plants. Most of the big steam plants have regenerative heating to improve their efficiency. If you apply this principle to the absorption machine, theoretically at least, the coefficient of performance may be enhanced by the rise of temperature of the supplied fluid. And so I think we will see in the future better absorption machines becoming available. One may have to go back to the Ammonia-water type machines instead of the lithium bromide-water type, because the lithium bromide-water machine, which is the common one used for air-cooling now, may suffer from crystallization problems as the temperature is raised. The ammonia-water system may be a disadvantage due to the danger of ammonia if it escapes from the system and already there are some countries in the world that do not allow the use of ammonia for air-cooling. There could be some relaxation for absorption machines because there are no open glands that would enable the ammonia to leak easily. So it may be possible to design absorption machines using ammonia that would be suitable for air-conditioning purposes. But I am no expert to say whether this would indeed be so. At the moment, certainly, I think, in Australia, ammonia is not allowed for air-conditioning purposes because of the danger in the air-conditioned space.

Question:

I would like to raise another area which is connected to this subject, that is cooking. At the moment in the Third World, the forest is depleting quite rapidly. There is a need to provide enough fuel, whether it is wood or cowdung, for cooking of meals.

Some attempts have been made in the past, of course, to introduce alternatives. I understand there was a suggestion to produce something like a million umbrellas with silver linings. However, they are not really practical, simply because every one in the tropics would prefer to cook in the shade rather than to do so in the sun and get broiled with the food! Is there another way around it? What about creating a bio-mass system to generate fuel so that there is no need for wood?

Dr. N. Sheridan:

Bio-gas would be able to solve the problem.

Comment:

But bio-gas is rather complicated.

Dr. N. Sheridan:

I do not think that is true. I am impressed with the bio-gas programme in India. I was given to understand that a company was going to sell 50,000 bio-gas plants in 1978. The plants are really very simple. The difficulty with the simple bio-gas plant is that it does require cowdung or other animal excreta to be introduced into the system, though some other materials may also be added in certain proportion. The advantage of the bio-gas plant is that the cowdung is not burnt, and it is still available for returning to the soil. Bio-gas is certainly a way to exploit bio-mass, if you have it available.

I share the concern at the deforestation that is occuring in the Third World countries, but this is surely a matter of educating people. If they cut down one tree, they should plant two in its place. Certainly every country in the world has been depleting its forest; I can remember the story of people coming to America, and finding all the virgin trees. The trees were gone in Europe, but they were there in the new land. But in the space of a few centuries, almost all the extensive deciduous forest in the United States has disappeared. They cut them down to grow corn and so on. This is what men do. They simply cut their trees down but do not think of replacing them, because they think they can use the land for other purposes.

I share your concern about the use of wood as a fuel, and I agree that other methods ought to be found.

Experimental Solar Dryer at AIT - Front view showing solar collectors, drying cabinets and chimneys

Experimental Solar Dryer at AIT - Back view showing drying cabinets and chimneys

Experimental Solar Refrigeration Unit at AIT - Front view showing solar collectors and mirrors

Experimental Solar Refrigeration Unit at AIT - Back view showing storage tank

WORKSHOP 4: SOLAR HEATING AND COOLING SYSTEMS

SOLAR HEATING SYSTEM ESTIMATING

Don Hampton,
Heating, Ventilating and Solar Specialist,
Department of Public Works, Canada;
Adviser on Mechanical and Electrical Systems in Buildings
for the Department of Building, Ngee Ann Technical College,
Singapore.

ACKNOWLEDGEMENTS

The information used in this report has been taken from the engineering reports by the staff and consultants concerned with the design and implementation of solar heating systems for the Department of Public Works, Canada. Particular reference is made to the work done by the Design and Construction Directorate and the Energy Conservation Group of the National Capital region. They have endeavoured to introduce solar heating projects with solar panels mounted close to the building profile to limit the cost of solar heating installations.

Because of my detachment from the Department of Public Works Canada, it has not been possible to submit this report to that department for policy review and, as such, this report cannot be directly connected to this department. The material was presented to that department earlier this year as part of a familiarisation seminar on solar heating applications in Ottawa, Canada.

SUMMARY

The Canadian Federal Government has embarked on a S$225,000,000.00 programme to help develop solar energy in Canada. The programme is being handled by the construction departments of the federal government who will be installing solar heating systems in federal buildings.

To help introduce the technique of solar heating to the implementing groups in the regions of Canada, a study was made of the trends in performance and cost of various solar heating systems.

The writer has formulated the theory that as the performance of solar heating systems for domestic hot water will not be dramatically affected by mounting angles, costly mounting frames for solar panels should be avoided for practical solar heating systems, particularly for domestic hot water systems. The panels should be mounted on existing building frames to reduce the overall cost of these systems. The result

Bill B. P. Lim (ed.), Solar Energy Applications in the Tropics, 321–346.

may be an increase in the area of solar panels installed, but the total cost will be reduced.

A method has been devised of assessing the performance of the systems with panels mounted at various tilts and azimuths. In addition, unit prices have been shown that reflect the cost of various methods of mounting solar panels.

The method created is similar to the estimating procedures currently being used for mechanical services in buildings, and is to be used in the preliminary stages of the planning to reduce the time spent on solar heating projects that are impossible to justify.

1. INTRODUCTION

This paper is presented to the Regional Seminar and Workshop on Solar Energy Applications in the Tropics with some reservations. Obviously the geographic location of Canada precludes it from the general forum. However, it is presented for consideration at the workshop sessions because it provides a method of estimating the cost and performance of an important part of the solar energy field in a manner suitable for mechanical engineers and designers familiar with the mechanical services in buildings who are faced with the implementation of solar heating systems.

The presentation is made possible by the courtesy of the Organising Committee and UNESCO. The writer is currently in Singapore under a UNESCO/UNDP project.

2. CANADIAN GOVERNMENT PROGRAMMES

Support for solar energy in the early 1970's came from demonstration and research programmes that were jointly initiated by the National Research Council, a federally funded group, and the provincial governments, particularly Ontario, the most industrialised province.

The federal government was approached by many groups, including the solar panel manufacturers, to create a more realistic programme to support the development of a solar energy industry in Canada.

In 1978, the Minister for Energy, Mines and Resources, in a joint release with the Minister of Public Works, announced a series of programmes for renewable energy, and one of these programs was the Purchase and Use of Solar Heating (PUSH) programme(1).

The organisation of this programme was undertaken by Public Works Canada (PWC). This is the largest construction group within the Federal government, and it is to work with the other construction groups, the Department of National Defence (DND), the Ministry of Transport (MOT) and the Department of Indian and Northern Affairs, (DINA) with each group working as an autonomous group and a section within PWC, the Solar Programme Office, responsible for coordination and information transfer.

3. ROLE OF THE CONSTRUCTION DEPARTMENTS

The construction departments are managed by professionals who use the consulting engineers and architects in the private sector to develop the designs.

As this field is new, there has been a tendency to design systems from first principles and to complete the designs before an estimate of the cost is made. This has often resulted in the design of a system that cost more for the design than the energy savings over twenty years.

Guidelines were needed to make preliminary estimates of the cost and the energy savings for the system. The experience that the professionals had in other work associated with building changes were not directly applicable to this field. However, their knowledge of building processes has helped in overcoming this.

4. PUSH

PUSH (Purchase and Use of Solar Heating) is a S$225,000,000.00 industrial development programme that will support the use of solar heating systems for federal buildings. It is to provide a temporary market for the manufacturers of solar panels until the industry and its technology are established and able to provide alternate energy systems that are competitive with the conventional systems. It is hoped that at the end of the program, in 1984 approximately, solar heating systems will be self-sufficient.

The immediate need has been to introduce the regions across Canada to solar heating. At the time that the programme was announced only one solar heating installation had been built (see also (2)).

The initial phases of the programme have been planned with the aid of those consultants who were involved in the early demonstration programmes - they were the only groups with local experience.

5. ROLE OF THE HEADQUARTER'S SPECIALIST

As the specialist in solar energy, I endeavoured to present the principles associated with solar energy to the regional offices of PWC and other departments, in a form that would complement their expertise.

Various rules of thumb have been established for all the mechanical services in a building, i.e. air-conditioning is valued at an overall cost per unit area, plumbing fixtures are estimated in terms of cost per fixture, etc. Then the question arises - can a unit cost be fixed for solar heating systems?

Obviously, the value of solar heating cannot be fixed in terms of the occupied area for all buildings, but some trends could be established for a solar heating system for a domestic hot water service.

The immediate prospects for solar heating are brightest in existing buildings as the planning time required for a solar heating system is short compared with the overall commercial building; also, there is more scope for solar heating with domestic hot water systems, as the low grade heat can be used as a preheat system - it is usually difficult to use low grade heat with building heating systems.

The estimating system evolved is the result of a survey made of the engineering studies by various regions concentrating on the use of solar heating of domestic hot water in existing buildings.

6. DESIGN APPROACH

Solar heating systems are not a "natural" selection. The capital costs are too high and the energy collected is not certain; a back-up system is always needed and the solar heating system's value is calculated from the energy saved (see also Appendix D). This energy has a smaller value in Canada than most of the industrialised world. It has been argued that the key to Canada's position as an industrial nation is its abundant and low cost fuel. Energy costs for homes is also low and this low energy policy makes the justification of solar energy systems very difficult.

The initial optimism that solar heating systems would become more viable with the introduction of mass production for solar panels must be questioned when the cost of the solar panels is approximately 30% of the total cost and that of the raw materials for solar panels is at least 40% of the selling price of the panels. This means that the cost of assembling the panels represents about 10% of the total project cost.

The next area that is often considered is the design of the solar system. Many groups have considered and developed alternate system but the changes made often seriously affect the durability of the system.

This leaves the question of improving the performance of the panel and the costs associated with the mounting of the panels. With the former, the Canadian Government through the Solar Programmes Office is working with the manufacturers, to develop improved panels under PASEM, Programme of Assistance to Solar Equipment Manufacturers. It is in the latter area, that associated with the cost of mounting panels, that some immediate progress has been made.

The mounting angle for solar panels would seem to be the most important criterion at first glance. However, to present the panel at the "correct" angle often means that building changes have to be made. (The correct angle has been determined as equal to the latitude for domestic hot water solar heating systems and as equal to the latitude plus fifteen degrees for space heating systems).

A check on the cost of the Confederation Height's solar system(2) reveals that the cost of the building changes were at least three times the cost of the panels, and this is typical of most demonstration systems.

The studies made for PWC have shown that the energy collected by solar domestic hot water systems would not be drastically reduced if the panels were:

(a) mounted almost flat on the roof - the panels may be covered by snow for some months, but this would have only a minor effect;

(b) mounted flat against the side of a building - the heat collected in the winter months would be almost equal to the heat collected during the summer months, and

---

Note: An important corollary was noted in connection with solar heating systems: the high capital cost and the low value of the heat collected made it imperative that all the heat collected be used if the system had any chance of being cost effective. This precluded the idea of formulating guidelines for space heating projects as the heat collected in summer would be wasted. However, the guidelines created on the mounting of solar panels applies equally to space heating systems, i.e. no solar heating system can afford heavy building costs.

(c) mounted at angles of up to 70° and 15° - panels will still produce approximately 85% of the heat available from the panels mounted at the ideal angle.

7. ALTERNATIVE SYSTEMS

The system used as the benchmark for this thesis is shown in Fig 1. Alternative systems can be considered. However, they must meet the criteria of durability. Reduced performance of alternative designs can be balanced against the reduced costs of the systems.

The cost reductions available through changes in the designs of the systems are marginal, whereas the cost of the building changes etc. will determine the viability of the project.

8. SYSTEM SELECTION

The system selected is the glycol filled heat exchange type with the forced circulation on both liquid paths of the heat exchanger.

The use of the heat exchanger type has been made as it provides the most proven method of freeze protection for air-conditioning and mechanical systems. The alternative of drain-down has been dismissed as uncertain, and the coupling of the technique of drain-down with glycol filled to allow for system malfunctions has been disallowed as the latter will mean that the glycol will be in contact with air. The sealed industrial heat exchange systems have a life expectancy of approximately six times the life of systems that are open to the air.

Propylene glycol has been selected as the heat exchange medium. It is less toxic than ethylene glycol and the products that are produced by this glycol when it is breaking down are not toxic. This also eliminates the need for a double-walled heat exchanger.

9. ESTIMATING THE PERFORMANCE

The tool used to predict the performance is the f-chart(3). This programme has been modified to include weather data for approximately ninety Canadian cities and has been made accessible to the private sector through PWC mini-computers.

The accuracy of this programme has been questioned. However, this is probably academic, as most systems fail to meet the predicted energy levels because of panel deterioration, heat losses and

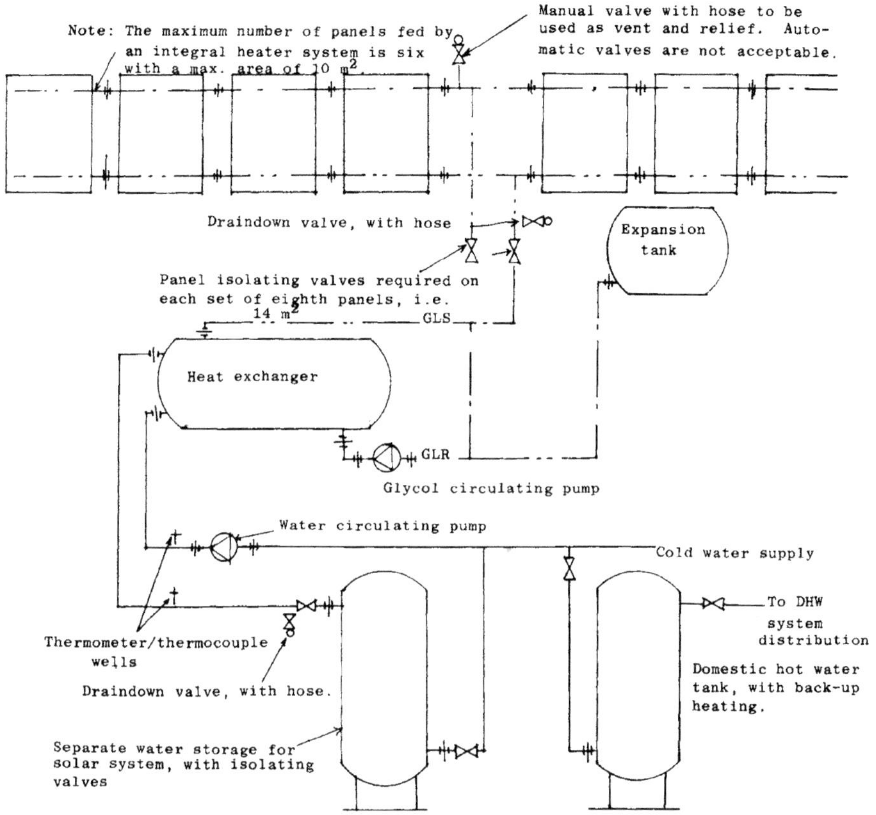

The system is to consist of a double pump system using propylene glycol as an antifreeze solution in the solar panel loop. The water and solar fluid are to be pumped through the heat exchanger. Piping in all loops is to be type L and all hot lines are to be insulated to a value of R_{SI}= 3 m. °C/W and the insulation is to be protected with aluminium sheeting: 0.9 mm thick.

All systems dealing with potable water must comply with the requirements of the local, provincial and federal plumbing codes.

Fig. 1 - Solar Heating System Estimating: System Schematic - Mininum standard mechanical deficiencies, and there is always some discrepancy between the model year and the actual.

A study made of the performance in the major cities in Canada shows that the total energy collected is roughly the same; the complete details for the development of the performance figures shown in Table 1 is given (see also Appendix A).

The system used in these calculations included a flat-plate single glazed solar panel with a selective coating. The factor F_R'. Tau.Alpha (see Ref 3 and 4) has been reduced by 10% to allow for deterioration of the panels during the service life.

It is PWC policy that all domestic hot water systems in federal buildings be set at 40°C as an energy conservation measure. This is part of the input to the f-chart (see Appendix A).

Appendix A also shows the derating of the f-chart results to account for snow build-up and for dirty covers on low angled solar panels.

The runs have been made to allow for collectors mounted at azimuth angles up to 45°.

The performance figures shown in Table 1 are based on the provision of 60% solar assistance and on the 40°C service water requirement. The 60% assistance has been selected to ensure that all the heat collected will be utilised, i.e. the solar fraction for each month is kept below 90%. If these values are reduced, the performance of the solar system will improve.

10. PAYBACK

The graph in Appendix E shows that the value of the energy collected will not repay the capital cost under current conditions in Canada within the lifetime of the system. Should the cost of fuel rose to international levels and if improvements could be made in the performance of the system, then a different conclusion could be drawn.

11. COSTS

The costs of the system have been divided into two areas:

(a) those directly associated with the solar panels; these cover the panels, the cost of assembling them in place, and the

Table I - System Performance

In gigajoules per square metre per year $GJ/(m^2.a)$

Collector Slope		0° to 30°				31° to 55°		
Azimuth Angle	0°	15°	30°	45°	0°	15°	0°	45°
January	0	0	0	0	0	0	0	0
February	0	0	0	0	0.03	0.03	0.03	0.02
March	0.15	0.15	0.15	0.14	0.18	0.18	0.18	0.17
April	0.17	0.17	0.17	0.17	0.18	0.18	0.18	0.18
May	0.21	0.21	0.21	0.21	0.21	0.21	0.21	0.21
June	0.23	0.23	0.23	0.23	0.22	0.22	0.22	0.22
July	0.24	0.24	0.24	0.24	0.23	0.23	0.23	0.23
August	0.22	0.22	0.22	0.22	0.22	0.22	0.22	0.22
September	0.17	0.16	0.15	0.13	0.19	0.19	0.19	0.19
October	0.11	0.10	0.09	0.08	0.15	0.15	0.15	0.14
November	0.04	0.03	0.03	0.02	0.07	0.07	0.06	0.06
December	0	0	0	0	0.03	0.03	0.02	0.02
Annual	1.54	1.51	1.49	1.46	1.71	1.71	1.69	1.68

Collector Angle		56° to 75°				76° to 90°		
Azimuth Angle	0°	15°	30°	45°	0°	15°	30°	45°
January	0.12	0.11	0.11	0.10	0.15	0.14	0.13	0.10
February	0.15	0.14	0.13	0.12	0.17	0.16	0.15	0.13
March	0.20	0.20	0.19	0.19	0.18	0.18	0.18	0.17
April	0.17	0.17	0.17	0.17	0.14	0.14	0.14	0.14
May	0.18	0.18	0.18	0.18	0.14	0.14	0.14	0.14
June	0.18	0.18	0.18	0.18	0.14	0.14	0.14	0.15
July	0.19	0.19	0.19	0.19	0.15	0.15	0.15	0.16
August	0.22	0.22	0.22	0.22	0.16	0.16	0.17	0.18
September	0.21	0.21	0.21	0.21	0.17	0.17	0.17	0.17
October	0.17	0.17	0.17	0.17	0.16	0.15	0.14	0.13
November	0.09	0.09	0.08	0.08	0.08	0.07	0.07	0.06
December	0.10	0.09	0.09	0.09	0.11	0.09	0.08	0.06
Annual	1.98	1.95	1.92	1.90	1.75	1.69	1.66	1.59

Note: The values shown use Ottawa International Airport weather. Monthly values for other major centres in Canada will vary slightly from this but have almost the same totals per year.

structural costs of changing the building to accommodate the panels; and

(b) those associated with the connections of the panels and the back-up heating system, etc. These have been called the "plumbing" and have been separated as the size will depend on the amount of energy collected and not the panel area.

This information on the costs is shown in Table II.

Table II - Costs

1. AREA DEPENDENT COSTS

(a) Solar Panels

This is only value that is readily available to the designer and it is anticipated that local values will be developed.

Collector Types	Price	Comments
Double glazed collectors, metal type with selective surfaces	$\$280/m^2$	Generally used for space heating applications
Single glazed collectors, same construction as above	$\$210/m^2$	Standard, i.e. recommended for domestic hot water solar heating systems
Plastic collectors	$\$100/m^2$	Generally used for outdoor swimming pool heating systems

Note: The values shown here are Canadian dollars. It is felt that these values cannot be realistically converted, anymore than true poetry can be translated; the values apply to Canada only. This does not eliminate the possibility of developing values for other countries based on these parameters.

(b) Erection Costs

This covers the cost of hoisting the panels to the support structure and coupling the collectors together. Extras would be applicable to collectors without headers and if couplings were not included with the collectors.

Basic price	$\$75/m^2$
Add for installation above 30 m	$\$10/m^2$
Add for provision of headers at site	$\$25/m^2$
Add for site supplied couplings	$\$10/m^2$
Deduct for ground level mounting	$\$10/m^2$

Multiply for areas above ground level without working platforms and/or guard-rails. 1.5 x

(c) Structural Addition to Buildings

This includes the modifications to the building to support the panels and to withstand the wind forces, flutter etc. and the space frame supports for the panels.

Mounting the panels on building components with all the forces supported by the existing building frame, allowance for simple clamps, angle frames, wooden sleepers for house roof, etc. $\$ 50/m^2$

Mounting panels on frames set on the ground or an adjacent beam supporting the weight of the panels with the frame clamped to the wall of the building. $\$100/m^2$

Provision of support structure on a flat roof to present panels at a high angle:

(a)	with new roof designed for duty	$\$ 75/m^2$
(b)	with existing roof, columns at less than 15 m spacing	$\$150/m^2$
(c)	with existing roof, columns spanned at more than 15 m	$\$300/m^2$

2. PERFORMANCE (POWER) DEPENDENT COSTS

Basic costs to allow for piping, heat exchanger, storage, controls, instrumentation, cut and patch are based on a seven day per week operation. $\$185/(GJ.a^{-1})$

Addition to allow for additional storage for a five daysper week operation	$\$40/(GJ.a^{-1})$
Addition for 30m to 50m runs from panels to heat exchanger/storage of existing heating system	$\$30/(GJ.a^{-1})$
OR	
Addition for over 50m runs from panels to heat exchangers, etc	$\$50/(GJ.a^{-1})$

12. DISCUSSION

The cost of solar heating systems is higher than conventional heating systems. This cost is often accentuated by the building modifications that are made to accommodate the solar panels.

Most of the expense for building modifications is made necessary by the high wind forces that result from mounting the panels at angles between 25° and 75° to the horizontal, and it has been shown that the improvement in performance within these angles is marginal.

The values shown here are for preliminary estimates only. The performance of the panels at angles well beyond the "correct" angle may be questionable, and the cost estimates for the building work are to be used as guides only.

Costs and, to a certain extent, peformance will be dependent on the local conditions, and the real cost and performance of solar heating systems will be proven by the local experience. However, the main points still persist at the moment:

(a) in Canada, particularly, solar heating systems cannot be cost effective, if any of the heat is allowed to be dumped, e.g. during the summer; and

(b) building changes are extremely costly and difficult to control and it is recommended that these be kept to a minimum.

The premise that the reduced performance is a minor concern may be a luxury that is permissible for low grade heating systems. With high grade heating systems and absorption systems for cooling, the angle of the collector is important. If this is so, then a lot more attention must be given to the wind problems associated with solar panels.

If panels are mounted on the building face etc. the result will be the use of more solar panels for each project and more projects. This will give more support to the Canadian Solar Panel Manufacturers, and they are the group that PUSH is supporting.

Appendix F shows the calculations that are used to determine wind forces. These show a conservative approach to the structural support problem, and more basic research is required to determine the real forces involved. Unfortunately, there is a limitless number of profiles, and the result will probably be the general acceptance of the approach shown.

REFERENCES

1. Minister of Energy, Mines and Resources (Canada), 1978 "To Support Strong Canadian Solar Effort", Development of Renewable Energy, Toronto Ont., Canadian Government Printers

2. Hampton, Don H. and Tom Lefeuvre, 1978. "Solar Heating of Domestic Hot Water at the Confederation Heights Cafeteria", Renewable Alternative (1978 Conference), London, Ont., Solar Energy Society of Canada, Inc.

3. Beckman, William A. et al., 1977. Solar Heating Design by the f-chart Method; John Wiley & Sons, Inc.

4. American Society of Heating Refrigeration and Air-conditioning Engineers 1977. ASHRAE Standard 91-1977, Methods of Testing to Determine the Thermal Performance of Solar Collectors, New York ASHRAE Publishing.

Appendix A

SOLAR HEATING SYSTEM ESTIMATING

f-chart input

Code	Variable Description	Value unit
1	Air System = 1 Liquid System = 2	2
2	Panel Area	Open (enough for 60% solar assist in this run)
3	F_R' Tua Alpha (Normal Incidence) (Taken from ASHRAE 93-75)	0.63 (Allows for deterioration of panels)
4	F_RU.L.	5.26
5	Number of Transparent Covers	1
6	Panel Slope	See Table 1
7	Azimuth Angle	See Table 1
8	Storage Capacity	350 $k^J/(°C.m^2)$ for systems using water seven days per week, and 500 $k^J/(°C.m.^2)$ for systems using water five days per week.
9	Effective Building, U.A.	0
10	Constant Building Heat Generation	0
11	Epsilon C Min/Effective Building U.A.	0
12	Hot Water Usage	Determine from site; see also Code 2
13	Water Set Temperature	40°C
14	Water Mains Temperature	5°C
15	City	Ottawa

16 Thermal Print out by Month = 1 Year = 2 1

17 Economic Analysis Yes = 1 No = 2 2

The results from these runs, the performance for each month, is then derated by the following:

(a) to allow for snow coverage, the energy collected for panels mounted below 30° is presumed to be zero during the months of December, January and February; between 31° and 55° the amount of heat collected is presumed to be zero for January and reduced to 50% during the months of December and February;

(b) due to dust accumulation on the panels, the performance for panels mounted at angles below 30° is reduced by 15%, and for panels mounted at angles between 31° and 55° there is a 7.5% reduction.

Appendix B

SOLAR HEATING SYSTEM ESTIMATING

Selection of Heat Exchangers in Solar Heating Systems

This is being developed as an Appendix for the paper on the estimating of Domestic Hot Water Systems.

1. GENERAL

This procedure has been developed to cover the selection of solar heat exchangers, as there are no established criteria for solar systems. The basic rules are no different for solar systems than for other systems. However, a more generous selection is required as the performance of these systems will be marred by the use of standard-sized heat exchangers.

For a solar system, the size and mass of the heat exchangers are not critical, except that the element should be located in a space that will allow cleaning of the unit.

To ensure that cleaning is possible, the heat exchanger is to be a removable header type. The performance of the heat exchanger is also diminished by scale build up. As the scale will form on the domestic water side of the heat exchangers, the domestic hot water must flow in the tubes with the inhibited solar fluid on the shell side.

2. SELECTION

The temperature difference for the collector should be designed for the general maximum rise of the design flow. The design flow recommendation is 15 $mL(s.m^2)$. The heat supplied is in the order of 1000 W/m^2, then:

Heat = Flow (L/3) x t°C x Thermal Capacity of Fluid
i.e. of Propylene glycol
solution 50% strength. (Cp)

i.e. Cp = 3560 J/(L. °C)

$$\text{i.e.} \quad \Delta t°C = \frac{\text{Heat (W)}}{\text{Flow (L/s)} \times 3560 \text{ J/(L.°C)}}$$

$$= \frac{1000W}{0.015\ L/s \times 3{,}560\ J/(L.^\circ C)} \quad \text{(Based on } 1\ m^2\text{)}$$

$= 18.73°C$ i.e. Theoretical maximum based on 100% solar heat utilization by the solar system

Presuming a 60% efficiency for the collector, the maximum temperature gain will be:

$$\Delta t^\circ C = 0.6 \times 18.73^\circ C$$

$$= 11.2^\circ C, \text{ say } 11^\circ C$$

The water side of the system has a smaller loop (generally) and the flow-rate can be increased to double the solar fluid flow-rate.

$$t^\circ C = \frac{0.6 \times 1000W}{0.03\ L/s \times 4190\ J/(L.^\circ C)}$$

$$= 4.77^\circ C, \text{ say } 5^\circ C$$

Assuming that the temperature of the water can be raised to the level of the outgoing solar fluid then*

$$t_i = 11^\circ C$$

$$t_o = 6^\circ C$$

$$LMTD = \frac{t_i - t_o}{\ln \frac{t_i}{t_o}}$$

$$= \frac{11^\circ C - 6^\circ C}{\ln \frac{11^\circ C}{6^\circ C}}$$

$$= 8.25^\circ C$$

---

* Heat Exchanger Effectiveness (m) $= \frac{11^\circ C}{16^\circ C} = 0.688$

Heat Transfer Coefficient for a Heat Exchanger, water to water type. (Taken from Holman, J.P. Heat Transfer, McGraw Hill Book Company New York, 1972) Table 10-2.

$$U = 150 Btu/(h.ft^2.°F) = \frac{225\ Btu}{h.ft^2.°F} \times \frac{5.678264\ W/(m^2.°C)}{1\ Btu/(h.ft^2.°F)}$$

$$\text{i.e.} \quad U = 1280\ W/(m^2 °C)$$

(Capacity of a heat exchanger = A.U.LMTD.F F varies with flow-rate use F = 0.6 (also from Holman)).

The area required for a small heat exchange system, suitable for a solar DHW system, using four panels with an area of $2m^2$ each, is

$$\text{Area} = \frac{600 W/m^2 \times 4 \text{ panels} \times 2m^2/\text{panel}}{1280\ W/(m^2.°C) \times 8.25°C \times 0.6}$$

$$= 0.76m^2$$

Using a 10" heat exchanger (four-pass type) with 5/8-in tubing, there are 62 tubes (see Chemical Engineers Handbook pp 11-13). This represents an area of $1.7 \times 10^{-3} m^2$ for each tube or $0.11m^2$ for the total, and with four passes the length of tubing required is 1.77 which may correspond to an Armstrong W 86-44 heat exchanger.

Appendix C

SOLAR HEATING SYSTEM ESTIMATING

Storage Requirement

The heat storage for solar systems is designed for short-term heat storage. Some systems have been developed in Ontario with yearly storage capabilities, but these must be considered as developmental only and should not be considered for PWC unless they are especially identified and approved, as seasonal storage systems are being installed for developmental purposes.

For domestic hot water solar heating systems the storage is capable of storing the heat for at least three days, which is particularly important for the commercial units that are being used in most Government buildings.

For solar heating systems providing heat to domestic hot water systems operating five days per week, the storage should be a minimum of 500 kJ/(°C.m^2) to ensure that a reasonable amount of heat is stored; this will apply to all systems regardless of the solar percentage assistance.

(a) The maximum average sunshine for any location is approximately 500W - the minimum average insolation allowed for the testing of solar collectors is 900 W - and this intensity could last all day.

(b) The temperature rise in the storage for the first day would be

$$\frac{500\ \text{W} \times 0.6\ \text{h/d} \times 3600\ \text{s/h}}{500\ \text{KJ/(°C.m}^2\text{)} \times 1000\ \text{J/k}^{\text{J}}}$$

0.6 allows for 60% efficiency of the solar heating system

= 30.24°C

(c) On the second day, presuming a warm weekend with no hot water usage over the weekend, - with the increased temperature the solar system will not perform as efficiently - presume 40% efficiency -

$$\frac{500\ \text{W} \times 0.4 \times 14\ \text{h/d} \times 3600\ \text{s/h}}{500\ \text{K}^{\text{J}}\text{/(°C.m}^2\text{)} \times 1000\ \text{J/k}^{\text{J}}}$$

= 20.16°C

(d) With the normal water intake of about 15°C and some heat pick up during the Friday, the temperature could rise to 80°C by the end of the weekend, which is warmer than is preferred. However, the size selected, i.e. 500 $kJ/(°C.m^2)$ is a reasonable compromise for a five days per week operation.

<u>Note</u>: It will still be necessary to have a safety release for excess heat and blending valve(s) on the supply to protect the users.

For domestic hot water systems that are generally used seven days per week the storage can be reduced to 350 $kJ/(°C.m^2)$.

Appendix D

SOLAR HEATING SYSTEM ESTIMATING

Examples

1. OPPORTUNITY

A major office block for the Department of Energy, Mines and Resources has a centralised domestic hot water service that serves the office area, the shower facilities for a fitness centre and water to the cafeteria. The water supply temperature has been reduced to 40°C throughout the building as an energy saving measure, with booster heating coils for the cafeteria dish-washer.

The building is in a park-like setting, with a river to the south. The building is ten storeys high.

The Minister for Energy, Mines and Resources is anxious to show the use of solar energy as a demonstration of the use of renewable energy.

2. ALTERNATIVE PROPOSED

(a) Mount the panels on the roof, run the lines in the service shaft of the building to the basement and the existing hot water service.

(b) Mount the panels between the first and second floor windows and reduce the cost of the system.

3. ANALYSIS

Aerial photographs of the building with roof-top panels will be dramatic. They will be attractive for magazine illustrations, a symbol for future energy programme promotions.

The view of the panels by building visitors will show the solar system to all, if the panels are mounted on the side of the building.

Obviously, there are two ways of viewing this and a cost analysis will be required.

4. UNIT COSTS

(a) Area Dependent Costs

Roof Scheme

Single glazed panels	$210/m^2$
Structural work for roof with 20m columns	$300/m^2$
(Extra cost of structural analysis)	$ 30/m^2$
Collector erection, allowing for high hoist	$ 85/m^2$
Total area dependent costs	$625/m2

Wall Scheme

Single glazed panels	$210/m^2$
Structural work for wall supporting	$100/m^2$
Standard erection schedule	$ 75/m^2$
	$385/m2

(b) Performance (Power) Dependent Costs

Roof Scheme

Plumbing connections allowing for long runs and a five-day operation per week $255/($GJ.a^{-1}$)

Wall Scheme

Basic installation cost with allowance for a five-day operation per week $225/($GJ.a^{-1}$)

5. PERFORMANCE OF PANELS

Roof Scheme

1.98 GJ/(m^2.a)

Wall Scheme

1.74 GJ/(m^2.a)

i.e. increased area required for the wall scheme for the same performance = 1.98/1.74
= 1.14 i.e. 14% additional panel area is required for the wall scheme

6. EQUIVALENT COST

Roof Scheme		Wall Scheme	
Area dependent cost	$= \dfrac{\$625/m^2}{1.98\ GJ/(m^2.a^{-1})}$	Area dependent cost	$= \dfrac{\$385/m^2}{1.74\ GJ/(m^2.a^{-1})}$
	$= \$316/(GJ.a^{-1})$		$= \$221/(GJ.a^{-1})$
plus performance costs	$= \$225/(GJ.a^{-1})$	plus performance costs	$= \$225/(GJ.a^{-1})$
Total unit cost	$= \$571/(GJ.a^{-1})$	Total unit cost	$= \$446/(GJ.a^{-1})$

7. PERFORMANCE

The system performance required = $Q_w.Da.(t_r-t_s) \times 0.6 \times (4.187 \times 10^{-6})$ GJ/(L.°C),

where Q_w = water requirements per day in litres
D_a = number of days per year the system will operate
t_r = temperature of water supplied to the solar energy system (from the mains)
t_s = temperature of water supplied to the domestic hot water system from the solar energy system

and the factor 0.6 allows for a solar fraction of 60%.

For the building in question, the following are used:

Q_w = 6000 L/d
D_a = 250 d
t_r = 5°C
t_s = 40°C

The system performance required for this building is

6 000 L/d x 250 d/a x (40°C - 5°C) x 0.6 x (4.187 x 10-6) GJ/(L.°C)

= 219.8 GJ/a

8. COST OF SCHEMES

Roof Scheme

$\$571/(GJ.a^{-1}) \times 219.8$ GJ/a
= \$125,500.00 (approx.)

Wall Scheme

$446/(GJ.a^{-1}) \times 219.8$ GJ/a
= \$98,000.00 (approx.)

A clear saving is made if the wall scheme is adopted.

9. PANEL AREA

Roof Scheme

$$= \frac{219.8 \text{ GJ/a}}{1.98 \text{ GJ/(m}^2\text{.a)}}$$

$= 111\ m^2$ (\$1,130/m^2)

Wall Scheme

$$= \frac{219.8 \text{ GJ/a}}{1.74 \text{ GJ/(m}^2\text{.a)}}$$

$= 126\ m^2$ (\$780/m^2)

Appendix E

SOLAR HEATING SYSTEM ESTIMATING

Payback

The attached graph shows the simplest economic review of solar systems. The curves show the years to payback for solar heating systems with three models shown.

Current "Demonstration" Installations

Installation cost	:	\$900/m^2
Performance	:	2 GJ/(m^2.a)
Description	:	Mounted on a flat roof, roof to be reinforced and space frames supplied with the panels; costs also increased by feasibility study.

Current "Simplified" Installation

Installation cost	:	\$600/m^2
Performance	:	1.75 GJ/(m^2.a)
Description	:	Mounted against the south side of a building with a series of clamps, no structural reinforcement required, minimized pipe runs, standard consultant's fees.

"Future" Installation

Installation cost	:	\$750/m^2
Performance	:	4GJ/(m^2.a)
Description	:	Using concentrating collectors and based on a simplified installation practice.

Also shown on the graph is the current value of energy in Canada with values from West Germany and Singapore.

Appendix F

SOLAR HEATING SYSTEM ESTIMATING

Wind Forces

The force from the wind on a solar panel is an unknown quantity and the structural designer has no alternative but to return to basic theory. The force due to the wind is derived from the equation of static pressure:

$P = 1/2. \; .v^2$, where v = the velocity of the wind

The actual force will depend on the profile of the building and the solar panels, and until now this has not been studied in any systematic manner. The only alternative available is a wind tunnel test of the actual shape of the building under the most adverse conditions.

In addition, the designer is dealing with a wind gust and this represents a shock load. To allow for this, the calculated force, from the above equation, is quadrupled.

Then the force on a single panel (area = 2 m^2) in a wind gust of 160 km/h:

$P = 1/2. \; .v^2 \qquad = 1.14 \text{ kg/m}^3$

$= 1/2. \; 1.14 \frac{kg}{m^3}. \; (45 \frac{m}{s})^2 \qquad 160 \text{ km/h} = 45 \text{ m/s}$

$= 1\,154 \frac{kg.m}{s^2} \; . \; 1/m^2 \qquad$ Note: $1 \frac{kg.m}{s^2} = 1$ Pa

$= 1\,154$ Pa $\qquad 1 \text{ N/m}^2 = 1$ Pa

The force on a panel two square metres is

$F = 1\,154 \text{ Pa} \; . \; 2 \text{ m}^2 \; . \; 4$ (allowing for the shock load effect)

$= 9\,232$ N (i.e. 2075 lb)

If the array consisted of fifty panels the force would be

$F_{50} = 50 \times 9\,232$ N

$= 461600$ kN (104 000 lb)

DISCUSSION

Chairman : Assoc. Prof. T. Y. Bong

Panel Members: Mr. D. Hampton
Dr. Chiam Joon Tong

Question:

Is electricity generated from mineral sources like coal or oil in Canada?

Mr. D. Hampton:

The simple answer to that is "yes", but these sources are unimportant for the costs shown in my paper. These are generated by Canada's extensive system of nuclear and hydro-stations. The country also supplies a great deal of electricity to the United States. It has been government policy to create cheap power for industrial development to help Canada compete, particularly with its neighbour. The United States has also developed a similar policy. Two myths add fuel to the policy: the first being that a high standard of living is coupled with high energy consumption and that electric companies must expand their sales to prosper. The economics of solar heating systems is distorted by the low electrical charge, but it should be realised that the industrial rate of 1.2 two cents per kilowatt hour is coupled with a demand factor. As the systems described here require a full back-up system, it must be assumed in the economic analysis that there will always be a requirement for full power at some time. If there is a reduction in demand, this can only be considered a bonus that has an improbability factor.

Question:

Concerning the connection of the collectors, the system that is normally used by manufacturers is direct coupling of panels in series. What is the advantage if the manufacturer proposes a system by which the panels are diagonally connected. This connection will increase the plumbing cost. Will there be any cost saving in the long run?

Assoc. Prof. T. Y. Bong:

No, it is a requirement. In our connection i.e. for solar cooling where we want to increase the temperature by something like 5°C or 10°C, this can be done by parallel connection. If it is necessary to increase the temperature very much more, say from about freezing point to about 20°C, then the collectors are connected in series. Otherwise, the temperature rise will not be enough. So it all depend on the application.

Mr. D. Hampton:

I think you were contrasting the individual connections to solar panels to the "through" header connection that I showed in the slides. The through header is different to the standard practice for heat transfer element connections; by "standard" I mean for general heating and air conditioning systems. The theory that allows this is that because there are thermal cycling effects, there will be self-balancing. If there is a hot spot on a collector, this will transfer more heat to the fluid, which will create a lower density and the extra flow will balance the system. The normal HVAC system needs a more positive balancing system.

Question:

I noticed that the system used to inter-connect the panels is some sort of a hose, whereas other manufacturers have standard copper coupling which perhaps may be more lasting in the long run.

Mr. D. Hampton:

Yes, this may be true for systems that are not subjected to high temperature ranges. In Canada, the coupling must accommodate the coldest temperature and the temperature reached under full sunlight. Under these conditions a mechanical joint may be optimistic. The advantage of a hose is that its embrittlement can be seen on inspection while the metallic joint may fail despite inspection.

Question:

Dr. Chiam showed a cold water system where water from the main tank is pumped directly through the panels. It goes down to a main storage tank, and is directly linked to an auxilliary boiler system which is tapped for consumption from the secondary tank. Would it be an improvement in the system if there was another linkage back from the main storage tank to the solar collectors to re-circulate the water for reheating, if necessary, especially when the radiator is at its minimum?

Dr. J. T. Chiam:

First of all, there is re-circulation from the main tank system to the solar collectors in my diagram. Are you suggesting that, when the solar radiation is low, the line running from the large tank to the secondary tank should be linked to the solar panels again?

Let us say, for example, the solar radiation is low, and the temperature on the solar panels is 30°C. If the water temperature in the main tank is also 30°C, this temperature difference will not activate the differential thermostat. Hence, the re-circulation pump, in turn, would stop operating. So, whether there was a pump on the line I have indicated in my diagram, or whether the pump was in the line connecting the main tank to the secondary tank would not matter, because the pump would not be activated. It would not serve any purpose. The reason why we have this line going from the main tank to the secondary tank is purely to ensure that there is no interference from the auxiliary heat in the main tank. To my mind, I see no use in connecting a second pump on this line to the collectors because if there was little radiation, no matter how much water was pumped through the panels, there would be no temperature gain. So it is better just to have one re-circulation pump.

Question:

There is a central boiler system in a hotel as shown in your diagram. Would it be better to connect the panels directly to the boiler tank instead of using a separate main tank?

Dr. J. T. Chiam:

Now, supposing there was only one big tank as you have suggested and the ambient temperature was low in the early morning, the solar heating system would not be activated. In other words, there would not be a sufficiently large temperature difference to activate the re-circulation pump. This being the case, the controls in the boiler system, sensing the low temperature, say, below 40°C, would automatically activate the conventional heating system. So, if the boiler was running, say, at about 6 o'clock, and assuming by 8 o'clock the temperature in the tank connected to the boiler reached 60°C, then even though the sun was shining by then, the solar panels would not be operating at the maximum efficiency.

Question:

If one proposed a central hot water system for condominium units in Singapore, what would be the ideal monitoring system for billing the residents?

Dr. J. T. Chiam:

This is a problem that comes up again and again when we talk to developers. And up to this point in time, no real solution has been found. Now, the problem here is that we only monitor the consumption of cold water at present. If there was a central solar heating system in a condomimium situation, the solar panels would be installed on the common roof with a common tank, and the hot water system would be connected directly to all in the flats below. In order to monitor the amount of water used, one simple solution, of course, would be to install a hot water meter in each flat. But who would read the hot water meter? The management corporation feels that it is not its responsibility. The Public Utilities Board does not want to do it. If the householder was allowed to do it, he might understate the reading. In fact, this is one of the stumbling blocks as far as developers and management corporations are concerned. Naturally, they would like to minimise their work. By installing the additional hot water meter, there would be extra work which no one would be prepared to do. My experience in Israel is that the residents apportion the total hot water consumed amongst themselves depending upon the floor area of their flats. There is a common water meter, and if all the flats have equal area, the cost is divided by the number of flats. But, for some reason or other, developers here feel that sort of system would not be applicable in Singapore, because the moment you told the householder that he would get the hot water where there was no monitoring system for each unit, there would be indiscriminate use of hot water. So it appears that, first, one needs to instill a sense of civic responsibility in Singapore.

Question:

Would it be feasible to incorporate an instant heater with a solar system so as to minimise the back-up system? It could be switched on if needed.

Dr. J. T. Chiam:

Yes, I think this is a very good point. In fact, in my own home, I have a system similar to what you have suggested. Now, as you know, in the normal units for homes, the solar tank is equipped with an auxiliary electric heater. This adds to the cost of the system. However, taking the case suggested, on a rainy day when there was no input of solar radiation to the solar hot water system, the cold water would merely run into an instant heater. Now, this idea is good because in most homes there is an instant or electric heater, and the normal auxiliary heating element is not necessary. I installed my own solar hot water heater earlier this year and, to my surprise, after operating it for about 10 months, I found that I hardly needed to operate the electric heater because the climatic conditions here are such that there is at least one or two hours of sunshine each day. In fact, I have recorded this, and over the last 10 months, have used the electric heater not more than 4 times. So actually, in practice, it is still possible to install a solar hot water heater without the heating element. Being without hot water four or five times a year would not hurt anyone, I think.

Assoc. Prof. T. Y. Bong:

Whether or not we actually need auxillary heating depends on the size and type of collectors to be installed, and the amount of hot water needed. To give you an example, at one time we did a study of the hot water requirement for a student hostel. You know that after sports events they want hot water. Now, if we assume that at any time of the night if anybody wants to turn on the tap, he would be able to get hot water at 40°C. According to our calculations, it was done by the student also, that it would still require about 60% of the heat. Early in the morning the collector will not be able to pick up any heat while the radiation is still low, but the water in the tank is already 40°C. Now, if there was an instant heater, then one would need to heat as much water as one would require. If a few students needed it at midnight, only one small heater would have to be on. That could definitely be a big saving. However, there is one shortcoming with this instant heater. It would require some pressure at the inlet to operate. If it did not have enough pressure, say from a tank, then there might not be enough flow. But one might get it directly from the line.

Comment:

From what we have seen, the performance of collectors is affected by the cleanliness of the surface outside and inside, especially in evacuated tubes. But from what I can see, whether it is experimental or commercially available, none of the collectors has any device incorporated for cleaning the tubes.

Please explain.

Dr. J. T. Chiam:

In Singapore, we are in a very fortunate position because the water here is extremely soft, i.e. the amount of calcium and magnesium carbonates is negligible. The reason for this is that we have convectional rain, and the water in our reservoirs does not come through long rivers where there are calcium deposits. Water in our reservoirs is, therefore, basically very soft, and I have seen collectors operating for several years here which are still perfectly clean. I cannot speak for other countries, but certainly in my experience in the last 4 to 5 years in Singapore we have not encountered problems of scaling, blockages in tubes, etc.

Assoc. Prof. T. Y. Bong:

We may have filtering problems with water in Singapore. The installation I showed you just now was commissioned about a month ago, and when I looked at my flow meter just now, it was almost impossible to see the floater. It was red. I think the best thing is to check what type of water we use. We may occasionally have sediment in the water.

Mr. D. Hampton:

I raised this problem with the National Research Council of Canada. I felt that it was essential to have filtering in solar heating systems. I was directed to some of the basic texts used by various authorities, in the U.S.A., on flat-plate collector development. The mathematical model they use makes no allowance for scale on the inside surface. I remain sceptical. Obviously some filtering is required or the system passages will become clogged.

Dr. J. T. Chiam:

Most of you are aware that the temperature at which we operate the solar collector is relatively low. Generally, the operating temperature range varies from 30°C right through to about 60°C in Singapore. At such low temperatures, scale is not readily formed. Generally, formation of phosphate and calcium scale occurs near the boiling point of the liquid. In boilers we have this sort of situation, and, therefore, we need to chemically treat the water chemically before feeding it into the boiler. In the case of solar hot water systems in Singapore, we are fortunate because the operating temperature is well below that of scale formation, and the scale formation substances are negligible. In the case Dr Bong mentioned of sediment, we also found that in old piping, for example GI pipes, which contain rust and corrosion, the problem can be easily solved if there is filter just before the pump.

Question:

Do you experience biological growth, furring and scale formation in the pipes of the solar collectors? If so, is heat transfer adversely affected?

Dr. J. T. Chiam:

In the case of solar heaters, the flow of the water is fast, and there is no time for any biological growth to take place, in my experience. With regard to furring and scale formation, in practice, we have not come across these in Singapore. In the absence of such phenomena, we have no serious problems in heat transfer. Scale formation is possible in boilers or other high temperature systems, but not likely in the case of solar collectors in the Singapore situation.

Assoc. Prof. T. Y. Bong:

I would agree with Dr. Chiam in this matter. With regard to biological growth, if you are not going to drink the water, perhaps calcium sulphate will do to stop the growth.

Mr. D. Hampton:

For domestic hot water systems, we are merely extending the basic supply plumbing system. We can generally assume that there is no growth in these systems; this is a portable water system. The standard solar heating system does not allow the water to come into contact with the air - it is a closed system. In the same way as there is no problem with conventional electric heaters, there will be no problems with solar hot water systems.

CLOSING WORKSHOP

THE FUTURE OF SOLAR TECHNOLOGY AND ITS IMPLICATIONS

Chairman : Prof. B. P. Lim

Panel Members: All Speakers

Chairman:

I would like to direct our attention to four areas: Architecture, Engineering, Finance and Policy. I would suggest that we think in terms of the role of an architect in solar technology. Is there going to be a new direction in architectural design? Obviously, there will be a new dimension in architectural design if solar energy is going to be a vital part in the operation of the building. What kind of problems will architects see in designing solar buildings? What kind of integration will take place between architecture and technology? Will there be new forms of architecture as a result of the solar input, and what will be the future of passive and active systems in building, particularly in urban areas? I would like to invite comments from the panel as well as from the floor.

Prof. J. Cook:

Well, this is certainly a subject about which I could spend some time talking. I think you have already implied that we should expect a new form of architecture which would integrate these new concerns. But not only do I look forward to a new kind of architecture, but also to a new description of an architect, if there is a suitable term for him. In my estimation, the kinds of cities that we have built and the kinds of buildings that we have promoted in the recent past are all that we can expect from architects who have received the education that has been charasteristic up until relatively recently. If we want to assume new professional responsibilities, we are going to have new professional orientations. We have to have new technical skills. Now it may be a vanity (Prof. Lim and I are architects in our professions) that architects are going to do this, and I would like to challenge you. It may be that there will be new responsibilities of integration of the new energies in a new kind of urban context. It is a responsibility that could be taken over by other professionals. We are seeing in the United States, for instance, that architects have not necessarily been the leaders in terms of solar applications to buildings, nor to energy conservation programmes which have been strongly affecting recent activities in the US. I think in

Bill B. P. Lim (ed.), Solar Energy Applications in the Tropics, 355–374.

South East Asia there is a similar kind of challenge. These responsibilities may also not be taken by engineers. So it really is an open question.

Comment:

For the past three days, we have seen a review of the experience in the industrialised world, in particular, America, Japan and Australia. So I would like to attempt to analyse how it can be applied to Singapore in particular. Now, what we see around us are many tall buildings, in point-block form, which are highly non-conducive to the application of solar energy and its appliances. If we are to incorporate solar technology in Singapore, I can see an end to point-block architecture because of the limited space for incorporating photovoltaic cells, solar collectors etc. So, perhaps architects in Singapore should think about the adoption of a slab-block form of architecture. Unless we change direction and design buildings that can incorporate these solar collectors and photovoltaic cells, there is very limited scope.

The other thing that we have seen is the process of incorporating solar collectors in buildings. In Japan and America, expensive steel structures have been assembled to support these solar collectors. Now, in our country, steel is an expensive material. If these collectors were incorporated into the roof structure, thereby eliminating roofing materials, there would be a great saving. However, there might be problems of leakage which would have to be solved by architects.

Comment:

What bothers me quite a bit about the responsibility that is being imposed on architects for a future solar Utopia is that one gets the impression that architects could transform that dream into reality.

My submission is that this is a very narrow perspective. If one takes Singapore as an example, one is talking about a city which one knows and which (as Prof. Cook has said) is the same as any other part of the world. Most cities are the same - because of the common problems that they have. Problems in cities are not really caused by the architects who design the buildings in them. Architects, like any other professionals, work within the framework of serving their clients. And, in the context of city architecture, the market system and shortage of the land are two factors which are completely beyond the architect's control. In the city, architecture basically depends on commercial developments. Any developer in a city wants to maximise land-use. That is the context in which the architect has to work.

One cannot have tower blocks and solar energy at the same time. So, one has to understand that in implying that architects can bring about solar architecture, we need to look at so many other factors which are beyond the control of the architectural profession. To encourage solar architecture, the government, for example, may play an important role in decisions over land-use and densities. One has to bring about certain changes before solar architecture can materialise.

Another aspect of solar architecture concerns education. For example, we should teach as much Structures as Solar Energy. Students in architecture schools should be made aware that Solar Energy is as an important to design as Structures or any other inputs. And I can only assume that certain buildings may be suitable for solar energy and others will not. And I think that there is a danger of walking away from this seminar thinking that one could end up with a Utopia. In city architecture, one cannot always identify whether a certain scene is Los Angeles or Singapore or Kuala Lumpur or Sydney. They look the same. My own feeling is that we can, for the time being, only make sure that there is something called Solar Energy, which can be developed, and include that input in architecture courses and other building courses. We cannot guarantee that this input will bring about solar architecture unless other factors which predominate or restrict the practice of architecture are considered.

Prof. J. Cook:

I would like to compliment the last speaker. I think he showed a very perceptive understanding of the ramifications of the problems. I suppose all of us, as professionals, want to be responsible, and yet architects at least are usually educated as public servants or as servants of their clients. That point of view is a sort of closed circle. How will that circle be broken? Obviously, those who pursue academic careers and who do not try to run a conventional practice have a unique opportunity to influence the future. But most conventional architects must serve their clients first. Our problem then is how to break the circle and how to intervene, and to bring to bear the new economics, the new perceptions of the problems that the world of the future is going to have to solve.

Prof. K. Kimura:

In attending the seminar for the last three days, I have been very much enlightened by the many presentations, questions and answers from the participants. I would like to add that there is a need to measure the environment. In large and small buildings, factories, hotels and so on, one has to examine how a comfortable environment may be achieved with the least amount of energy, within the limits of the present capability of machines and facilities. In our estimation of energy requirements in heating and cooling, about 10% of the energy now used could be saved if one lowered or raised the temperature 1°C on an annual basis, in terms of air-conditioning for heating or cooling. So, this is the first thing to consider. Then, if one measures the total energy requirement, one has to find out what kind of energy components are utilised within the building, such as lighting and external cladding.

In the U.S., much energy is used in lighting. If the heat from lights is re-used, some energy may be saved. More than 10 years ago, I worked out how to extract heat from lights, how to use it for heating in winter time and how to discharge in summer time. In my laboratory, the students are working on a project for admitting direct solar radiation into buildings. It could save electric power for lighting and also save energy for heating in winter, although it would result in some heat gain in summer. But on an annual average, we could save about 20% of lighting energy now used. Now this condition requires the automatic control of lighting levels. The next component is outside air intake, and this is taken care of by a total heat exchanger: we call it total heat exchanger. That means total sensible plus latent heat. The discharged air has a lower enthalpy than outside air.

In Japan, the absorption air-conditioner is very expensive. But if this absorption unit worked more efficiently at a higher chilled water temperature so as to pre-cool the in-coming outside air, then the total efficiency would be much higher.

As for the responsibility of architects, I would like to recommend that all architects should learn skills in how to design low energy buildings. It may be difficult to find the right clients. But, if the potential clients could foresee future energy costs, then one might be able to interest them. If the architect could combine his ideas with the client's ideas, build one house or one building and then measure the results for 2 years afterwards, he could build a convincing case for solar architecture. This would gradually develop into an environmental issue.

Assoc. Prof. R. Exell:

In the field of architecture, I myself am a layman but I do claim to keep my eyes open. Perhaps, I can make an observation about what I see in Bangkok. Although in Bangkok there are tall buildngs going up in the centre of the city, there is around Bangkok a very large suburban development for an expanding middle class in 2-storey houses over very wide areas and the architecture in this development is generally improving, and there are more interesting designs appearing. About a decade ago, the architectural design was rather boring and dull, but I think nowadays there are definitely much more pleasant looking building developments.

With the generally rising quality of life in domestic homes and the increasing price of electricity and fuel, I think that in this particular field there is tremendous scope for the Utopian solar suburbia. This could develop in Bangkok and I should imagine also in Singapore away from the central parts.

Comment:

Regarding the roles of architects, I think it is a commonly accepted fact that in any project the architect is the team leader in most projects, unless the project is too advanced in cases like an airport where specialists are needed as team leaders.

Now, being an architect, I think it is our great responsibility to lead the way in showing how we can integrate solar appliances into buildings. Going back to my colleague from the floor who raised the point just now about the intense use of land - I believe, from my experience in Singapore, we have such design concepts and the tools to assist us. In our approach to a solution, say, a high rise complex in Singapore, we are first of all working under constraints like plot-ratio control. The architect has to consider the aesthetic and other values of point-blocks or slab-blocks, as well as the application of solar appliances in such blocks. However, it does not mean that one must go high-rise all the time, because a high-rise building may block all the other buildings which would then not be able to incorporate solar appliances. So here again is the great contribution architects can make. Of course, we are limited in our ability because of the high technology involved, and we need specialist consultants in solar technology to assist us to work together as a team to see how it can be done. So I still maintain that the architect should lead the way. Unless we are aware that we can incorporate solar appliances, we may still carry on designing buildings that in the end cannot be used at all.

So I think we should now perhaps encourage all architects to go back for a refresher course in the N.U.S. to become acquainted with this new technology.

Mr. D. Hampton:

I would appreciate the opportunity of making comments on four factors that affect solar energy systems.

Firstly: this area probably has more scope for the architect than in Europe or North America. It is generally accepted in North America that most of the buildings we have will still be there in the year 2000 and that within that time there will be a lot of innovations adopted for solar energy and for energy conservation. A grandiose scheme to show how architects will control future development of solar energy systems is rather immaterial, unless we are building a new city. The most we can expect is that less than 50% of the accommodation needed for the year two thousand is yet to be built - even in Singapore. So the role of the architect will change to a more subordinate one of implementing solar systems developed by engineers.

Another factor is the enclosure of buildings: the development of a sealed building. The technology for this is not taught to architects, and builders in this area, including those in Singapore, have not paid attention to this detail in the past, simply because it is not important except when air conditioning is built in.

The third factor is the relevance of heat exchangers, as raised by previous speakers, for energy conservation. Economically speaking, the use of these is limited in Canada and there the potential for heat exchange is enhanced by the high temperature difference between the discharged air and the fresh air, with average differences of 20°C prevailing for many months. In the tropics the average outside temperature is rarely higher than 30°C, and, with room temperature of 26°C, there is little potential for heat transfer. There is also a maintenance problem.

And lastly, I would like to point out the need for education in the use of solar energy systems. The user can be excused for thinking that all the energy supplied through a solar energy backed system, such as a solar heated domestic hot water system, is free and that extravagant use of hot water from that system could be encouraged. The fallacy of this approach is revealed if the back-up system is considered, as this will automatically make up any deficiency from the solar heater. As this is a difficult concept to realise, it is also a big task to convey the need for energy conservation in connection with "free" energy.

Comment:

Mr. Chairman, I feel slightly out of place here. I happen to be a plumber and I am out-numbered by architects. We did a study for a 400 bedroom hotel in Spain, simply for the hot water requirement, and it would require an area of solar collectors equivalent to 21 tennis courts to supply the hot water. The question of plumbing has been brought home, and perhaps we should bring this to the attention of the institutes which train plumbers to help upgrade the skills and knowledge in this area, which I believe is vital, as has been proven in this conference, in order to make this system work. Even if only one valve does not work, the clients will say that our systems do not work. That is where the problem really lies.

Comment:

Mr. Chairman, at this point in time, I would like to really go slow on solar applications in Singapore. The reason is that in the last few days, most of the illustrations we have seen are not really applicable in Singapore. That being so, I would say that if architects in Singapore could design buildings to keep out the sun (because there is plenty of sun here) and at the same time allow maximum ventilation, we would save millions of dollars spent on electricity per year.

Chairman:

I think that is a part of solar architecture. What the Americans have been calling "passive systems" is really good design. I think the speaker has made a point of importance on conservation of energy which really must go hand in hand with solar architecture.

Prof. J. Cook:

Yes, perhaps I could underline other areas of responsibility which have been hidden here. One may think of buildings in terms of 3 kinds of responsibility. There are certainly technical questions; there are economic questions; there are aesthetic questions. Most of this conference has been devoted to the examination of technical questions and it seems to me that not all architects are very good in technical areas. Not if you look around.

Perhaps, we should not expect them to be. But there is one very important role, and that is to humanize the technical solutions. Somehow we have to make them acceptable. We have to make them attractive, saleable, purchasable, lovable. This is the responsibility of architects. It is unique and cannot be assigned to developers, engineers, or building inspectors. It is difficult to decide the basis on which decisions are made, and why one is acceptable and another is not. It is hard to evaluate why some technical solutions become popular and others do not. But I do think the architect has a responsibility to maintain aesthetic values and to find appropriate solutions which are atrractive to the general public.

Chairman:

Thank you. May we turn to the engineering problems. What has emerged at this conference is that we have made something of a breakthrough on the technology front. Is that the way we expect technology to go? In other words should we wait till R & D move in the direction where we can take advantage of it? Or is the engineering problem in solar application one of taking advantage of what is available right now, and making the best use of it, using the engineer's creativity, imagination and ingenuity?

Dr. J. T. Chiam:

On the point of whether we should wait for more advanced technology and better economic solutions from mass production in the solar thermal field, the technology is well tested, proven and established. This technology has been with us in fact for the last 20 years. Many countries in the world have gone ahead with large scale commercial applications of flat-plate solar collectors. Countries like South Africa, Australia, Israel, Japan and the USA have been successful in developing and marketing flat-plate solar collectors on a commerical and nation-wide basis. Of course, one can still improve on the design of the flat-plate solar collectors, but that does not mean that we have to wait for more advanced technology before starting to apply this solar thermal technology. As I said in my talk, in the context of Singapore and this region where there is abundant sunshine, the application of solar thermal technology is viable. And I think with a little help from the government and with a little encouragement from the architects and engineers involved in the building projects, the wide-scale application of solar thermal technology will get off the ground. While on this subject, I would like to endorse a statement that was been put forward from the floor a few moments ago, that, in the context of Singapore, the architect plays a very important part in attempts to introduce solar thermal technology in Singapore.

Let me just give you two examples from my experience. I was discussing with a certain architect the possibility of introducing solar thermal technology in a condominium project. At that time, this architect was unfamiliar with solar thermal applications. But he had an open mind. Further discussions were carried out not only with the architect but also with the developers. The architect and developers adopted an attitude that permitted me to look objectively into the engineering and economic viability of solar heaters. In the end, this condominium project (those of you who are Singaporeans will know that this is the U.O.L. project at Faber Hill) got off the ground. We installed 32 units in the top flats of the condominium project. This condominium project has only 3 storeys. When the developers first priced the flats in this condominium project, they considered that the top flats (they are walk-up flats) less attractive, and so these were priced lowest. As a further incentive, solar heaters were installed in the top flats as a means of attracting buyers. The top flats sold out first, and the premium flats, on the lower floors, did not initially sell so well. This shows, among other things, that if solar heating systems are installed in buildings, they have the advantage of attracting buyers if it is realised that the monthly PUB bill will be reduced significantly.

Now there was the case of another architect who was also involved in another development project. Basically, this was a case where the man had a closed mind. He said he did not know anything about solar heaters and indicated, in no uncertain way, that he was not interested. He suggested that I had a talk with the M & E consultant, who suggested that I should speak with the developer. The developer, in turn, asked me to talk to his architect. So we got nowhere.

The point is that solar thermal technology is practicable and applicable today. It has been tried out and has been successful so far. As I mentioned in my talk, efficiency is very high, varying from 50 - 60% in the required hot water temperature range. This efficiency is certainly higher than photovoltaic cells which have an operational efficiency of about 10%. Thermal electric power stations are operating at about 30% efficiency. The internal combustion engine - the basic design has not radically changed since Henry Ford invented his car - operates with an efficiency of about 25%. No one has for a moment suggested we wait for improved overall thermal efficiency in power stations or the internal combustion engine. We just continue building power stations and manufacturing motor cars despite the relatively low efficiency, and, in the process, improve as we gain more experience. So, consider the case of solar flat panels. We have, if you like, a machine which is operating at 50 - 60% efficiency. But people still want to wait for more improvements in the technology. I see no sense in this. If it is technically and economically viable, let us use it. If we are not interested in introducing this technology in our buildings, then, of course, we can make all sorts of excuses for not doing so.

Dr. N. Sheridan:

Mr. Chairman, perhaps I should make a few observations. It is certainly true that energy conservative design is something that should go ahead. And I would say, as some have pointed out, that this has just got to be applied to existing buildings, which are going to be around till the end of the century. The main approach to energy conservative design is to look at the envelope of the building. And there is certainly no reason for hesitation. I would think that no matter what system is to be used, the energy saved means a saving of primary energy that must be imported. As far as multi-storey blocks are concerned, I am not sure if we are going to make much impact on the total energy consumed by such buildings if a few collectors are mounted on the roofs.

I am also not sure that we should presume that solar energy is going to be applicable only to decentralised systems. We should not really say that the current generations of collectors that we mount on buildings are the only devices for utilizing solar energy. If we want to look forward to new technology, we should keep our options open for centralised solar technology. I would think that there have been enough developments in centralised solar technology that will produce some results in the not too distant future, such as in centralised electricity generation. Some of these might not be applicable to the climate of Singapore, and there must not be enough land available. But one has to keep looking at what is happening and not say that the current approach is the only way that solar energy is going to develop.

Another possible development in the future is the use of hydrogen as a means of transmitting energy. Hydrogen is already used. In Brisbane, where I live, the gas that is distributed in the city is reformed petroleum gas which is 65% hydrogen. Hydrogen is not as dangerous as some people might think. But the problem is the source of the hydrogen. It is no good looking for a large contribution of hydrogen from reforming petroleum gas or using oil, because this is not going to be available in the future. We should look to the sun, therefore, as our source of hydrogen; there are many ways that hydrogen can be produced by using solar enegy. Of course, it must be produced from water; water is not going to be destroyed in the process, it is going to be recycled. The hydrogen will be formed from the water and eventually it will get back to water, so we are not going to waste the water, because it is a recycleable material.

Hydrogen gives a means of transporting solar energy from a place where it is plentiful to a central business district where you can utilize it as we use electricity. Hydrogen gives a means of storing solar energy over long periods and overcomes one of the biggest disadvantages of the use of solar energy at the moment. Water heaters on the roof for air-conditioning require extensive thermal storage which is really quite difficult to install. But with hydrogen, storage is provided in a fairly convenient form, and there is the means of transporting this energy probably more effectively and efficiently than by electricity.

And so we should not be restricted in our outlook by saying that we have only one view of what a solar city might look like, with every building covered with collectors. Perhaps it is not going to be that way at all. There may be a mixture of things - some absorbers where the advantage is to collect the thermal energy for direct use in buildings, for example. But for running engines and lights, we could use hydrogen.

From the engineering point of view, we have to look more to the system design. I was very pleased to hear Dr Chiam's paper which is concerned with the design of a system. I think it is an eye-opener to a lot of people to see his particular system in which he proposes to pre-heat the water for an electrical water heating system. If we want to have successful solar systems, we have to be very careful with our system designs. It is not good enough just to go and buy a few components, toss them together, and hope that they will work. We must optimise that system, and carefully match all the components, or otherwise the system will not work as well as it could.

As I look at some of the manufacturer's exhibits, it is obvious that some solar components have not been designed to fit into systems; there has not been sufficient consideration given to the design of simple interconnections and supporting members, which would facilitate site installation. I think this will reduce the number of failures. One cannot expect the plumber to rectify the faults of badly designed components. This is especially important with the higher temperature systems. All the connections ought to be done at the stage when the components are being manufactured. One should not expect the plumber to do all that thinking. It is the engineer's job to think out how the system is going to be built and he should not expect the tradesmen to rectify design mistakes.

Question:

Mr. Chairman, I have a question to ask and a few comments to make. I must first qualify what I have to say by saying that I do not know anything about thermal technology except what I heard at this seminar and from discussions with friends. I have one question for Dr. Chiam. He says that he has installed over thirty solar hot water heaters in the U.O.L. project. Why have they only been put in for the top flats, the highest level, and not for the others? Because, in my opinion, there should not be any problems involved in providing solar hot water heaters for the lower levels as well. Perhaps, it has to do with the design of the roof. Now, because of soaring energy costs, a lot of developers would like the idea, especially if they were keeping the building as an investment. The running costs should also be kept low. So, if they could make use of solar energy and save on energy costs, I think they would go in for it.

But I would say that, in the Singapore context, there is a limit to the amount of energy that might be extracted from solar collectors. Surely there would be a limit or constraint on the height of buildings or on the total area of buildings which might be served by solar energy.

Also, I think there would be a need for some control on the height of all the buildings to avoid shadow-casting. And if the roofs were not big enough for the installation of solar collectors of the size required for the total energy consumption, then there would be the problem of supplying sufficient heat for the systems.

You could also have a problem with the metering because, at the present moment, the Public Utilities Board only supplies one meter to register the consumption of cold water. There is no way of registering the consumption of hot water. The other question is storage. If you have to supply hot water for the residents in a building, you will have to find a big enough storage space for the hot water tank.

Dr. J. T. Chiam:

Thank you for those questions. I think they are very well thought through. I will try to answer those questions to the best of my ability. Now the first question concerns the housing project. Why is it that we have only installed solar hot water heaters on the top floor? It is not because of the roof. There is sufficient roof area. But when we talked to the engineers and the architects, we proposed the concept of having the panels on the roof, with the tank also sited on the roof. The common tank would suppply hot water to all the flats below. As this is only a three-storey building with a ground floor, for a given roof area it is only required to supply hot water to four flats. From the technical point of view, there is no problem. It is only in the implementation of this scheme that we ran into some snags. The problem of metering was mentioned. This is a real problem which I also brought up yesterday at the workshop, because it would involve a separate hot water meter for each flat. Now we all know that the P.U.B. will only read the cold water meter, so there is no problem in metering the total output of hot water. All one has to do is to meter the total cold water coming into the tank which equals the hot water leaving it. The problem is to meter each of the flats to which hot water is supplied. The P.U.B. will not read the hot water meters of these flats. We talked to the management corporation, and basically they wanted to do as little as possible, so they were not very keen. So this scheme fell through.

The second alternative was to divide the cost of hot water consumption equally between the flats. But this would also create a problem because it appears that there would be indiscriminate use of the hot water. So, because of these problems we decided that we would only install solar heaters on the top floor.

Now to this question of whether there is sufficient space on the roof of a point-block flat for solar collectors. If we take the middle-class flats, with a roof area, say, of $100m^2$, assuming that only $50m^2$ was available for the siting of the solar panels, each panel of $2m^2$ would supply the requirements of one apartment. So, theoretically, 25 apartments in a point-block building could be supplied with solar hot water. In the tropics, people do not use as much hot water as they do in temperate regions. For a start, we do not usually take baths, most of us use showers. Just to illustrate this point, I have a solar collector of $2m^2$ and we have found that this is sufficient for a family of five, and we have no problem with the hot water supply - in fact, we seldom use the electrical back up system. I mentioned in the workshop yesterday that over a period of ten months, we monitored this situation, and the number of times we put the electric heater on could be counted on the fingers of one hand - about four times. I hope that answers the question of whether we could install solar heaters in point-block projects for residential purposes.

The case of storage really relates to the consumption of hot water. A storage capacity of 150 litres is sufficient for a family of four or five. So really, there is no need for such a large storage capacity since we design our storage capacity to supply one day's demand only. So, in the tropics here we try to minimize the storage capacity by studying critically the consumption pattern for a given situation.

I just want to strengthen the point which Dr. Sheridan has just made, especially with the large number of engineers here, that engineers think through the systems they design. Let me give you a case in point, from my experience. I know of a particular club, which bought a number of solar collectors. A plumber was engaged to install these collectors on the roof. Now the plumber had no idea how to fix these collectors. So all he did was to connect the cold water from the mains into the solar collectors and ran the system as a once-through process with the output supplied directly to the consumers. Well, this was just disastrous. For two months, the club was monitoring this system and nobody got any hot water because they did not have a recycling system. So, it is very important that engineers think about the design of the system.

We should also make a study of the demand pattern, i.e. how much hot water is used. The results could be surprising. We monitored one particular clubhouse and the amount of hot water used by the members of the club is surprisingly low. It worked out, on the average, to be only about 25 litres per day per person. This is a very small amount, because most people in this particular club only use the showers for a relatively short time.

Comment:

Following on from Dr. Chiam's suggestions, I am glad to say that we have decided to install a solar system in our new factory building which is going up in Jurong. The factory is manufacturing food products, like potato chips. A solar system is supposed to be used for air-conditioning the building. We have a total of about 80 tons of refrigeration and we have about $1000m^2$ of roof area. The system which we have looked into is the absorption system. We have also examined the dehumidification system. But after hearing all the different comments on the dehumidification system, I don't think it would be viable in the tropics now. We will go in for the absorption system, the modified version which has a better COP. We had a computer run on this, and found that it would be about 0.6 to 0.7 COP, which showed that it is viable, but the economics may not be so very attractive. So, we are trying to make the system less solar-dependent and use the process to supplement the solar system. Then we will expect the system to be more economical. We also have a stand-by boiler which will cut in when the solar system does not operate on a cloudy or rainy day. We expect this factory to be ready by June next year and those of you who are interested are welcome to have a look at it when it is finished.

Chairman:

I would like to turn our attention to ASEAN. We have invited official delegates from our neighbouring-countries. We really should not just think in terms of Singapore, although we have been paying a lot of attention to urbanised areas. I would like to invite three speakers from three neighbouring countries to address us very briefly and report to us on the progress in their own countries. Firstly, from Indonesia, Dr. Mostavan.

Dr. Mostavan:

The problems in Indonesia are rather different from the problems in developed countries. At the moment, there is a lot of agricultural production. That is why the need in Indonesia is more for dryers. The latitude of Indonesia is between 6°N and 11°S. The longitude is between 95°E and 141°E. The distance between north and south is 2,000 km, and between east and west, 5,000 km. The territory is made up of 13,677 islands of various sizes. Thus, we have a problem of communication, of transmitting energy from the source. Usually, Indonesia has a high daily temperature. The average temperature is 25°C to 26°C. The minimum annual temperature is 20°C and the maximum 33.7°C with some variation. It is said that the sun shines, on average, for more than 6 hours daily, and the number of hours of sunshine annually is 2,400. Indonesia needs a lot of instruments to measure meteorological data, and it is not surprising that Dr. Rao does not have sufficient solar radiation data for Indonesia.

Despite the lack of data, Indonesia has good potential with regard to the application of solar energy. We have several institutions working on solar energy. At the Institute of Application and Analysis of Technology, working directly under the Ministry of Research and Technology, at the moment there are solar panels used for pumping water for irrigation and for the refrigeration of fish. Secondly, there is Bogor Institute of Agriculture where studies are being done on solar air-conditioning and refrigeration. Thirdly, there is Gajah Madah University where solar concentration and hydrogen from the sun are being developed in co-operation with the Australian National University. And fourthly, there is I.T.B., where we have two Departments which are active in solar energy research. One is the Department of Mechanical Engineering, dealing mostly with bio-mass. The other is the Department of Physical Technique which is my laboratory; in this case we have no research activities financed by the government, but students are engaged in solar energy work. Applications, such as solar panels, are ordered by the Tele-communications Department, the Army and the Railway Department. Small solar water heaters can be found in some hospitals in Bandung and in Jakarta. Demonstrations of solar dryers have been done by I.T.B. Solar radiation data is published by the Department of Meteorology.

Recently, we discussed the creation of a National Association of Solar Energy. It will be called Forum Energy Suria (suria means solar in Indonesian). We are planning the secretarial office in Bandung. To conclude, I can say that more and more institutes are becoming aware of the importance of solar energy in Indonesia. However, as Indonesia is a large area, introducing something new takes a long time.

Chairman:

May I now invite Miss Angelita Panis to address us briefly on the work in the Philippines.

Miss A. Panis:

Thank you, Mr. Chairman. The research and development in solar energy application in the Philippines are in the following fields: solar air-conditioning, solar water heating, solar pumping and solar stills. A 2-3 ton capacity solar air-conditioning system was installed in the Philippines by the Yazaki Corporation. Government funds have been used to buy the system. So far, we have not encountered any major problems in the system. We are trying to get some relevant data so that we can compare solar air-conditioning under conditions in the Philippines with the conventional system. I would also like to point out that there is only one solar air-conditioning system in our country. We are trying to convince investors to buy them but they are rather reluctant to do so because of the very high cost.

Another application of solar energy systems in the Philippines is in water heating. Considering the tropical climate of the Philippines, we have recognized that solar water heaters have greater potential for extensive use in industrial and commercial establishments rather than in domestic applications. We have been trying to convince the industrial sector, hotels, country clubs, restaurants, gyms and hospitals to change to solar water heating. So far, we have had some favourable reaction. One of the biggest solar water installations in the Philippines is in a local hotel. The installation provides hot water for all guest-rooms and the kitchen. The Yazaki blue panels have been used for the system, and it is already in operation. There is another solar water installation at the Magnolia Dairy Plant. This is also industrial in scale, and it is in operation and also uses the Yazaki blue panel. The hot water is used for the sugar hydration process. This is to name but a few of the industrial solar water heater installations we have in our country. We have also distributed domestic types of solar water heaters nation-wide in order to popularise the technology.

Another application is solar pumping. The Philippines, together with Mali, Sudan and India, was chosen by the United Nations Development Programme as one of the sites for country demonstrations of small scale solar-powered pumps. These solar pumps have already been installed and are in operation, and currently some of our staff are monitoring these projects.

Another application of solar energy in the Philippines is solar distillation. At present, a 100 square metre area of solar stills has been constructed at the Bureau of Animal Industries. Its high initial cost and the large land requirement for the set-up are the negative factors.

Solar dryers are the most common application of solar energy in the Philippines. A simple local solar dryer has been designed in the Philippines. The dryers are good for drying marine products and fruit. Data gathered shows that products dried in this dryer have a longer shelf life compared to products subjected to the usual practice of direct exposure to the sun. Besides this, the drying time is also lessened. There is one on-going research project on solar crop drying. This project has been designed to establish a systematic method of developing solar crop dryers as a means of increasing the benefits derived from solar energy, while simultaneously minimising cost production losses of crops. On a smaller scale, a panel of solar cells is used to run a television and is installed in our office for research and demonstration purposes.

However, there is a joint-project between the Philippines Ministry of Energy and the Federal Republic of Germany's Ministry of Economic Cooperation and Ministry of Research and Technology to test the applicability of the portable type of system to provide basic power needs to our remote villages. The German Government has contributed a certain amount to finance the design, construction and installation of a 10-15 kilowatt portable type of solar power plant. At this moment, the silicon cells for the portable type are being made in Germany. Telefunken has been chosen as the principal contractor for this system. A site in the Philippines has already been chosen, and the system will serve 50 households.

In conclusion, one of the major problems we have in the Philippines is that, although we have the technology, the dissemination of information is difficult. However, we are trying hard to persuade people to change to solar energy or other alternative sources of energy. They are rather reluctant, nevertheless. They prefer the existing system because it is more convenient to use.

Chairman:

May I invite Mr. Oran from Thailand to address us briefly on the development in Thailand.

Mr. Oran:

Before speaking about the solar energy situation in Thailand, I would like to give some idea of the overall energy situation in Thailand. About 99% of the fuel consumed in Thailand is imported. So, the fuel imports in 1979 were about 75% of the gross national product. Therefore, one can expect that in the next few years the country will be in debt. So, it is not too surprising that people there are interested in solar energy utilisation. For example, there are more than 10 private companies dealing with solar hot water applications and the utilisation of photovoltaic cells. For this year, they have signed contracts for more than 1,000 square metres of solar collectors. The Department of Telecommunications has awarded a contract of about 18 million bahts (about US$40,000) for photovoltaic applications in telecommunications. In the government sector we have a number of agencies involved in solar energy utilisation. For example, there is the Electricity Authority of Thailand, the National Energy Administration, The Institute of Applied Science and Technology, Department of Science, Chulalongkorn University, and the King Mongkut Institute of Technology. The Electricity Authority of Thailand has awarded a contract to the universities to study the feasibility of electricity generation by solar energy. Chulalongkorn University has been working on a system to produce about 100 kilowatts of thermal energy, and the King Mongkut Institute of Technology will develop parabolic solar concentrators necessary for the production of about 10 kilowatts of thermal energy. The National Energy Administration is considered a co-ordinating agency. These agencies have also been concerned with the implementation of the project. For instance, from 1978 to now, ten paranometers have been installed to measure the total solar radiation, with some cooperation from the Meteorological Department. On the demonstration side, since May 1980, a 2-ton air-conditioning system has been installed, and also a 50 m^2 solar still has been constructed to produce pure water for laboratory purposes.

As far as research and development are concerned, universities or higher institutions are mostly carrying out both basic and applied research, ranging from a small cabinet dryer and solar pump evacuated tubes to a modified technique for producing solar cells. Finally, with regard to for the next five years, the government has been placing more emphasis on solar energy, especially for rural areas, such as in the use of solar dryers and solar pumps. There will be many more demonstration projects within the next few years. We have the same problem as the Philippines - it is difficult to try to disseminate information on solar energy and to interest people in using it.

Chairman:

It seems to me that we in Singapore (and, to a large extent, in other countries of the region) have come to understand the need for energy conservation, though perhaps not as well as we ought to have done. We have introduced sufficient legislative measures so as to ensure that our buildings do not become monstrous energy consumers. To that extent, we have done something in the right direction. We should now educate people more and more with respect to energy conservation at all levels, and ensure that buildings and industry at large will minimise their consumption.

Our end task then is to look into the area of alternative energy sources. It seems to me, then, that solar energy is the first thing we should look at. To me, Singapore is the last frontier. If solar energy can be adopted here, it can be done almost anywhere else. This is because there are many factors in Singapore which are against solar energy application. We do not have the sunny climate of Arizona, though we are right on the Equator. Because of our maritime climate, we only have about 5.6 sunshine hours per day. We are highly urbanised. The national grid is effective and convenient, and it is difficult to get people to do anything else. The modern development of solar energy, of course, does not originate from this part of the world; it started in temperate regions where there was not enough sun and people wanted to have it. The ideal of autonomy and self-sufficiency in energy began in these countries. So, the technology that came over here is a different kind of technology. It is heating and hot water. These ideals are alien to us. Economically, it is not attractive to have long-term returns. Therefore, solar energy would present problems in Singapore because of this long-term problem of financial gain. The present pace of development means that the total global solar energy utilisation, as estimated, would not be more than, say, one percent by the turn of the century.

So we are facing many difficult problems as far as solar energy utilisation is concerned. But I think one should make a start, if it is ultimately for the good of mankind. We should begin to ask the right questions. This is the reason for the organisation of this conference: to stimulate some understanding and awareness of solar energy. We are not saying that it will cure all our ills, and obviously it will not, as has been expressed by many speakers in this conference. But we have also heard from many other speakers that, on a project to project basis, it may pay. As Dr. Chiam and other advocates of solar energy have told us, there are many opportunities for us to develop good systems even on a short-term basis. Dr. Sheridan has also reminded us that there are many varieties of solar applications we could use in buildings. As architects, engineers and executives, we will have to think about the best ways for us to minimize the use of energy in a particular building and to supplement the requirement by solar energy.

We have also seen that technology is improving and that there are now many varieties of appliances we can use - even collectors have improved over the last 5 to 10 years. Attitudes are also changing; and, as we have heard from the ASEAN delegates, although they are facing similar problems of people finding solar energy a little inconvenient, attitudes may also improve as the years go by.

I believe that in the ASEAN context, the major energy consumption will be in the cities and not in the rural areas. If we could solve the problems in urban utilisation of energy, particularly in Singapore, then the same strategy could be used in Bangkok, Manila and the other major cities of the developing world.

The prospect is a good one.

INDEX OF NAMES

INDEX OF SUBJECTS

9 789027 715067

GPSR Compliance

The European Union's (EU) General Product Safety Regulation (GPSR) is a set of rules that requires consumer products to be safe and our obligations to ensure this.

If you have any concerns about our products, you can contact us on ProductSafety@springernature.com

In case Publisher is established outside the EU, the EU authorized representative is:

Springer Nature Customer Service Center GmbH
Europaplatz 3
69115 Heidelberg, Germany

Batch number: 10815974

Printed by Printforce, the Netherlands